Integrated Public Lands Management

Integrated Public Lands Management

Principles and Applications to National Forests,
Parks, Wildlife Refuges, and BLM Lands

John B. Loomis

Columbia University Press
New York

Columbia University Press
New York   Chichester, West Sussex
Copyright © 1993 Columbia University Press

Library of Congress Cataloging-in-Publication Data

Loomis, John B.
 Integrated public lands management :
principles and applications to national forests,
parks, wildlife refuges, and BLM lands /
John B. Loomis.
  p.   cm.
 Includes bibliographical references and index.
 ISBN 0-231-08006-9
 1. Public lands—United States—Planning.
 2. Public lands—United States—Management.
 I. Title.
 HD221.L66   1993
 333.1′0973—dc20   93–17218
                                     CIP

                                     ∞

Printed in the United States of America
c 10 9 8 7 6 5 4 3 2 1

To JCL^2 and KML: my wife, Jayne,
and daughters, Jennifer and Kelly

Contents

List of Illustrations

List of Tables

Objectives of the Book

This book on public land management originated from two concerns. The first was the lack of a unified treatment of the analytical procedures used by federal land management agencies in planning and managing their diverse lands. There were excellent books on public land laws and on forest management, but no single book dealing with management of National Forests, National Parks, National Wildlife Refuges, or Bureau of Land Management lands. The second motivation stems from my concern that advances in valuing environmental benefits provided by natural resources have not been incorporated adequately into public land management. This is coupled with a belief that if such advances in nonmarket valuation are incorporated into public land management decisions, these decisions would be more sound, both environmentally and economically. I hope to demonstrate in this volume that application of nonmarket valuation techniques to public lands management is relatively straightforward and that an economic way of thinking about public land management trade-offs can help improve the environmental and economic benefits provided from public lands.

About half the book (chapters 2 and 9–13) focuses specifically on the four main federal land management agencies (Bureau of Land Management, Fish and Wildlife Service, Forest Service and National Park Service). However, much of the first half of the book is devoted to presenting general principles applicable to management of many other types of public land. Specifically, chapters 1 and 3–8 present concepts and techniques that are equally applicable to other federal public land management agencies such as the U.S. Army Corps of Engineers, Bureau of Reclamation and Tennessee Valley Authority as well as state forests and state parks agencies.

The primary intent of this book is to teach-upper division and graduate students about how public land planning and management is performed (i.e., how many of the theories and techniques learned in many other classes are applied to managing natural resources of National Forests, Parks, and Wildlife Refuges). Some basic knowledge equivalent to introductory economics and statistics courses is desirable but certainly not mandatory, as nearly all topics are introduced at an elementary level.

This book also offers a constructive criticism of the current state of public land management and provides concrete suggestions for improving planning and management of our public lands. This criticism is in no way meant to belittle the extraordinary efforts made by the people who work for these four federal public land management agencies. From my own years working with three of the four federal land management agencies I know only too well what a remarkable job these professionals do despite the odds facing them. They overcome inadequate funding, unrealistic time deadlines, and often conflicting directions to manage natural resources in what they consider to be the public interest. It is my goal in this volume to provide suggestions for better utilizing the talents of these dedicated professionals to improve natural resource management on public lands.

Acknowledgments

I am indebted to many people and institutions who facilitated my completion of this volume on public land management. The first draft of this book was written while on an "in-residence" sabbatical from University of California-Davis. Thanks to a mini-grant from University of California-Davis Teaching Resources Center, Ray Sauvajot (Graduate Group in Ecology, UC Davis) reviewed the first drafts of the core chapters and Brian Yonts proofed the final chapters and compiled a much-needed acronym list. Their comments and insightful suggestions will save future readers much confusion. Students in my Public Lands Management course in the fall of 1990 and 1991 provided numerous suggestions for clarifying the text. Finally, Catherine McCarthy (Graduate Group in Ecology at UC Davis) painstakingly critiqued several lengthy chapters and provided substantiative comments, references, and suggestions. Sy Schwartz (UC Davis) provided detailed suggestions and clarifications on the benefit-cost chapter. John McKean, of Colorado State University, and Gregory Alward, of the U.S. Forest Service, significantly updated the chapter on input-output models. Joe Cooper, now with USDA Economic Research Service, provided assistance in running the sample linear programming model used in chapter 9. Richard Behan, of Northern Arizona University, also provided several useful suggestions and references.

The Division of Environmental Studies at UC Davis was in many respects a suitable habitat to write such an interdisciplinary book as this. Discussions with natural scientists, such as Charles Goldman, Jim Quinn, Alan Hastings, Susan Harrison, Ted Foin and Christine Schoenwald-Cox, kept me acquainted with important ideas in conservation biology and landscape ecology. Social scientists such as Paul Sabatier, Sy Schwartz, Bob Johnston and Ben Orlove ensured that the institutional settings in which these agencies operate were firmly kept in mind. Janet Kendrick provided secretarial support on several of the lengthy tables. Most of the better-looking graphics used in this book were initially developed by Hanna Fischer and Karen English-Loeb of Instructional Media for my public lands management courses in fall of 1990 and 1991. Their cheerful assistance was much appreciated.

Rob Lilieholm, at Utah State University and David King, at University of Arizona, provided not only excellent suggestions but also invaluable support in the publication process. No doubt this book would not have been published without their statements of interest.

I am also indebted to several people in the federal land management agencies studied in this book. Daryl Trotter and Bob Milton (both of the Bureau of Land Management) supplied many of the BLM planning documents used in the case study and provided the updates on how planning was proceeding at BLM. Daryl reviewed the BLM chapter so that it reflected the ''unwritten guidance'' field personnel were relying on as well. Numerous people in the U.S. Forest Service have aided me over the years in understanding their planning process and models. They include members of the staffs on the Siuslaw National Forest, Gallatin National Forest, Eldorado National Forest, Tongass National Forest and, the Rocky Mountain Forest and Range Experiment Station. Al Dyer and Richard Walsh of Colorado State University, as my former professors, and John Hof and George Peterson of the Rocky Mountain Forest and Range Experiment Station will recognize my synthesis of their philosophies toward natural resource management in this volume. Drs. Dennis Schweitzer, Linda Langer, and Chris Hansen have provided me with much material on the future direction of the Forest Service planning process. Spencer Amend, Curt Laffin, and Charles Houghton of the U.S. Fish and Wildlife Service all provided valuable reviews and input on their agency's chapter. Stephen Veirs and Kathleen Davis of the National Park Service corrected many terms and interpretations of National Park management policies, for which I am very grateful.

Executive editor Edward Lugenbeel, and the staff at Columbia University Press (especially Laura Wood and Leslie Bialler) were most supportive and helpful in guiding the final preparation of the manuscript. Ronald Harris' editing nicely smoothed out a great deal of rough text and removed many inconsistencies. Virginia and John Cullen did an excellent job cross-referencing the index words and final page numbers in the final weeks of book preparation.

As always none of the preceding people are responsible for my personal interpretations, and they should be absolved of any guilt by association. Finally, keep in mind that no book is absolutely perfect for every purpose. Each reader will wish that additional topics were covered or less emphasis was given to a particular point of view. Suggestions such as these are of course quite welcome, because a book on a rapidly evolving topic such as public lands management can never be finished completely. It must change as society changes the laws and their expectations of what public lands are to provide.

List of Acronyms

ACEC	Area of Critical Environmental Concern
AMS	Analysis of Management Situation
ANILCA	Alaska National Interest Lands Conservation Act
ANWR	Arctic National Wildlife Refuge
ASQ	Allowable Sale Quantity
AUM	Animal Unit Month
BCA	Benefit Cost Analysis
BCR	Benefit Cost Ratio
BLM	Bureau of Land Management
CMR	Charles M. Russell
CVM	Contingent Valuation Method
EIS	Environmental Impact Statement
FHI	Fish Habitat Index
FLPMA	Federal Land Policy and Management Act
GAO	General Accounting Office
GIS	Geographic Information Systems
GMP	General Management Plan
GYCC	Greater Yellowstone Coordinating Committee
GYE	Greater Yellowstone Ecosystem
HEP	Habitat Evaluation Proceedures
HMEM	Habitat Management Evaluation Model
HSI	Habitat Suitability Index

ICO's   Issues, Concerns, and Opportunities

ID   Interdisciplinary

IFIM   Instream Flow Incremental Methodology

IRR   Internal Rate of Return

LP   Linear Programming

MAU   Multiple Attribute Utility

MB   Marginal Benefit

MC   Marginal Cost

MMR   Minimum Management Requirement

MSA   Management Situation Analysis

MUSYA   Multiple Use Sustained Yield Act

MVP   Marginal Value Product

NAWMP   North American Waterfowl Management Plan

NEPA   National Environmental Policy Act

NFMA   National Forest Management Act

NHQ   Natural Habitat Quality

NPS   National Park Service

NPV   Net Present Value

NRA   National Recreation Area

NRDC   Natural Resources Defense Fund

NWR   National Wildlife Refuge

PLLRC   Public Lands Law Review Commission

PPF   Production Possibilities Frontier

ORV   Off-Road Vehicle

RNA   Research Natural Area

RMP   Resource Management Plan

ROD   Record of Decision

RPA   Resources Planning Act

RVD   Recreation Vistor Day

SIA   Special Interest Area

SPG   Supplemental Program Guidance

SPNM   Semi Primitive Non-Motorized

SRMA   Special Recreation Management Areas

TCM   Travel Cost Method

USDA   US Department of Agriculture

USFS	US Forest Service
USFWS	US Fish and Wildlife Service
USLE	Universal Soil Loss Equation
WSA	Wilderness Study Area
WTP	Willingness to Pay
YNP	Yosemite National Park

Integrated Public Lands Management

1

Natural Resource Use Interactions: The Key to Modern Public Land Management

A Long-Term Global View

Early conservation efforts in the United States were aimed at preventing the nation from running out of such natural resources as game animals, which provided food and clothing, and timber, which provided fuel and shelter. The concern over adequate long-term timber supplies was one of the reasons for establishing the National Forests in 1891. However, conservationists also worried about preserving unique natural environments, not for possible consumption but for aesthetic and spiritual values. This motivation led to the establishment of the National Parks in 1872.

By the early 1960s it became apparent that in meeting our growing material wants, we were polluting the air and water in the cities, where we lived. Although society has made some progress in controlling air and water pollution, the validity of Boulding's "spaceship earth" metaphor became very apparent in the late 1980s. Most people now recognize the magnitude and subtlety of the changes in the earth's overall environment caused by millions of localized human actions. The 1980s brought widespread recognition that human alteration of local environments has changed the world's atmosphere, forests, and oceans for generations to come. In the United States this recognition was epitomized by *Time* magazine's selection of the Earth as "Man of the Year" in 1988.

Recognition of these new environmental problems amplifies the ecologist's maxim that "in nature everything is linked to everything else." Unlike readily identifiable point source water pollution from industrial and municipal polluters, many of these new environmental threats arise from thousands of interrelated natural resource decisions all over the planet. Resource use in one country affects the atmosphere of the entire planet. The most compelling examples of such global interdependence are depletion of the ozone layer and the buildup of greenhouse gases in the atmosphere.

To illustrate the global linkages of newly recognized environmental problems let us examine one potentially far-reaching new environmental threat, the greenhouse effect. The greenhouse effect is the gradual warming of the earth in response to a net gain in carbon dioxide and other gases. If some scientists' worst-case predictions prove correct, the greenhouse effect would melt the polar ice sheets and expand the ocean mass, thereby raising the sea level and flooding most of the world's major coastal cities. The warming trend may reduce our storable water supplies as less precipitation falls in the form of snow. Many current agricultural areas could become too hot and dry to grow crops. There may also be substantial changes in current wildlife and fisheries habitats that may exceed the ability of species to adapt. As with most environmental issues, scientists are uncertain about whether the greenhouse effect has started already and what the rate of warming is likely to be over the next several decades.

If and when enough evidence of global warming has accumulated that world leaders decide to respond, the solutions will be more complex than dealing with past environmental problems. Unlike point source water pollution, with its readily identifiable polluters, the greenhouse effect has many causes. The major source of atmospheric carbon dioxide is the burning of fossil fuels in power plants and automobiles. At the same time that humans have been adding more carbon dioxide to the atmosphere, nations around the world have been rapidly cutting down one of the major ecosystems that could consume some of this additional carbon dioxide: our forests. Deforestation through timber harvests and outright burning of tropical rainforests is adding greenhouse gases and removing forests that normally sequester carbon dioxide as part of their growth process (Harmon, Ferrell, and Franklin 1990). This suggests that forestry decisions should be linked with other global strategies to reduce the greenhouse effect. Preserving existing forests (Harmon, Ferrell, and Franklin 1990) and replanting deforested areas can help reduce the buildup of carbon dioxide in the atmosphere. Recognizing the relationship between natural resource management and environmental quality is a major part of integrated natural resource management and a major theme of this book.

The benefits of using an integrated approach do not end with this one effect of better forest management on the global atmosphere. Reducing the greenhouse effect is not the only reason to cut back on deforestation. Integrated resource management suggests that reduced deforestation may have additional *joint* benefits. These occur simultaneously with maintenance of forests and include protection of water quality, natural beauty, and outdoor recreation, as well as preservation of other plants and wildlife species that depend on forest ecosystems. In the Amazon region thousands of species depend on the rainforest for their survival. In the states of California, Oregon, and Washington the northern spotted owl depends upon old-growth Douglas fir forests. In the southeastern U.S. the red-cockaded woodpecker requires an old-growth forest. Thus while living forests are fixing carbon dioxide, they are providing wildlife with a habitat and ensuring water quality. Society benefits in many ways from not cutting forests.

The fact that preservation of forests and reforestation of previously cut areas throughout the world will make a small contribution to slowing the greenhouse effect does not, by itself, mean that every nation should stop cutting trees and start planting forests. However, it does mean that controlling timber harvest and accelerating reforestation on public lands may play a small part in solving the greenhouse problem.

The proper role of forestry practices in reducing the greenhouse effect can be determined in theory by comparing the cost of a reduction in atmospheric carbon dioxide achieved by (1) reducing emissions of carbon dioxide from fossil fuel use versus (2) increasing consumption of carbon dioxide by planting more trees and cutting fewer trees. It may be less costly to society in terms of the value of other goods and services given up to reduce emissions. For example, energy conservation in the form of reduced electricity usage and fewer automobile miles driven may be less of a sacrifice than giving up wood products derived by cutting virgin forests.

This example illustrates the potential role for resource planners and analysts in public land management. Their role is to evaluate and communicate the environmental and economic consequences of alternative public land management strategies to decision makers and society. In this way society can make informed trade-offs involving publicly owned natural resources, the management of which affects the environment at many levels (e.g., local, national, and global). One of the goals of this book is to provide a framework for evaluating the desirability of alternative strategies for managing natural resources on publicly owned lands.

Natural Resource Planning and Management Issues Closer to Home

Integrated Natural Resource Management: Balancing Competing Resource Uses

The criteria that natural resource agencies use in decision making are becoming broader. Gone are the days when agencies wishing to build hydroelectric dams did so without concern for fish and river-based recreation. Gone too are the days when the only decisions a forester on public or private land had to make was when to cut the trees and the cheapest way to bulldoze a road to reach the timber.

Today laws and regulations require most public land managers to recognize that their job is *not* to maximize a single use of a natural resource but to optimize the many uses of natural resources occurring on public lands. What we mean by *optimizing* is attainment of the highest level of overall benefits from all the resources considered together. At an existing dam this may mean maximizing the benefits attainable from all uses of a river, such as fishing, rafting, swimming, water supply, and hydroelectricity production, not just benefits from hydroelectricity production by itself. Providing some river flows for fish and recreation may more than outweigh the electricity revenues foregone.

Thus one aim of this book is to explain resource planning and analysis techniques/methods that make it possible to determine how much of each resource to produce from a given type of public land (e.g., how much recreation, timber, minerals, and wildlife to produce with the available public land resources and facilities).

Integrated Natural Resource Management Involves Trade-off Analysis

Resource management may sometimes involve choosing between quite different bundles of resource uses. When society is deciding whether to build a new dam on a free-flowing river, it must choose between two different bundles of natural resource uses. With the dam there will be hydropower production, water storage, and lake recreation. Alternatively, the river, if not inundated by the dam, would continue to provide rafting and stream fishing. In this case a given area can generate only one of these two sets of natural resource services.

Determining which one of these mutually exclusive bundles of natural resource services to produce is a dilemma commonly faced by society and governmental agencies in charge of public lands. The same issues arise in decisions over wilderness designation. If an area of land is designated as wilderness, then it might provide, for example, backpacking, trophy elk hunting, trout fishing, protection of old-growth-dependent wildlife, and protection of water quality. If the same tract of land is logged, it can produce timber and provide road-oriented recreation and deer hunting but less trout fishing. The issue once again is which competing bundle of resource uses is of greatest benefit to society.

This book will present resource evaluation techniques and planning methods that allow comparison of the benefits of different combinations of multiple resource uses. The goal is to learn how to perform multiresource evaluations that identify the desirable alternative resource decisions in the face of these trade-offs.

The key word in all the preceding examples is trade-offs. Society cannot preserve all its forests and still satisfy current demands for lumber and firewood. But this does not mean we should not preserve any forests. Old-growth forests also produce services that many people in society find just as valuable as lumber. The goal is finding the most beneficial mix of resource uses. One of the challenges of integrated resource management is determining which land and water resources should be developed and which should be preserved.

Another challenge is to manage both the developed and preserved areas to maximize the overall benefits to society. The developed areas should not be exploited to the extent that society is giving up $100 worth of trophy elk hunting just to get another $50 worth of timber or livestock.

Defining What Is Integrated Natural Resource Management

To better understand the concepts embodied in integrated natural resource management we must define each part. We will start by defining the term *natural resources.*

What Are Natural Resources?

Natural resources include plants, fish, wildlife, and other living organisms as well as nonliving matter, such as minerals, water, and air. Often this distinction between living and nonliving resources is made by referring to *renewable resources* and *nonrenewable resources.* The key distinction is that renewable resources have the potential to replenish themselves, whereas nonrenewable resources are fixed and can only be depleted. The replenishment of renewable resources depends on the physical and biological environment. Natural limits impose a ceiling on the production from both renewable and nonrenewable natural resources in two ways. First, at any given time, a fixed amount of natural resources is available. Second, even renewable natural resources have a finite growth rate, which management can increase by only a limited amount. This contrasts with manufactured goods, where production can be increased several-fold over a few years.

People use individual or groups of natural resources as inputs to produce goods such as food, clothing, and housing. Groups of natural resources (even entire ecosystems) are often directly used by people when they sightsee in a scenic forest.

Most natural resources are *scarce,* by which we mean there is less available than people would like to use, either directly or indirectly. Both renewable and nonrenewable resources, such as wood, water, and even clean air, are scarce. (When water becomes very scarce, we often call the resulting condition drought.).

Scarcity is relative, however. What is or is not a scarce resource at any one time is constantly changing, because it is defined relative to people's use of resources. For example, bauxite was not a scarce resource until the process for manufacturing and demand for aluminum arose. Seawater is not currently a scarce resource in most places of the world, since, given our current technology, there is more seawater available than people want. If society develops an inexpensive method to desalinate seawater into fresh water, seawater may become scarce. Wilderness and free-flowing rivers were not considered a scarce or valuable resource in the 1600s. Today, with only a small fraction of the world's original wilderness remaining and humanity's increased desire for wilderness recreation, wilderness is a scarce natural resource.

Anthropocentric Versus Biocentric Views of Natural Resources

As the reader probably noticed, the preceding definition of *natural resources* is human centered: what matters for analysis is what people care about. Before we continue, it is important to acknowledge the major assumption made here. This viewpoint is often referred to as anthropocentric, or "man centered." But do resources really only matter to the extent people care about them? Given most (but not all) of our current environmental laws and public land legislation (reviewed in detail in the next chapter), the answer is yes. That people are the key is not too surprising. After all, they make and enforce resource decisions, not plants and animals. Thus the approach adopted in this book and for that matter most social sciences, such as economics, planning, or policy analysis is human centered.

This is not as narrow-minded as it might first sound. People can care about natural resources in many ways. For example, I enjoy eating salmon for dinner. I also enjoy seeing ducks fly across the sky, and I would not want to eat one. I care about condors, even though I have never seen one in the wild, nor likely ever will. But I get satisfaction from knowing that condors exist in the wild. I may not care much about brine shrimp. However, if I were told that they make up 50% of the diet of such shore birds as gulls, grebes, and snowy plovers, then I would have a "derived" concern for them because of my concern for the birds. In addition, I may care about a species because of a desire to protect them, so my children or some unspecified future generations can enjoy viewing them.

Although this human caring about species can be quite broad, there is clearly still an ethical judgment in believing that resources matter only if people care about them. This is not an overly restrictive view, as not all people must care about the resource for it to matter. If even a few people care a great deal about it, this value information is useful for resource management decisions. A market analogy may be useful here. Cauliflower has an economic value to some people, but many people never eat it and some even dislike it. However, enough people like it and are willing to pay the farmers' costs to make it a profitable vegetable for society to produce and consume. The same logic holds for grizzly bears, timber wolves, desert tortoises, or pupfish. If enough people derive satisfaction and enjoyment from the knowledge that these wildlife species exist, then they have a value to society. This is still a humanistic or anthropocentric view because it is centered on people only.

A legitimate alternative philosophy is called a biocentric or deep ecology view. Although *biocentric* means "biology centered," it is in many respects a different human view of how people fit in with nature. With this philosophy all creatures in the ecosystem are viewed as equally important. The biocentric view urges people to learn to work with nature, not subdue it for their own personal use. One goal is to achieve an earth where all biosystems can be sustained and where people live in balance with nature. A brief summary of the biocentric, or deep ecology, view is given in more detail at the end of this chapter (see "Another View").

At this point we should note that if most people's behavior became consistent with the deep ecology view, this would not alter the need for evaluating trade-offs in integrated natural resources management. Such a change in philosophy would change the relative values people place on different uses of the natural resources.

Since the techniques presented in this book are in large part based on people's preferences, the net result of more people behaving consistently with deep ecology would be more land and water resource preservation and less development. Until most people are willing to develop the preferences associated with deep ecology, realistic resource management strategies must recognize that what motivates most people's current and near future behavior is their concern for resources that they enjoy consuming, viewing, or simply knowing they exist.

However, the fact that people may not care directly or indirectly about a specific species does not mean we will ignore it when performing natural resource management or multiresource analyses. In general, enough people do care about biological diversity and maintaining viable populations of all species that countries have laws aimed at preventing extinction of species. For example, this concern about maintaining species is manifested in the Endangered Species Act of 1973. As we shall discuss in more detail in chapter 2, public land laws contain both explicit and implicit concerns about species or environments that we must incorporate into our planning and analysis. In the case of the Endangered Species Act, specific plant and animal species about which people may not directly be concerned still enter our analysis. They do so as a constraint that limits the range of our management actions to ones that do not threaten the continued existence of the species. Now that we understand how natural resources are defined and how this is influenced by human perceptions, we turn to defining the next part of integrated natural resource management, *management*.

What Is Management?

In the context of this book, *management* is the organization or coordination of natural resource uses. It is the human input of labor, capital (man-made objects such as machines and buildings), and most important, knowledge. Knowledge takes the form of information about the uses of natural resources that people care about and how strongly they feel about each use relative to the other. (Remember, we must make trade-offs between competing uses of a resource.) The knowledge also takes the form of methods to harvest or extract resources. Another important type of knowledge is information about how to enhance the growth of renewable resources, whether domesticated (e.g., cattle), cultivated (e.g., wheat), or wild (e.g., antelope). Lastly, knowledge involves understanding the interrelationships between how use of one resource influences other resources.

Management is also about coordination, control, and scheduling of resource uses. For example, the timing of livestock grazing by cattle ranchers on federal rangelands can be modified so that it does not interfere with hatching of ground-nesting birds in the spring. In addition, the livestock can be withdrawn from the federal rangelands early in the fall so as to leave enough vegetation for big game animals that will use those rangelands as winter range.

What Does Integrated Mean?

Integrated means ''unified or brought together''; thus integrated natural resource decisions are not made separately, in a piecemeal fashion. Since resource uses are usually linked together ecologically, we must make a comprehensive plan of action that considers how all resources will be affected by resource management decisions. Some of the resources will be directly changed by management actions (e.g., timber harvesting), and these actions will affect other resources (e.g., wildlife habitat).

Integrated Natural Resource Management Defined: Putting the Pieces Together

Integrated natural resource management is the process of organizing the different human uses of natural resources in such a way as to produce the greatest value of goods and services from those resources over a given period of time. A comprehensive plan or accounting of the direct or intended resource effects plus all the indirect or unintended resource effects must be described, and where possible, quantified and put into common units that allow comparisons between alternative uses of natural resources.

This is, of course, quite a challenge, given our current knowledge of ecological systems. Where the resource commitments cannot ever be changed once they are made (i.e., they are largely irreversible, as in damming a free-flowing river), such an exhaustive (and exhausting!) analysis is warranted. For less-long-lived and smaller-scale resource actions, the analysis must be tailored to the magnitude and value of the natural resources involved. However, treating each resource in a piecemeal fashion and ignoring the effects on other resources is what caused many past environmental problems and continues to generate many new ones.

Role of Planning in Achieving Integrated Natural Resource Management

Since one of the primary methods by which federal land management agencies determine their integrated natural resource management actions is through planning, it is useful to provide a few general planning principles. General definitions and principles of planning underlying the basic approach taken by all federal land management agencies follow. Chapters 9, 10, 12, and 13 discuss in detail the planning process for each of the four major federal land management agencies.

A Definition of Planning

Planning has many definitions, the shortest of which is a ''design or scheme of how to attain a given objective.'' Thus there are plans for building everything from houses to skyscrapers. But there are also plans for scheduling one's courses so as to meet graduation requirements within four years, for getting oneself elected governor, or for becoming financially independent. Thus a more complete definition might be "an integrated system of management that includes all activities leading to the development and implementation of goals, program objectives, operational strategies and progress evaluation" (Crowe 1984:1). This definition clearly states the key components of all planning: statement of objectives (what to attain), operational strategies (how to attain the objectives), and progress evaluation (whether the strategies are being implemented and are attaining their objectives).

Another way to look at planning that matches what federal agencies do in planning is to answer the following questions (adapted from Crowe 1984:3):

1. Where are we?
2. Where do we want to be (i.e., the ends)?
3. What alternative actions will best get us there (i.e., the means)?
4. Did we make it?

The first step involves an inventory and assessment of the quantity and condition of the natural resources the agency administers, including its own agency resources (i.e., personnel, budget, equipment, and so on). The second involves a determination of whether the resource conditions (such as wildlife populations, water quality, recreation facilities, and so on) are acceptable. Specifically, are the natural resources in the condition that Congress specified in public land legislation, as desired by the public, or as expected by a resource professional? If the answer is no, then step 3 is to develop alternative solutions to attain the desired resource conditions. In many respects planning is nothing more than structured problem solving. This view is useful to keep in mind as we move through the details of each agency's planning process.

This general approach to planning is sometimes called the *rational-comprehensive* approach (Garcia 1987; Cortner and Schweitzer 1983) and is just one of five approaches to planning (Garcia 1987). Rational-comprehensive planning is the most common approach used by federal land management agencies and will be the primary focus of this book.

How Integrated Natural Resource Management Can Solve Past Problems and Avoid New Ones

Many environmental problems today arose from failure to perform a comprehensive multiresource analysis prior to making decisions. In part, this stems from historic single-resource-use agencies or single-resource-use laws. For example, the allocation of rangeland forage to livestock with little regard to the needs of wildlife or erosion control partly resulted from the single-purpose Taylor Grazing Act of 1934 and its implementing agency, the Grazing Service. This act and agency had primarily one job: to allocate the available forage among competing ranchers and to exclude transient sheepherders (Clawson 1983:37).

Even though the multiple-use Bureau of Land Management (BLM) has replaced the Grazing Service, rangeland forage continues to be largely allocated to livestock. Focusing primarily on forage requirements of livestock and giving only secondary consideration to other resources have resulted in a deterioration of a majority of rangelands. In addition, inadequate forage has been available for big-game populations. It was not until the Natural Resources Defense Council lawsuit in 1974 that the BLM was forced to perform a thorough multiresource analysis of forage allocations.

Several other single-purpose laws, such as the Mining Law of 1872, resulted in the mining of certain mineral ores being given precedence over other resource uses. Often the value of the minerals gained was quite small compared to that of other resource uses given up. The resulting environmental destruction from uncontrolled and haphazard staking of claims and mining is well documented (Leshy 1987). Society will be working for many decades to repair the environmental damage resulting from maximizing production of one resource without concern for the consequences for related resources.

Integrated natural resource management can help us avoid these problems by explicitly including the effects on other resources in the decision-making framework. Some resource uses will always be forgone to gain others, but we want to make this an explicit choice involving the input of all members of society, not simply one group imposing its view upon the rest of society. In addition, environmental restoration costs need to be included in the analysis to ensure enough benefits in the proposed activity to repair the damage created by that activity.

An Example of Conflict Resolution Through Integrated Natural Resource Management

The techniques of integrated natural resource management can often be a conflict resolution device when attempting to balance competing natural resource uses. Often one resource agency mandated by the federal government to develop resources for producing commodities will find itself competing with another agency required by the same federal government to preserve that same environment.

Such conflicts can often be resolved by these agencies jointly performing or having a mutually agreed-upon third party perform an integrated natural resource analysis. This analysis would determine which uses of the natural resource are most valuable to society and how much of each use there should be. The agencies agree upon the evaluation criteria for determining the desirability of different alternatives and the measurement techniques or models to be used to measure the consequences of each; then they agree to accept the outcome. This can shift the conflict, so that instead of being an agency "turf battle" or a political test of wills, it becomes rational, goal-directed behavior. Often the agencies find they share much of the same decision criteria in terms of what they define as best. But the key lies in performing a comprehensive assessment of intended and unintended consequences of the resource action to calculate which is best overall.

A good example of using multiresource analysis as a conflict resolution device can be seen in the controversy surrounding increased hydroelectric production at Glen Canyon Dam. As shown in figure 1.1, this dam is located on the Colorado River, just upstream from Grand Canyon National Park. Water releases from the dam for hydroelectric production can dramatically change the flow of the Colorado River through Grand Canyon National Park. Uses of the river include some of the most-sought-after white-water rafting in the world and trophy trout fishing.

The Bureau of Reclamation, an agency within the U.S. Department of Interior, operates the dam. Unlike coal and nuclear power plants, hydroelectric dams can easily produce extra power when demand for electricity is at its peak. Such power is called peaking power and is more valuable to society than baseload power, since it can be produced when people most want it.

Releasing water to produce electricity just at peak times results in large fluctuations in the river through Grand Canyon National Park (GCNP). During off-

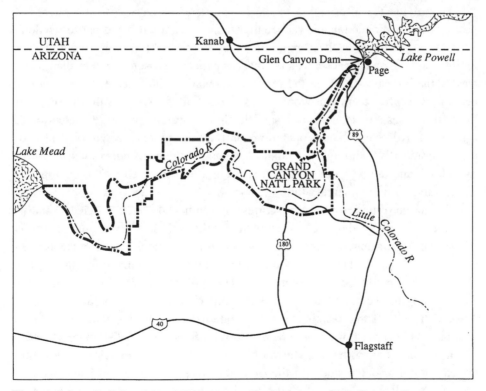

Figure 1.1.   Relationship of Hydropower Operations at Glen Canyon Dam to Colorado River through Grand Canyon National Park

peak electricity production hours, there may be too little water downstream for rafters and anglers. During peak electricity production hours, the volume of water in the river and velocity of the river flow would be ten times more than that when the flow is lowest. Such fluctuations are very disruptive to fish, anglers, and rafters. For example, at low river flows certain rapids become impassable because of exposed rocks. At very high flows, the waves in several of these rapids become so large that rafting is unsafe.

The Bureau of Reclamation had proposed increasing the peaking power fluctuations at Glen Canyon Dam so as to produce more peaking power. The National Park Service (also in the Department of Interior), which administers rafting in the GCNP, was very concerned (to say the least) about the Bureau of Reclamation's new peaking-power plans, as were the State of Arizona Fish and Game, which administers the trout fishing along the river, and the rafters and anglers. As it became clear that neither public concern nor related natural resource agency concerns were being addressed by the Bureau of Reclamation's engineers, resource analysts were brought in to deal with the controversy. The first question the head of the Department of Interior asked was whether the impact on all the downstream natural resources of increasing peaking-power operations was fully understood. The Bureau of Reclamation's answer was no.

To provide the information to the Department of Interior on the effects of peaking-power production and reduce the controversy, all affected parties (federal and state agencies as well as representatives of the public) were assembled as multidisciplinary analysis teams and advisory panels. In essence, rather than being part of the bureau's problem, the resource specialists and affected parties would become part of the team that would resolve this conflict. As a result, the scope of the analysis was broadened from just evaluation of peaking power to evaluation of a range of hydroelectric dam operations on the ecology and recreational uses of the Colorado River through the Grand Canyon. Multiyear studies on fish, birds, hydrology, and rafting were jointly designed and implemented by the interagency Glen Canyon Environmental Studies team.

The moral of this story is that serious environmental effects on GCNP and its resources were avoided by incorporating the results of these analyses into operation of the dam. The agencies worked jointly to optimize natural resource use at Glen Canyon Dam and Grand Canyon National Park rather than to maximize just hydroelectric power production. Partly in response to preliminary results from the Glen Canyon Environmental Studies, the U.S. Department of Interior announced in November of 1991 that peaking-power operations would be substantially reduced and more stable river flows would be provided (U.S. Bureau of Reclamation 1992).

It is worth noting that the Bureau of Reclamation's participation in this integrated natural resource study on the Colorado River has been consistent with its refinement of its mission. In 1989 the commissioner of the Bureau of Reclamation stated the agency had an expanded mission to be a "total resource manager," not just a water and power agency.

Integrated Natural Resource Management Requires an Interdisciplinary Approach to Planning and Management

True integrated natural resource management requires the blending of natural sciences with social sciences. Resource management must be consistent with the physical and biological forces in an area. The erosivity of soils, the length of the growing season, and the precipitation will strongly affect the amount of erosion from off-road vehicle use or the abundance of wildlife. Beyond a simple inventory of existing resources is the need to predict the biological response of the ecosystem to management actions. Predictions of the amount of soil erosion from clear-cutting, fish population responses to alternative river flow levels, and wildlife populations' response to alternative livestock grazing rates are all critical for assessing the consequences of resource utilization.

However, it is not enough to know the potential quantities of different natural resource uses the ecosystem can produce. Integrated natural resource management requires that we know which of these sometimes mutually exclusive natural resource uses society prefers in a specific geographic area. Combining information on natural sciences with social sciences, such as economics, can provide both types of information. Therefore it is critical that an agency use the combined expertise of those working in a wide variety of disciplines, including hydrologists, soil scientists, foresters, botanists, wildlife biologists, recreation planners, and economists. Most important, these individuals must work together as a team toward a common goal of balanced natural resource use. To do this they must agree on a common set of criteria for judging the desirability of different combinations of resource uses. In essence, they must share a common frame of reference.

This text provides such a framework and set of criteria that integrates the information from natural and social sciences into a comprehensive management framework within the bounds of existing natural resource laws and regulations. Although an anthropocentric view is adopted, which may be uncomfortable for some natural science disciplines, integration of natural and social sciences requires a common ground. Whether specialists draw their intellectual justification from laws, from the publics they serve, or from a concern about future generations, we are often forced to recognize that in one form or another, preferences of people do matter in public land management. People may not be all that matter, but it is they who influence soil erosion, water quality, fisheries, and wildlife. Recent history has shown us that ecologically sound plans that ignore the desires of people may never be successfully implemented. Voluntary compliance with the plan's regulations by people requires their support and understanding of the plan's goals.

Evidence of Public Demand for Integrated Resource Analysis in Public Land Management

In many ways the public increasingly perceives itself as an owner of many natural resources. Whether on public or private land, the public is asserting its implied rights under the Public Trust Doctrine for fisheries and wildlife protection. In essence, this doctrine assigns the rights to most fish and wildlife not to the land-owner, but to the citizens of the state. The public outcry over below-cost timber sales in many National Forests in the intermountain West and on the Tongass National Forest in Alaska has been reported in many newspapers and magazines.

The public is using the court system more frequently to scrutinize timber plans and mineral leasing. And the courts, more often than not, are upholding the public's interest in natural resources, especially when those resources are on publicly owned lands. For natural resource managers there is no turning back to the "good old days" of managing the trees and to heck with the people. It does no good to plan to graze cattle more efficiently in Wilderness Areas if the public does not want cattle there. Natural resource management no longer involves only the application of the principles of natural sciences; it requires trade-offs between competing uses of natural resources. Although a solid grounding in natural sciences is important, so is an understanding of planning, economics, and decision sciences.

This text provides the background in planning, decision sciences, and economics necessary to integrate a concern for people with one for natural resources on public lands. The first way in which people's concerns influence public land management is through laws and regulations. The next chapter provides a review of those laws and their implications for the planning and integrated natural resource management of the major federal land agencies in the United States.

The Appropriate Roles of Multiresource Analysis in Resolving Public Land Resource Conflicts

Although a multiresource analysis may identify which alternative resource management action is most socially desirable, it is important to remember that the *results of the analysis do not make the decision!* People make the decision—field-level managers of agencies, heads of agencies, governors, Congress, or even the president. The analyst is responsible for performing what is often called a *positive analysis.* This is a factual description of what will happen if action A is selected over action B. In essence, a positive analysis is a statement of "what is" or "what would be" if a particular choice is made. This compares with a normative statement of "what ought to be" or "what should be." This latter statement about what ought to be involves some implicit value judgment about what outcomes are preferred or what state of the world is desirable and what is not. Making these value judgments is the responsibility of agency heads or elected officials, not of the analyst.

In some cases the analysis will make the socially preferred alternative obvious. But sometimes analysis alone may not be not enough to identify the superior course of action. There are at least three cases when this is true.

First, determination of which alternative is most socially desirable requires agreement on *criteria* about which effects are desirable and which are undesirable. It also requires agreement on the relative importance of different criteria. For example, everyone may agree that both preservation of salmon fisheries and low-cost hydropower are important, but which should receive the greatest weight in the decision when these uses conflict with one another? The purpose of analysis in this case is to ensure that all the potential effects associated with each criterion have been identified, and quantified where possible. Then it is up to the agency or political decision makers to judge which criteria are most important.

Second, the analysis should clearly identify not only which alternative yields the greatest overall benefit to society as a whole, but how the benefits and costs are distributed. For example, a group of 1000 farmers in a particular watershed may grow a crop that currently requires a highly toxic herbicide. The runoff of this herbicide results in water pollution of rivers used by several cities for their water supply. It might cost a $100 million a year to remove the pesticides from the water. The cost per city dweller might only be $100 a year, whereas the gain in income from growing the crop might be $30,000 per farmer. If these were the only effects, the agency might be unwise to approve a permit for continued use of this herbicide: the farmers' gain is only $30 million (1,000 × $30,000), whereas the downstream cities' costs would be $100 million. But the farmers might argue the trade-off at the individual level: a loss of $30,000 per farmer's family versus a loss of $100 per urban family. The farmer would argue that a small loss to many is better than large losses to a few. As will be shown in chapter 6, an economic efficiency analysis of these data would suggest that the cities could pay the farmers to grow some other crop that did not require the herbicide. The point, however, is that neither the analysis nor the analyst can objectively state whether growing or not growing the herbicide-using crop is more fair or equitable. Greater concern for large losses to a few individuals over small losses to many individuals is inherently a subjective value judgment that is within the authority of elected officials, not the analyst. The primary contribution of the analysis in this case is to quantify accurately the costs and benefits as well as to identify who bears those costs and receives those benefits.

Lastly, some individuals or groups may have near veto power over which alternatives are politically feasible. The use of low-sulfur coal as an inexpensive means of cutting acid rain from sulfur emissions is a good example. Numerous analyses showed that it was much cheaper for power plants to burn western coal with a low-sulfur content than to install and operate stack scrubbers to remove excess sulfur from high-sulfur eastern coal. But the incentive to use western low-sulfur coal was continually weakened by a few eastern senators, led by former Senate majority leader Robert Byrd of West Virginia. Byrd did not want to see a reduction of coal mining jobs in West Virginia, which would occur as western low-sulfur coal was substituted for high-sulfur eastern coal, much of which is mined in his home state of West Virginia. It took a change of Senate leadership to implement what the analysis had shown to be an overwhelmingly large cost savings for use of low-sulfur coal. In this case the analysis plays the important role of informing about the high cost to society of Senator Byrd's position. The analysis was there waiting for national leadership to break the near veto power of a few politically powerful but parochial interests.

Although a comprehensive or integrated analysis will not make the decision, it often will influence the nature and scope of the final decision. The influence of the analysis may not always be readily visible or instantaneous, but a thorough analysis often has several important effects. An integrated analysis will often lay out the true costs and benefits, thereby stripping away self-serving claims of extraordinary benefits by proponents and extraordinary costs by opponents. The analysis may reveal that a hotly debated issue may not be a substantive issue. A long-lasting effect of a comprehensive analysis is that it often changes the way the decision makers and the public view and understand an issue. This can help avoid unnecessary conflicts in the future.

Why Should People Interested in Natural Resource and Environmental Management Study Federal Lands?

As the introduction of the next chapter will detail, *federal lands alone make up nearly one-third of all the land in the United States. Therefore the management of these lands can have a profound impact on local, national, and global environmental quality.* In some western states, public land represents over two-thirds of the land area of the state. In these states, federal land management decisions have a pervasive influence not only on the environment of the state but also on its economy. The four main federal land-managing agencies that will be the primary focus of this book manage over 90% of all federal lands and over 25% of all lands in the United States.

Although the primary focus in this book is on federal lands such as National Forests, Parks, and Wildlife Refuges, the framework, principles, and techniques are equally applicable to state and county forests, parks, and wildlife areas. Many of the principles presented in chapters 3 (rationale for public ownership of land), 4 (criteria for public land management), 5 (modeling), 6 (benefit-cost analysis), 7 (input-output models), and 8 (linear programming models) are applicable to and used by other smaller federal land management agencies, such as the Army Corps of Engineers and the Bureau of Reclamation. Most of the principles and techniques discussed in chapters 3–8 are also applicable to management of state lands, such as state forests or state parks. There is considerable flow of management techniques and information between state and federal public lands. In some cases states such as California have adopted analysis and planning techniques very similar to what the Forest Service uses. Thus understanding the techniques used to perform integrated natural resource management of federal public lands will be of great assistance in providing more balanced natural resource management on similar types of state and county lands.

Another View: Biocentric or Deep Ecology View of Natural Resources

The idea that people are just a part of the natural order is an old one and can be found in the writings of Henry David Thoreau, John Muir, and Aldo Leopold. To many of these thinkers, all species are citizens of nature. All organisms have a right to live. Certainly survival requires that some individual organisms must die so that others may live. But each species' consumption is limited to the needs of survival, not accumulation of resources beyond one's needs. The emphasis is on being in balance with nature. The appropriate goal is a sustainable environment that can meet the needs of future generations as it met those of the current generation.

Biocentrism, or deep ecology, is a more recent label given to the view that people should live in harmony with nature. The goals of deep ecology are to preserve the integrity, stability, and diversity of the ecosphere. Because ecosystems are at the same time complex and self-regulating, attempts to manipulate or control the ecosystem are likely to fail over the long term. It is this concern with the very long-term view (in terms of centuries, not decades) that requires us to pause in our current rate of development. Deep ecologists believe the current rate of development is not sustainable, because the world's entire stock of resources is being depleted by one generation, with little left for the future.

Deep ecology is a movement of hope, not despair. Adoption of the philosophy of deep ecology should not only begin with individual life-styles, but also be advocated at the national and international level. At the individual level, each person should live according to the theme of voluntary simplicity. Resources need not be rapidly depleted if everyone lives simply. People's material *needs* are few, relative to their current wants. With an eye toward more equal distribution of resources among current members of the society and a more equal distribution to future generations, the following saying sums up much of the deep ecology movement: "Live simply, so that others may simply live."

2

Laws and Agencies Governing Federal Land Management

Size and Scope of Public Lands in the United States

Although most of the human population lives in urban and suburban areas, urban land uses make up only a small fraction of total land use in most countries, including the United States. Table 2.1 illustrates that urban land uses make up only 2% of the land area in the United States. By comparison, agriculture uses 17% of the land area. However, forests and rangelands are the dominant land types in the United States, representing nearly two-thirds of the U.S. land area. As table 2.1 illustrates, natural resource management on rangelands and forests (about half of which are publicly owned) is a substantial undertaking. Because of the dominance of these two land types, forest and range management are likely to have a profound influence on the environment.

The federal government owns nearly one-third of all the land in the United States. Table 2.2 illustrates the distribution of federal land ownership in the United States. The many reasons for this land ownership pattern include national defense (military lands), preservation of migratory and other wildlife (wildlife refuges), and protection of special areas of scenic or historic importance (national parks, monuments). Each of these special land use designations will be discussed in greater detail later. The bottom line of table 2.2 is that public land management involves land use decisions on one of out every three acres in the United States and in all fifty states. As figure 2.1 illustrates, resource management decisions on public lands are even more important in western states, where over half the land is federally owned. (In such states as Alaska and Nevada, this percentage rises to as much as 80%.) Even in many eastern states, where the federal government had to acquire land through purchase, the percentage of federal land is frequently 5% to 8% of the state.

Table 2.1
*Categorization of Land Uses in the
United States*

Land Use	Acres (Millions)	% of U.S.
Urban/suburban	45	2.0
Agriculture	383	17.0
Forests	731	32.4
Rangelands	770	34.2
Other lands (such as deserts, tundra)	325	14.4
Total	2,254	100

Table 2.2
Federal Landownership in the United States

Agency	Acres (Millions)	% of U.S.
National Forests	192	8.5
Bureau of Land Management	270	11.9
Fish and Wildlife Refuges	90	4.0
National Park System	80	3.5
Other Dept. of Interior	7	.3
Military and other federal	34	1.5
Total	673	29.7

SOURCES: BLM Public Land Statistics, 1989; National Park Service, 1987:13.

The agencies that manage these large areas of federal lands are major enterprises, rivaling some multinational companies in terms of resources, employees, and budgets. Because we discuss these agencies in more detail later, it is useful to keep in mind that both the Forest Service and the Bureau of Land Management (BLM) would easily qualify as Fortune 500 firms in the U.S. economy (Clawson 1983:11). Boskin and colleagues (1985) estimate that rural federal lands themselves are worth $63 billion and that onshore minerals associated with some of them are worth $146 billion. Together the Forest Service, BLM, and Fish and Wildlife Service employ about forty-six thousand resource professionals and support staff. The Forest Service has a budget of about $2 billion, the BLM's budget averages around $650 million, and the Fish and Wildlife Service's budget averages about $500 million annually. One of the key questions of public land management is whether the American taxpayers are receiving a good return (broadly speaking, not necessarily in a financial sense) on the funds spent by these natural resource agencies.

The BLM and Forest Service provide significant quantities of commodities to the U.S. economy each year. BLM and Forest Service land produce about one-quarter of the lumber consumed in the United States. The public lands provide about 150–160 million barrels of oil and 60 million tons of coal annually (Clawson

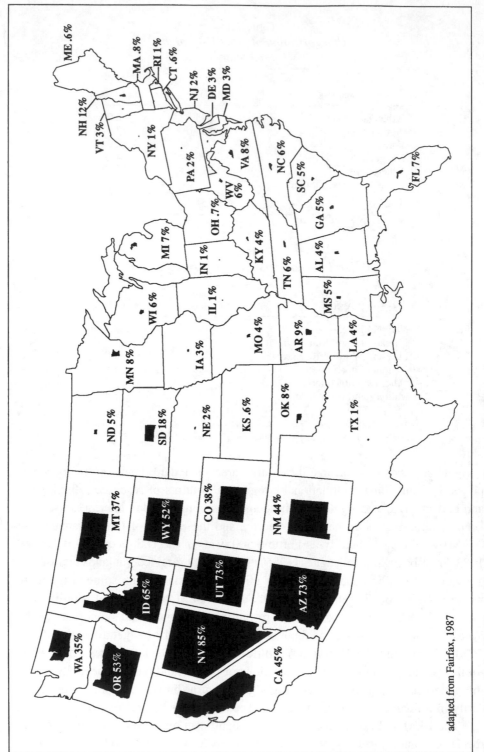

Figure 2.1. Federally Owned Lands Shown as a Percentage of State Area

adapted from Fairfax, 1987

Table 2.3
Recreation Use on Federal Lands

Agency	Visitor Days (Millions)
Forest Service	242
Bureau of Land Management	38
Fish and Wildlife Service	7
National Parks Service	115
Total	402

SOURCE: Developed from U.S. Army Corps of Engineers, 1990:2.

1983). National Forests provide about two-thirds of the U.S. production of certain minerals, such as lead and molybdenum (Forest Service 1989:53). The public grazing lands support 7% of the U.S. livestock grazing, although this percentage rises to 34% in the southwestern United States (Joyce 1989:48).

Public lands also affect the quality of life for hundreds of millions of people throughout the world. The four main land-owning federal agencies shown in table 2.3 provided about 400 million visitor days of recreation in 1988. These federal land agencies provided about two-thirds of all federally provided recreation (with the U.S. Army Corps of Engineers and Bureau of Reclamation being the other main federal suppliers). The recreation provided by these four federal land agencies represented about 40% of all public (federal and state) rural (i.e., nonurban) recreation in the United States in 1988. For some recreation activities such as nonresidential wildlife observation, the National Wildlife Refuges alone account for 16% of total use, with the other federal lands accounting for an additional 22% (Flather and Hoekstra 1989:55). In addition to recreation, the quality and quantity of water flowing to many cities in the western United States depend upon land management practices of the Forest Service. The last remaining habitat for hundreds of wildlife and fish species occurs on federal lands in National Forests, Wildlife Refuges, National Parks, and lands administered by the BLM. As one can see, management of federal lands is a substantial part of overall natural resource management in the United States.

A Brief History of the Evolution of Federal Land Ownership Patterns with Implications for Management Today

Acquisition Era

Nearly all the land of the original thirteen colonies was privately owned. Only in the 1900s has significant amounts of land been retained under federal ownership as National Forests and National Parks.

The westward expansion of the United States came through several federal government actions, including the Louisiana Purchase, Oregon Compromise, the Mexican cession, and the Alaskan Purchase. These major purchases plus numerous other smaller acquisitions added 2.313 million acres to the United States.

Disposition Era

The federal government's intention was to transfer the land it was acquiring to private citizens, who would then cultivate or develop it in one form or another. To facilitate the orderly transfer and development of these acquired lands, the federal government established several acts, the imprint of which is still very evident on maps today. The resulting mixed public and private ownership is the source of considerable conflict in land management.

Homestead Act.   The most famous land disposal act was the Homestead Act of 1862. This act allowed citizens to obtain, for a minimal fee, up to 160 acres of land under the condition that they live on it and farm it. As westward expansion to less fertile soils progressed, it became apparent that more than 160 acres was needed to form a viable farm or ranch. The acreage limit was raised in later acts to 320 acres in 1909 and 640 acres in 1916 (Clawson 1983:23). Before the formal close of homesteading in 1976, 288 million acres, or 12% of the entire western land acquisitions, were transferred to individuals.

Transfers to States.   The largest transfer of federal lands was to the states that were formed within the federal territories and acquisitions (such as the Louisiana Purchase). The federal government granted a total of 328 million acres, or 14%, of the federal lands to states. Nearly half of this went to midwestern and southeastern states. The eleven western states received 71 million acres, a great deal of which is in a checkerboard pattern associated with "school sections." In general, two different sections of land in each large block of land, called a township, were transferred to most western states. These school sections were to be managed by the state to provide revenue for public education. Additional lands granted the states included those for the establishment of agricultural colleges (hence the label *land-grant college* in many western states [Clawson 1983:23; Dana and Fairfax 1980:19]). Alaska received grants of federal lands totaling more than 100 million acres. This represents 32% of all grants of federal lands to the states.

Transfers to Railroads.   Another major disposal of federal lands that continues to influence modern land management was that to railroads. To encourage railroads to build and hence "open up the West," the federal government made grants of alternating sections of lands on either side of the railroad. If some of these sections were unavailable, "in lieu" sections farther away from the rail line could be substituted. In total, 94 million acres, or nearly 10% of all disposed-of lands, were transferred to railroads, again in a checkerboard pattern.

Continued Management Conflicts from the Pattern of Disposition

The legacy of the intermingled private homesteads and state lands in public land states influences land management today in many ways. Typically, the fertile valley and river bottoms in many western states are privately owned, whereas the surrounding high country remains publicly owned. However, by controlling the valley bottoms, private landowners frequently control and therefore limit public access to the public lands above their property. Assuring public access to public lands is a significant challenge to land managers in the western United States. In addition, commodity production activities or intense development on private lands can negatively affect wildlife habitat, wildlife migration, or recreational enjoyment of the adjacent public lands. Examples include the building of a major religious headquarters and settlement on land adjacent to the Gallatin National Forest and Yellowstone National Park in Gardiner, Montana.

However, the conflict sometimes works in the other direction. Activities on public lands can spill over to private land. Elk do not recognize the differences between public and private lands, and an unusually cold and snowy winter can push them down to the private valley and river bottoms in search of food. Elk grazing of private hay and alfalfa fields adjacent to public land is also the source of much conflict between private landowners and public land managers. Finally, intense commodity production on public lands can seriously detract from the aesthetic qualities on adjoining private lands. This too is a source of conflict between adjacent landowners and public land managers.

The intermingled tracts of state and retained federal land often make land management difficult for both governments. Much state land is associated with school sections, mentioned earlier. Consequently, state land managers are often charged with maximizing revenue from these school sections. Therefore commodity development on these lands is often encouraged. However, Congress may have directed the federal agency to manage the surrounding federal land for other uses, such as wildlife, wilderness, or recreation. Again conflicts develop when the state wants to allow road construction to its state lands, but such roads must cross wilderness areas or critical wildlife habitat. Dealing with these conflicts is another challenge for land managers.

In many western states, railroad sections contribute to the problems associated with checkerboard land ownership mentioned earlier. In some states, railroads have successfully swapped lands so that they have consolidated large tracts. These lands are managed to maximize profits. Such business objectives may be incompatible with management objectives of surrounding federal land. In Montana, for example, intensive timber harvesting on Burlington Northern Railroad lands so fragmented elk habitat and increased sedimentation in trout streams that the U.S. Forest Service had to cancel its planned timber sales to maintain compensatory habitat on the National Forest.

To offer a preview of the federal agencies and departments involved in public land management it is worthwhile to mention that much of the land disposal activity was handled by the General Land Office that was a part of the U.S. Department of Interior. As the name implies, the Department of Interior was the federal department that dealt with the "interior," or domestic, concerns of the United States as opposed to the Department of Defense (defense against foreign nations) or Department of State (foreign relations). The importance of the Department of Interior with respect to federal land management will become more apparent as the chapter progresses.

Selected Retention of Federal Lands from 1870 to 1900s: Emergence of National Forests and National Parks

The lands remaining in federal ownership today are largely a product of two factors. Some of them reflect an explicit federal action to withdraw the lands from consideration under the various disposal acts discussed earlier. Such lands include National Forests and National Parks. Much of the remaining federal lands (particularly those arid western lands largely administered by the Bureau of Land Management) were literally the "lands nobody wanted" specifically, lands that, as of 1946, neither homesteaders nor railroads or other groups were interested in acquiring.

We want to trace briefly the history of the origin of these explicit decisions by the federal government to retain ownership of its land. The main focus, however, is not to dwell on this history but to understand the motivations for retention of these lands and the resulting laws that govern their management. The history is important to provide the context for how these lands were intended to be managed. In addition, the early years of federal management of these lands have left a legacy that today's land managers must deal with. Early uses of many National Forests and Parks, such as horse pack trips, summer cabins, roads into fragile areas of National Parks, and grand lodges in Yosemite, Yellowstone, Grand Canyon, and other National Parks, have become in the public's mind part of the park experience. However, these established uses add to the difficult balancing act in the increasing emphasis of the National Park Service on maintaining naturalness of park environments.

Reasons for Initial Federal Land Retention Efforts

Three key factors led to the beginning of retention of some land in federal ownership. First was a concern over the abuses and fraud associated with the different land disposal programs. Second was the desire to preserve some of the unique scenic and geologic wonders of the United States, such as Hot Springs National Park in Arkansas in 1831 and Yellowstone National Park in 1872. Third, and perhaps most important, was the growing perception of the shortsighted, destructive influence of human activity on the land and the potential for shortages of critical natural resources such as timber.

People had watched as those with the "cut-out–get-out" mentality had clear-cut several forests, first in the Northeast and the South, then in the Great Lakes states; and the West appeared to be next. In the economy of the 1800s, wood was a vital natural resource for building, heating, and cooking; the forests also provided game animals for meat. People also perceived that water supply and flood control were linked to forest management. Cutting over of lands without reforestation left bare soils that were more prone to landslides. Such bare ground also caused water runoff to be concentrated in floodlike surges rather than in the more gradual runoff of water from natural forests. During this same period, the concept of scientific management of forests was being transplanted from Europe to the United States.

The publication of George Perkins Marsh's book *Man and Nature* in 1864 drew attention to the destructive influence of man and the importance for maintaining a balance with nature. John Muir's writings planted the idea of preserving natural areas as a place of respite for the "overcivilized," harried masses of the industrial revolution.

In these seeds of discontent over timber/water management and protection of nature, two significantly different responses began to emerge: National Parks and Forest Preserves.

National Park Reservations.   The first response was to preserve federal land as National Parks (e.g., Yellowstone in 1872). This response took the views of resource management associated with John Muir: preservation and protection, not utilization. This view permeated many discussions of federal land retention and management in the 1870–90s. In 1890 the federal government set aside Big Tree National Park (what is now Sequoia National Park) and General Grant National Park (both in California) and took back Yosemite from the state of California to make it a National Park. The land preservation movement was certainly alive and well, and in some people's minds, federal reservation of land meant federal preservation of land, with no commodity utilization allowed.

Origins of Efforts to Establish Forest Reserves.   In the 1870s, the second response to the growing concern over timber and water led to some congressional interest in introducing bills to appoint a commission to study utilization of forest resources. After little success in passing a separate bill, an amendment to the U.S. Department of Agriculture's (USDA) appropriation bill was approved to establish a Division of Forestry in USDA (Dana and Fairfax 1980:50). This inauspicious beginning set the stage for the future administration of the U.S. Forest Service (and later the National Forests) within the Department of Agriculture. The first director of the Division of Forestry in his required study for Congress "strongly endorsed the concept" of reserving some forestlands in federal ownership (Dana and Fairfax 1980:51).

The Division of Forestry's activism for retention of forestlands and emphasis on scientific management of forests accelerated with the appointment of Bernhard Fernow as chief in 1886. Fernow was trained in Germany as a professional forester and spent several years there practicing his skills. He adopted the principles of German forestry, which held that wood was scarce and necessary for life. Consequently, his view of forestry was far more utilitarian than preservationist. When Fernow became chief of the Division of Forestry, he infused these utilitarian ideas and his training on the division. Some authors question the appropriateness of these views toward scarcity of wood in the United States in the 1890s (Dana and Fairfax 1980:53), but this German viewpoint toward the objective of forestry continues to have some influence on many foresters even today.

Fernow's reorientation of the Division of Forestry based forest management on scientific principles, considered economic factors, and used these principles in forest regulation and management. These factors were woven together into the concept of providing a sustained yield of wood for society. Specifically, sustained yield involves harvesting only an amount each year that is equal to the annual growth. The resulting level of harvest can then be sustained indefinitely into the future. Their emphasis was on wise use of resources, not preservation.

In addition to establishing the Division of Forestry, some congressmen dealt with concerns about forest and water management by withdrawing forestlands from disposal and retaining them in a reserve. For fifteen years, starting in 1876, bills were introduced each year to establish some form of Forest Reserves. As an individual bill, none of the many versions ever became law. Surprisingly, the power to create a system of Forest reserves emerged as a section of a conference committee report that settled differences between the House and Senate versions of general land law reforms. This section become known as the Forest Reserve Act of 1891 (Dana and Fairfax 1980).

Forest Reserve Act and the Birth of the National Forests

The Forest Reserve Act of 1891 gave the president of the United States the continuing authority to reserve "public lands wholly or in part covered with timber or undergrowth, whether of commercial value or not, as public reservations" from homesteading and private entry. The withdrawn lands were administered by the Department of Interior (not the Division of Forestry in the Department of Agriculture).

President Harrison first used this authority to create the Yellowstone Forest Reservation adjacent to Yellowstone National Park. In the next two years, Harrison made fourteen additional reservations in the West, resulting in 13 million acres of Forest Reserves (Dana and Fairfax 1980:58). By 1893 President Cleveland added 4.5 million acres, bringing the total to 17.5 million acres.

Dana and Fairfax claim there was confusion over the initial purposes of the Forest Reserves: were they to be "used" or preserved? The Department of Interior felt they were to be preserved (e.g., withdrawn from all uses [Dana and Fairfax 1980:58–59]). Dana and Fairfax (1980:65) as well as Miller (1973:290–298) contend that Congress may have intended the Forest Reserves to be "parklike preserves." In 1895 a "carefully wrought compromise to authorize the Secretary of Interior to regulate limited use of the reserves without altering their basic purpose, which was watershed protection," passed both houses of Congress, but the absence of the bill's sponsor from the conference committee charged with resolving differences between the House and Senate bills resulted in the bill's death (Dana and Fairfax 1980:59).

As efforts continued to define appropriate uses of the Forest Reserves, President Cleveland, in one of his last official acts, established thirteen new reserves that more than doubled the acreage of the Forest Reserve system to 39 million acres. The size of the new withdrawals, the lack of consultation with affected states, and the fact that these new reserves actually included existing towns and potential agricultural land brought an outcry from all levels of government.

To deal with the issue of purpose of the Forest Reserves and the process for creating and revoking Forest Reserve designation, Congress and the new president (McKinley) passed what is known today as the Organic Act of 1897.

1897 Organic Act Establishes the Purpose of the Forest Reserves

The Organic Act of 1897 stated the purposes of Forest Reserves and authorized the secretary of the interior to establish rules for use of these reserves. These purposes were (1) to preserve and protect the forest within the reservation; (2) to secure favorable conditions of water flow; (3) to furnish a continuous supply of timber for the use and necessities of the people of the United States (Dana and Fairfax 1980:62). Two themes are evident here. First, the confusion over preservation versus utilization is answered with utilization. Thus Forest Reserves were not synonymous with National Parks. This is an important distinction that carries through to management of these two different types of federal land today. Second, the act did not spell out broad multiple use; in fact, it emphasized just two key resources: timber and water. The act provided for timber harvesting but specified harvesting of only "dead, matured or large growth trees." These trees must be "marked and designated." Interpretation of these conditions in the 1973 *Monongahela* court case determined that clear-cutting was inconsistent with the Organic Act.

Implementation of this act established the basic structure of National Forests, with each Forest Reserve having a forest supervisor and under him a group of forest rangers. However, the administration of the lands was based in the General Land Office in the Department of Interior. However, most of the country's few trained foresters were in the Division of Forestry in the Department of Agriculture.

Expanding into the Present-day National Forest System

Transfer of Forest Reserves to Department of Agriculture

Although many conservation-related events of interest took place during Theodore Roosevelt's tenure as president, two are of particular importance to Forest Reserves. First, the alliance of Gifford Pinchot, the new head of the Division of Forestry in the Department of Agriculture, with Roosevelt allowed Pinchot to attain one of his long-term goals: to transfer administration and management of the Forest Reserves to the U.S. Department of Agriculture. Pinchot was another German-trained forester in a mold similar to that of the former head of the Division of Forestry, Bernard Fernow. Pinchot argued that the Division of Forestry had the forestry expertise lacked by the Department of Interior's Land Office and that the Department of Agriculture was a more natural home for management of the Forest Reserves. In 1905 Pinchot's goal was attained with the Transfer Act, which transferred management of the Forest Reserves to the Department of Agriculture.

Secretary of Agriculture Wilson sent Pinchot a letter (believed to be written by Pinchot himself [Dana and Fairfax 1980:82]) detailing the principles of management to be followed. These included the following:

1. Wise long-term use of the reserve so as not to conflict with their "permanent value." The Forest Reserves were to be devoted to their most productive use.
2. Key resources to be managed were water, wood, and forage.
3. Local, decentralized management of the resources.
4. A concern to accommodate industry and to prevent abrupt changes in management that would be detrimental to industries. Along with this charge was the instruction to use scientific management principles carried out by professionally trained foresters in managing the Forest Reserves.

In 1907 the Agricultural Appropriations Act made both cosmetic and substantive changes to the management of the Forest Reserves. The cosmetic change involved changing the name the Forest Reserves to the present one of *National Forest*. The more important change for long-term county–state–federal relationships involved Congress dedicating 25% of the gross timber-harvesting receipts from the forests to county schools and roads. This created an incentive for local counties to press for high timber harvests to increase local funding, regardless of the merits of cutting more timber.

Acquisition of Eastern Forest Lands

It was not until the Weeks Act of 1911 that the National Forest system became truly national. In the interest of watershed protection on deforested private lands in the east, the Weeks Act authorized purchases of private lands for National Forests. Most of the National Forests in the eastern United States, from the Applachians in the south to the White Mountains in the north, originated from purchases related to watershed management under the Weeks Act. The Clark-McNary Act of 1924 broadened the Weeks Act to allow the Forest Service to purchase private lands that were potentially valuable as timberlands. Acquisitions under the Weeks and Clark-McNary Act added 20 million acres of eastern lands to the National Forest system. The last significant additions to the National Forest System in the continental United States came in the late 1930s with the purchase of the Depression-era "dustbowl" prairie lands now, called National Grasslands (Culhane 1981:48).

In the continental United States, these laws set up the National Forest System as we know it today. With the land base in place, the Forest Service began the challenge of managing these lands to meet growing and changing demands for natural resources. We shall return to the agency's movement toward what we now call multiple-use management after we discuss the next few laws and organizational structure of the Forest Service.

One of the acts that was intended to improve the management of the National Forests was the Knutson-Vandenberg Act. This act required individuals and firms harvesting timber from National Forests to deposit money with the Forest Service for reforestation of the cut-over lands. This funding mechanism has received increased scrutiny as a result of O'Toole's (1988) hypothesis that it is the incentive to maximize the amount of money in this fund that drives the Forest Service to offer large volumes of timber for sale, even if the returns to the Treasury are negative.

In part to foster community stability in towns dependent on National Forest timber, Congress passed the 1944 Sustained Yield Forest Management Act (Dana and Fairfax 1980:167). This act permitted combined management of private forests and National Forests on a sustained-yield basis. The issues of sustained yield and community stability continue to influence forest management even today.

Organizational Structure of the Forest Service

In much of the following discussion it is worth remembering the basic organizational structure of the modern Forest Service. At the top is the national office of the Forest Service in Washington, D.C., where the chief of the Forest Service and associated staff for budgeting, planning, and legislative affairs are located. The next tier down is the Regional Office. The Forest Service has nine regions, shown in figure 2.2. The Regional Office serves as a link between the Washington office and the individual National Forests for policy, budget, and general administration. Each Regional Office is headed by a regional forester, who oversees several forest supervisors. The forest supervisor is specifically responsible for an individual National Forest. The forest supervisor is located at the forest headquarters and has a staff that performs planning and budgeting for the forest itself. This staff also provides technical expertise and assistance to the ground-level managers (called the district rangers). It is at the district ranger level that most recreation visitors, livestock users, firewood cutters, and so on generally interact with Forest Service personnel.

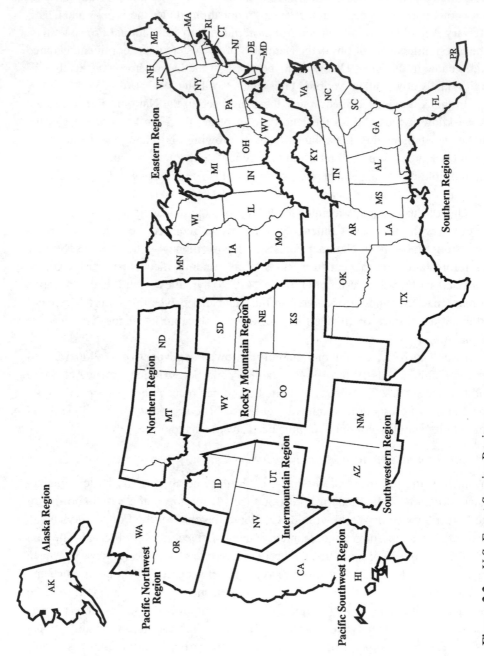

Figure 2.2. U.S. Forest Service Regions

Management of National Forests Since the 1950s: Laws, Social Forces, and Trends

The Forest Service Moves Toward a Multiple-Use Agency

Recreation as a Use of the National Forests

Although the Organic Act of 1897 emphasized that the purposes of the National Forests were protection of watersheds and supply of timber, the Forest Service also had a long interest in recreational use of the National Forests. An early indication of this interest is the 1918 Forest Service report entitled "Recreation Uses on National Forests." The report discussed the current and potential recreational uses of National Forests.

Recreational use of National Forests grew during the 1920s. However, the Forest Service was repeatedly unable to get funding for recreation facilities and management, because Congress and the newly formed National Park Service felt funding for recreation was appropriate for National Forests.

Primitive Recreation and the Emergence of Wilderness Land Use

Recreationists, made mobile by automobiles and a growing network of roads within the National Forest, were soon everywhere in many forests. Some foresters with an ability to look ahead, such as Aldo Leopold in the Southwest and Arthur Carhart in the Northeast, soon saw the need to maintain parts of their respective National Forests as roadless or wilderness areas for primitive forms of backcountry recreation (Dana and Fairfax 1980:132).

In 1929 the Forest Service issued what it called L-regulations, which established a land classification called primitive areas. These would remain without roads and such improvements as cabins and other buildings. These areas were to provide for packhorse- and backpacking-type recreation. Special management plans were required to ensure that the areas remained in their natural state. Regional foresters had the authority to identify and mark lands as primitive areas. According to Dana and Fairfax (1980:133), the permanence of these administrative designations was unclear, even to Forest Service personnel. It was also not certain if other forest uses, such as grazing, could occur. In general, there was considerable uncertainty about what this new land classification permitted and precluded.

Ten years later, when the Forest Service issued the U-regulations, the agency was far more specific about wilderness land classifications and management recommendations. These U-regulations created three new land classifications, called wilderness, wild, and recreation (Dana and Fairfax 1980:157). The wilderness and wild lands were to have no roads, timber harvesting, or motorized access, but existing grazing was allowed to continue. The primary distinction between these two categories was that the wilderness label applied to areas greater than 100,000 acres; the wild areas were those with less than 100,000 acres. Recreation lands would be kept in natural condition, but limited roads and timber harvesting would be permitted. Wilderness areas were to be designated by the secretary of agriculture; wild lands were to be named by the chief of the Forest Service (Dana and Fairfax 1980:157).

Growth in Demand for Forest Resources: Post–World War II Era

Twin Forces of Increased Commodity Demand and Rising Recreation Use

Many studies of public lands indicate that two forces had a dramatic effect on the nature and intensity of public land management: increased commodity demands from the National Forests and increased recreational use of the National Forests (Culhane 1981:51; Bowes and Krutilla 1989). Rising demand for wood for the postwar housing boom coupled with decreasing inventory on private lands resulted in increased pressure to harvest more timber in National Forests. The Forest Service began to accommodate some of this increased demand by moving from its custodial management of the forests to intense production of timber from National Forest lands. Timber sales that had rarely exceeded 1 billion board feet during the 1920s through 1930s were increasing nearly 50% each year, year after year. After World War II, the timber cut from National Forests went from about 2 billion board feet to 12 billion board feet by the mid-1960s (Clawson 1983:75).

At the same time, recreational use of National Forests was rapidly increasing. Recreation represented both an opportunity and a conflicting land use. Although opportunities for roaded recreation increased with timber harvests, many recreation activities, such as backcountry camping, hiking, and fishing, were adversely affected by timber harvesting. Clear-cutting of recreationists' favorite areas in what had been undeveloped areas intensified the conflict. Wilderness interests were concerned that the Forest Service would strip away wilderness or wild designations from existing areas to permit logging.

Because of the different groups clamoring for more timber, more recreation, and more wilderness, the Forest Service found itself in a position that was new to it: not all interests could be satisfied. No longer could all the demands of each group be met by "implicitly zoning" the forests into areas for recreation, wilderness, and timber harvesting. The growing timber demand would soon require that primitive areas that had been set aside administratively under the U-regulations be opened up for timber harvesting. However, the Forest Service had a legal mandate in the Organic Act of 1897 only for timber and water, not specifically for recreation and wilderness.

Broadening the National Forest Mandate to New Purposes: Multiple Use, Sustained Yield Act (MUSY) of 1960

The Forest Service sought congressional authorization for recreation and to provide the agency with the authority to balance competing demands for uses of the National Forests. The Forest Service answer to both these problems was the concept of multiple use of forestlands. Although this concept will be defined in detail later, its essence is captured in the Forest Service slogan: "Land of Many Uses." Dana and Fairfax (1980:201) state that the Forest Service basically wrote the MUSY Act and then lobbied for its passage by Congress.

This law has several important characteristics. First is the definition of uses permitted on National Forests: outdoor recreation, range, timber, watershed, wildlife, and fish. Although this broadening of purposes was critical to the Forest Service, traditional users, such as timber interests, insisted that the MUSY Act was a supplement to and not a reversal of the original purposes for which the National Forests were established in the Organic Act of 1897. Thus timber and water, although equal, remained the historic priorities of the Forest Service.

The second major part of the MUSY was the notion of balancing competing uses in relation to their relative values (Culhane 1981:53). This recognized that not every use could be supported on every acre and that a determination of which areas were best suited to which uses could be made. This is the key concept underlying the statement that National Forest lands are managed for multiple uses.

Third, the act formally defined multiple use as

> the management of all the various renewable surface resources of the national forests so that they are utilized in the combination that will best meet the needs of the American people; making the most judicious use of the land for some or all of these resources over areas large enough to provide sufficient latitude for periodic adjustments in use to conform to changing needs and conditions; that some land will be used for less than all of the resources; and harmonious and coordinated management of the various resources, each with the other, without impairment of the productivity of the land, with consideration being given to the relative values of the resources, but not necessarily that combination of uses that will give the greatest dollar return or greatest unit output.

The definition of *multiple use* stresses that any one resource use is not to dominate. It also emphasizes that multiple use does not require every use on every acre. Rather, the particular subset of uses on each acre would be determined by the suitability of the land and relative values of those uses. The mix of resource uses should be adjusted periodically as society's values for different resources change. Thus the act provided both guidance to current management and the urging to the Forest Service to update its land use decisions to keep them current with a changing society.

The last key concept included in the MUSY Act is the concept of sustained yield. Sustained yield was defined as the " achievement and maintenance in perpetuity of a high level annual or regular periodic output of various renewable resources of the national forests without impairment of the productivity of the land." In essence, the Forest Service was not to harvest more timber via logging or more forage (via livestock) than is grown each year. This sustained-yield requirement helps to ensure that current harvest levels will not be so high as to preclude future supplies of these renewable resources.

Perhaps what Congress did not say is even more important than what it did say. By being silent on a number of timber management and other issues, it left the exact management decisions to the discretion of the Forest Service. The agency professionals were left to decide what uses would be permitted on what tracts of land (Culhane 1981:53). Congress did not prescribe what to produce, how to produce it, or the timing and distribution of the five multiple uses.

Special Federal Land Classifications Passed During the 1960s

The Wilderness Act of 1964

Motivations for the Wilderness Act.   Even though the Forest Service had developed its own administrative wilderness land classification with the U-regulations (discussed earlier), wilderness was not explicitly mentioned in the listing of multiple uses in the MUSY Act. However, a different section of the act stated that establishment and management of areas of wilderness were consistent with it (Dana and Fairfax 1980:203; Culhane 1981:54). Nonetheless, the issue of permanent legislative protection of Forest Service administratively designated wilderness and wild lands continued to be a priority of many conservation groups. Conservation groups also wanted the wilderness concept applied to National Parks and Wildlife Refuges administered in the Department of Interior.

After nearly nine years of congressional debate and the introduction of sixty-five different bills, the Wilderness Act passed over the opposition of the Forest Service and timber industry in 1964 (Culhane 1981: 54). The Forest Service opposition centered around its perception that Wilderness might become a single-use classification that would preclude the other multiple uses and hence undercut the agency's discretion over management of these lands.

Allowed and Disallowed Uses of Wilderness.   As defined in the Wilderness Act of 1964, Wilderness is compatible with many other multiple uses. It is defined as lands that retain their "primeval character," with no permanent developments or human habitation. Lands designated as Wilderness allow for primitive (i.e., nonmotorized) recreation, including hunting, fishing, backpacking, and rafting as well as opportunities for solitude. Of course, Wilderness areas protect habitat for many species of wildlife, and the forest cover ensures a high-quality flow of water. If livestock grazing was an established use prior to designation, it is allowed to continue in the same manner and degree as before designation. The Wilderness Act does preclude timber harvesting, motorized access, and building of roads and permanent structures. Staking of mining claims was allowed only until 1983.

Wilderness Designation Process.　The Wilderness Act established that all National Forest lands that were classified under the U-regulations as wilderness or wild would become official Wilderness areas as part of the National Wilderness Preservation System. This amounted to more than 9 million acres of National Forest Wilderness upon passage of the act. Although the Wilderness Act did not grant instant designation to any National Park or U.S. Fish and Wildlife Service lands, it did legitimize Wilderness as a land use classification in management of National Forests, Parks, and Wildlife Refuges. Once designated, these lands were provided with a more permanent protection that could not be as easily undone as administrative designations.

Wilderness Study Process.　Another far-reaching feature of the Wilderness Act was establishment of a Wilderness study process for the National Forest "primitive areas" created under the old U regulations. In addition, the National Park Service and U.S. Fish and Wildlife Service were instructed to study their roadless areas and make recommendations to Congress regarding suitability for Wilderness. Although this study process is quite detailed, the key criteria involve a determination of whether the candidate area is truly roadless and natural (i.e., the imprint of human activity is not apparent), offers outstanding opportunities for primitive and unconfined recreation, and would add to the diversity of ecosystems represented in the National Wilderness Preservation System. This study process escalated into one of the major public land battles during the 1970s and is still an issue in most National Forest plans today. The Wilderness Act also created a National Wilderness Preservation System for all federal land-managing agencies (except BLM until 1976). How each agency implements the Wilderness Act will be discussed in more detail under each respective agency.

Wilderness Management.　Once an area of land is designated as Wilderness it is still managed by the same agency as before. Thus there are National Park Wilderness Areas, National Forest Wilderness Areas, and Fish and Wildlife Wilderness Areas (and after 1976, BLM Wilderness Areas). Given the common source of authority in the Wilderness Act, these areas are managed very similarly, but not identically. There has been an effort to develop a core of Wilderness management principles that all agencies would share. This effort culminated in the joint publication by the U.S. Departments of Agriculture and Interior of *Wilderness Management,* by Hendee, Stankey, and Lucas (1978), all of whom were wilderness researchers with the Forest Service. For an interesting comparison of the subtle differences in agency approaches to wilderness management see Allin (1990).

The 1968 National Trails System Act and National Wild and Scenic Rivers Act

The National Trails System Act and National Wild and Scenic Rivers Act passed within weeks of each other and are similar to the Wilderness Act in design and intent. These acts set up two national systems, one for trails and one for free-flowing rivers. They established "instant" National Trails (Appalachian and the Pacific Crest trails) and instant Wild and Scenic Rivers (such as the Middle Fork of the Salmon River in Idaho). The acts also provided a framework to add waters or trails to their respective national systems. The legislative land use classification systems are both applicable to all four federal lands agencies. Wild and Scenic Rivers continue to be managed by the agency that had administrative jurisdiction over the land in the first case. Thus the Forest Service, Bureau of Land Management, National Park Service, and U.S. Fish and Wildlife Service all manage Wild and Scenic Rivers when they occur on their lands. The primary intent of the Wild and Scenic Rivers Act is to protect some of the remaining free-flowing rivers from dams.

The Wild and Scenic Rivers Act created three river designations: Wild, Scenic, and Recreational. Wild rivers are those rivers or sections of rivers that are free of impoundments and generally inaccessible except by trails, with essentially undeveloped shorelines. Scenic rivers are those rivers or sections of rivers that are free of impoundments with mostly undeveloped shorelines but that are accessible in places by roads. This road access is one of the key distinctions between the Wild and Scenic designations. Recreational rivers are those rivers or sections of rivers that are readily accessible by road or railroad, that may have some development along their shoreline, and that may have had some impoundments or diversion in the past.

The second major component of the acts was to establish a study process whereby new trails and rivers could be added as National Trails or Wild and Scenic Rivers. Evaluation of federal lands and waters for their potential inclusion into these two categories is an important planning issue with many National Forests, Parks, and Wildlife Refuges.

Long-Term National Planning for Forests: 1974 Resources Planning Act

With the implementation of the Wilderness Act, Wild and Scenic Rivers Act, and National Trails Act, much federal land was restricted from timber harvesting, although much that was withdrawn had marginal timber value (Irland 1979). Nonetheless, the timber industry wanted assurance from Congress that increased funding for timber management on existing areas would be forthcoming. In 1969 industry started promoting a "Timber Supply Act" that would mandate minimum cut levels and provide a long-term national plan (Dana and Fairfax 1980:324). The purpose of the plan was to make comprehensive policy about land use designations such as Wilderness within a framework recognizing the implications to the national supply of timber.

The notion of long-term national planning for national forests fit then Senator Hubert Humphrey's inclination toward national central planning; he saw such planning as a way of providing a clear long-term direction to the Forest Service (Shands et al. 1988). The Forest Service saw an opportunity to tie long-term resource objectives to five-year budgets to ensure stable funding. Rather than annually debate budget levels for renewable-resource programs that are long term by their very nature, why not budget for the long term? More important, these five-year budgets would be based on an assessment of renewable-resource supply and demand to determine a program within the capability of resources (i.e., supply potential of National Forests) as well as within the confines of what the public demanded.

In 1974 such long-range national planning was passed in the form of the Resources Planning Act (RPA). The two key features of the act are the preparation of, first, a national Assessment of the demand for and supply of resources (reflecting all land ownerships in the United States) and, second, consistent with the findings of the Assessment, a Program for general management direction of National Forests for the next five years.

RPA requires the preparation of the Assessment every ten years. It is to include "(1) an analysis of present and anticipated uses, demand for, and supply of renewable resources of forest, range and other associated lands with the consideration of international resource situation, and emphasis on pertinent supply and demand and price relationship trends; (2) an inventory of present and potential renewable resources and an evaluation of opportunities for improving their yield of tangible and intangible goods and services" (Forest Service 1981).

The Program is to be completed every five years and is intended to provide Congress with a menu of what the Forest Service could do to meet the needs identified in the Assessment. That is, it presents alternative mixes of renewable resources that the Forest Service could produce and the associated budgets it would take to produce them. Although the Program emphasizes management of the National Forest System, it also addresses forestry research and cooperative state and private forestry efforts. The president transmits each five-year program to Congress. The president and Congress are supposed to use this Program as the basis for developing the Forest Service budget over that five-year period.

Since RPA emphasized national planning, it was initially implemented by the Forest Service as "top-down planning." That is, the Washington, D.C., office of the Forest Service, the executive branch, and Congress determine the long-term direction of the National Forests. Output targets in terms of board feet of timber to cut, acres to reforest, recreation facilities to supply, roadless acreages to preserve as Wilderness, and so on, are passed down to each Regional Office. Each of the eight Regional Offices then apportions these targets to each National Forest within its region. Each forest must meet these targets so that the National Forest System can meet its obligation to Congress.

The primary problem with this has been that targets set for each National Forest often exceed a particular forest's productive capability or do not make sense, given the array of other multiple uses that are jointly produced. That is, certain combinations are not just practical: most forests cannot produce more elk, more fish, and more trees, but they can produce more trees and less fish.

Coordination Between National- and Forest-Level Planning.   It is important to stress that this top-down process is how the first three RPA Programs were constructed. By the fourth (1990) RPA Program there were enough new forest-level plans (associated with the National Forest Management Act discussed in the next section) that the RPA Program included an alternative that was built "from the bottom up" using locally developed forest plans. In this case, the national direction would be arrived at by adding up the outputs each forest would produce under its approved plan. These outputs would be within the productive capacity of the National Forest and would be the mix deemed most beneficial at the regional level of the Forest Service.

The difficulty here is that the simple total of independently arrived-at National Forest output might not add up to the nationally desired levels of output for certain goods or services. That is, although each piece of the puzzle may, in isolation, be quite pretty, the overall picture these pieces make may not be that desired at the national level.

The potential of course exists to synchronize the ten- to fifteen-year cycle of preparation of individual forest plans under the National Forest Management Act (discussed in more detail later) with the preparation of decennial RPA Assessments and Programs. In this way, the alternatives in the RPA Program could be the sum of the alternatives in the forest plans. Certainly, this synchronization would be a demanding process for the agency, but it would greatly improve the linkage between national direction of forest management and local implementation of the national direction at the individual forest level. The Forest Service may be moving in this direction. Many of the forest plans prepared during the 1980s contained an RPA alternative that required that the forest be managed so as to meet the RPA targets for that forest. In the 1990 RPA Program, one alternative for the national direction of the National Forest System was to implement the individual forest plans.

Analysis Required Under RPA.   RPA requires several types of analysis, but quantification of demand and supply of outputs and benefit-cost analysis are prominent among the types of analyses emphasized by RPA (Krutilla and Haigh 1978; Fairfax and Dana 1980:325). This quote from Section 8 of the RPA legislation best summarizes the spirit of the new law:

> The report shall contain appropriate measurements of pertinent costs and benefits. The evaluation shall assess the balance between economic factors and environmental quality factors. Program benefits shall include but not be limited to environmental quality factors such as esthetics, public access, wildlife habitat, recreational use of wilderness, and economic factors such as the excess of cost savings over the value of foregone benefits, and the rate of return on renewable resources.

As the Forest Service began to perform the first RPA Assessment and Program, a storm was brewing over clear-cutting at the individual National Forest level. From this storm, the previously mentioned National Forest Management Act of 1976 to deal with planning, not at the national level but at the individual forest level, was born. It is to this critical legislation that we now turn.

The National Forest Management Act of 1976 (NFMA)

Origins of NFMA

The 1970s brought a convergence of several forces that would change forest management as it had been practiced during its first fifty years. First, much of the most productive timberland on gentle slopes had already been cut in the post–World War II era. The remaining old-growth timber was on steeper slopes in more remote areas. The second event was that the technology to harvest and manage timber on steep slopes was becoming available. Roads could be built in steeper areas. More elaborate and powerful yarding systems to haul logs were available. This led foresters in Montana's Bitterroot National Forest to log on steep slopes and bulldoze terraces into hillsides to facilitate regrowth of trees for continued intensive forestry in these areas (Shands et al. 1988). But such intensive forestry resulted in visible scars symbolic of a "forest management as solely timber management" mentality. Third, nearly 2,000 miles to the East, a series of large clear-cuts on West Virginia's Monongahela National Forest was eliminating wildlife habitat and popular recreation areas (Shands et al. 1988).

Although both timber operations may have been silviculturally sound to meet the single objective of maximizing the number of board feet of timber per acre of land, they displaced many of the other multiple uses. The final event that began to reverse the silvicultural dominance of the Forest Service was the strengthening of the environmental movement and the passage of several environmental laws in the late 1960s and early 1970s. No longer would the public allow elimination of the other multiple uses by timber harvesting. The public could use the National Environmental Policy Act (discussed in the next section) to lodge legal appeals and demand that recreation, wildlife, wilderness, and water quality be given equal consideration with timber. In the midst of the environmental decade of the 1970s, the Forest Service accommodation of increased timber harvesting in increasingly visible areas brought its management under a tremendous amount of scrutiny.

The University of Montana's School of Forestry took the Bitterroot National Forest to task for concentrating on timber management rather than on forest management. Its researchers recognized what the Forest Service field offices had missed: the Forest Service was out of step with changes in society's evolving environmental values. Although this recognition of misplaced emphasis was also coming from within the agency (Dana and Fairfax 1980:228), the agency had clearly lost a great deal of credibility as a professional resource managing agency.

The Monongahela National Forest was not so lucky as to receive just an academic scolding. In 1973 conservation organizations filed suit against three planned clear-cuts totaling more than 1,000 acres. These groups cited a little-noticed clause in the Forest Service's founding legislation (Organic Act of 1897) that allowed cutting of only dead, mature, or large-growth trees that were marked and designated before harvesting operations. The federal district court in West Virginia agreed to order the Forest Service to halt clear-cutting as inconsistent with the law. With this decision, the potential for stopping clear-cutting on all National Forests became apparent.

The specter of cessation of clear-cutting (and with it much of the National Forest timber program) as well as the clamor for restriction of Forest Service discretion led to the formulation of the 1976 National Forest Management Act (which is technically an amendment to the 1974 RPA). The act went well beyond what industry and the Forest Service would have liked. These interests simply wanted the offending phrase in the 1897 Organic Act changed to allow clear-cutting. However, environmentalists saw the clout of their court victory as an opportunity to restructure the way the Forest Service did business. They wanted to eliminate certain timber-harvesting techniques, such as clear-cutting, and greatly limit the Forest Service discretion over forest management techniques.

However, the final version of the bill that passed left much discretion to the agency. The Secretary of Agriculture was ordered to prepare regulations to address suitability of different timber-harvesting methods. Clear-cutting could be used as long as it was the best method for meeting the objectives of the land management plan for a specific area. In addition, the clear-cuts were to be blended to the extent possible with the terrain, giving attention to the size of the cut and carrying out the cuts in a manner consistent with the other multiple uses. Thus clear-cutting remained a legitimate use, and the 25-acre limitation of earlier cutting guidelines was not mentioned in the bill (Dana and Fairfax 1980:329).

Other issues addressed in the NFMA include biological diversity and determination of suitability of lands for timber production. Here again the emphasis is on process rather than on specific prohibitions. For example, the Forest Service must evaluate both biophysical and economic factors when determining suitability for timber harvesting. These double criteria significantly broadened those the Forest Service had used previously for determining timberland suitability. This is just one of several instances in NFMA of continuing the trend, started in RPA, by which Congress increasingly emphasized economic criteria.

Rather than dictate specific land management techniques and how much of each of the multiple uses the forests were to produce, Congress prescribed a decision-making or planning process. This process would set the "rules of the game" for Forest Service evaluation of which specific land management techniques were appropriate for particular National Forests and how much of each multiple use to produce on each National Forest. A formal public review and consultation role was provided in the NFMA legislation. This process orientation is wise, given the great variety of resources found on individual National Forests through the United States. A national prescription of land management techniques that might be appropriate for many National Forests would be inadequate for certain other forests because of differing climate, topography, or vegetation structure.

Main Features of NFMA

The main thrusts of NFMA are threefold. First, NFMA reaffirmed the multiple-use—sustained-yield approach of the Forest Service with two minor changes from the 1960 act: (1) that instead of five multiples uses there are now six, with the sixth being explicit inclusion of Wilderness as one of the multiple uses; and (2) sustained yield is constrained to be one of "nondeclining even flow," which, as the name implies, requires the sustained yield to be either constant or increasing over time.

The second major thrust was the requirement that each National Forest prepare a forestwide fifty-year multiple-use plan for its National Forest. These plans would be updated every ten to fifteen years. Most important, they would be far more comprehensive by simultaneously planning for all multiple uses for the entire forest, rather than the previous separate individual plans for each resource. In this way resource interactions would be accounted for from the start.

The third major thrust was the explicit requirement to perform interdisciplinary planning. That is, planning involving biological, physical, and social sciences. The approach involves integrating the viewpoints and principles of the different disciplines so that management actions undertaken make sense from the perspective of all the disciplines involved.

Translation of NFMA into Forest Planning Regulations

As is typical with most broad-based legislation passed by Congress, the administering agency is delegated the task of translating congressional intent into specific, detailed regulations. Although this was certainly the case for NFMA, the Forest Service had an unusual partner in this endeavor: a Committee of Scientists that would provide advice on writing the NFMA regulations (Wilkinson and Anderson

1987:43). Three years after the passage of NFMA, the Department of Agriculture issued final planning regulations. By 1980 many Forest Service regions had chosen a "lead" forest to begin the experiment of interdisciplinary comprehensive planning under NFMA. The lessons learned in implementing NFMA planning on the lead forest could then be applied to planning efforts on the other forests in the region.

Although Congress wanted the plans to be finished in 1985, the newly elected Reagan administration soon put an end to such hopes. The new administration's "Task Force on Regulatory Relief" identified the rules that guide land and resource management planning in the National Forest System as a high priority for review (Forest Service 1982:43026). This resulted in a decision to revise the newly released NFMA regulations (and as will be discussed later, BLM's planning regulations as well). Nearly three more years passed before the current NFMA regulations were issued in September 1982.

Final Planning Regulations.   What did these final planning regulations emphasize? Several features are worth noting. The first of these are elements present in the 1979 planning regulations that were carried forward to the 1982 final planning regulations. This includes fourteen principles of forest planning:

1. Multiple-use and sustained yield.
2. Consideration of the relative values of all renewable resources.
3. Recognition that the National Forests are ecosystems and consideration of the relationships among plants, animals, soil, water, air, and other environmental factors.
4. Protection of and, where appropriate, improvement in the quality of renewable resources.
5. Preservation of important historic, cultural, and natural aspects of our heritage.
6. Protection of the inherent freedom of American Indians to exercise their traditional religions.
7. Provisions for the safe use and enjoyment of forest resources by the public.
8. Protection, through ecologically compatible means, of all forest and rangeland resources from depredations by pests.
9. Coordination with the land-planning efforts of other federal agencies, state and local governments, and Indian tribes.
10. Use of interdisciplinary planning.
11. Early and frequent public participation.
12. Establishment of quantitative and qualitative standards for land and resource planning and management.
13. Management of National Forests in a manner that is sensitive to economic efficiency.
14. Responsiveness to changing conditions of land, other resources, and changing social and economic demands of the American people.

These fourteen principles emphasize the importance of the forest environment and the need to be responsive to changing demands for and values of the forest resources. In contrast to the 1960 MUSY, ecological values (numbers 3 and 8), economic principles (numbers 2, 13, and 14), and public participation receive explicit attention.

The other main carryover from the 1979 regulations is the continued reliance on mathematical optimization models as the main analytical tool for selecting the preferred multiple-use land-management options (within the constraints set by law or manager discretion) for the forest (Bowes and Krutilla 1989:35).

One of the most frequently discussed provisions in the new planning regulations was the explicit objective of managing the National Forests "towards the desired result of maximizing net public benefits" (Forest Service 1982:43026). *Net public benefit* is defined in the regulations as "the overall long-term value to the nation of all outputs and positive effects (benefits) less all associated inputs and negative effects (costs) whether they can be quantitatively valued or not" (Forest Service 1982:43039). Net public benefit includes an economic accounting of benefits minus costs plus a description of other positive and negative effects that cannot be quantified in dollars, such as social stress or such environmental indicators as biological diversity. As will be discussed in more detail in chapter 9, the Forest Service calculates dollar values for more than just such commodities as timber, minerals, and livestock. Included in measurement of the economic portion of net public benefits is the value of water, outdoor recreation (hiking, camping, skiing, boating, and so on), wildlife-related recreation (fishing, hunting, wildlife viewing), and wilderness recreation. The techniques used to measure the economic value of these multiple uses will be discussed in chapter 6.

Figure 2.3 illustrates the relationship between net economic value and net public benefits. (Technically, net economic value is the net present value of economic benefits minus economic costs. Present value will be explained in chapter 6.) As

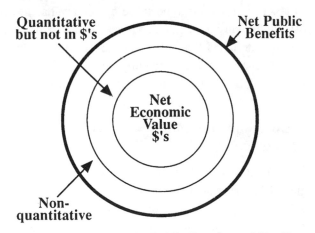

Figure 2.3. Relationship Between Net Public Benefits and Net Economic Value

one can see, *net public benefits* is an all-inclusive term measured by the area of the entire circle, with net economic value being a subset. In most forests, net economic value makes up a large part of net public benefits. A great deal of effort is going into expanding the positive and negative effects that can be included in the net-economic-value sphere, so that fewer resources or environmental effects are not quantified in commensurate dollar terms.

In chapter 9 we will investigate how the Forest Service calculates net economic values using its mathematical optimization model and what some of the other factors are beyond net economic values that make up net public benefits. In that same chapter we will examine the steps in the Forest Service planning process and how comprehensive, interdisciplinary planning is being implemented. Chapter 9 also will provide a case study of forest planning on the Siuslaw National Forest in Oregon.

As of 1992, the first round of National Forest planning was largely completed. Chapter 11 presents an evaluation of this first round of forest planning. As the Forest Service begins this second round of planning in the 1990s it is considering a revised set of guidelines specific to this second-round effort. These too are discussed in chapter 11.

Recent New Policies in National Forest Management

Although it does not have the legislative authority of multiple uses listed in NFMA, the Forest Service has recently issued policies that recognize two new resources on National Forests: preservation of old growth and maintenance of biological diversity. Both the Forest Service and the public have begun to realize that retention of some remnants of old-growth forests not only in the Pacific Northwest region but also in the intermountain West is important for some wildlife species. In addition, several old-growth ecosystem types are not well represented in the "rocks and ice" alpine Wilderness areas typical of most National Forests. Conservation of biological diversity has also taken on growing importance as society has become aware of the importance of maintaining interconnectedness of ecosystems. The loss of biological diversity on private lands (typically managed as monoculture agricultural or tree-growing operations) has increased the importance of public lands as the best opportunity for preserving biological diversity.

1969 National Environmental Policy Act (NEPA): Implications for Public Land Management

The National Environmental Policy Act (NEPA) applies to all federal agencies and thus will be discussed in terms of its broad implications for all four federal land management agencies rather than for just the Forest Service. In particular, the act played an important role during a critical time in the Bureau of Land Management's history (to be covered next), so it is important to discuss it at this time. However, it is worth noting that this act has a pervasive influence on all federal agency decisions, including the Departments of Transportation, Energy, and Defense (particularly the civilian works programs of the U.S. Army Corps of Engineers). The act has also served as a model for similar state legislation.

NEPA, after being broadly interpreted by the courts, has become a central feature that permeates all federal agency decision documents, whether they are land use plans, timber-harvesting permits, or Wilderness designations. The intent of NEPA is to ensure that environmental issues receive consideration in federal agency decisions. However, it is important to remember that NEPA by itself does not require selection of the most environmentally beneficial alternative.

NEPA Process. What has become known as the NEPA process requires that every federal action be evaluated in terms of its potential impact on the natural and human environment. A federal action in this context means either direct action by an agency itself (e.g., building a campground) or issuance of a federal permit to a private or public entity for its action (e.g., a utility's construction of a power line across federal lands). To begin this NEPA evaluation, federal agencies normally start with an environmental assessment (EA). An EA is a cursory evaluation of the likely magnitude and significance of environmental impacts of every proposed action by the agency or any use the agency might permit by other public or private parties. If the EA results in a "Finding of No Significant Impact," then the process stops there. If a finding of significant impact results, then the agency (or group requesting the permit) must prepare a full environmental impact statement (EIS). This process can take several months and is quite involved.

This EIS process imposes several important requirements on agency decision-making processes. First, the agency's proposed action must be written down in sufficient detail to allow agency staff to predict associated environmental effects. Second, a wide range of alternatives to the proposed action must be evaluated, along with the proposed action. The effects of implementing each alternative on the biophysical and human environment must be presented. A key requirement of the EIS is that a written justification for selecting the proposed action over the other alternatives be made.

Today, the preparation of an EA or EIS to accompany a public land management plan is required in all federal land-managing agencies. In addition, specific agency actions such as timber sales, range or wildlife improvement projects, construction of campgrounds, and management plans regulating the use of facilities or resources require at least an EA. Common federal actions, such as construction of dams, highways, and so on, frequently require an EIS.

Public Involvement in NEPA. Another important feature of NEPA is to expand greatly the opportunities for public participation in agency decision making. The general public, private industry, local governments, conservation groups, and other public agencies often provide input to and comment on a federal agency's draft EA or EIS. Public meetings and formal public comment periods are a standard part of the NEPA process. The federal agency must respond to these comments in preparing its final environmental assessment or EIS. This was of course a radical change for all federal agencies, but particularly the Forest Service, which viewed itself as a professional and scientific management agency. But failure to heed the public was far more hazardous with the advent of NEPA, for NEPA made possible additional grounds for suing an agency for failure to follow the procedures spelled out in the regulations. This was a common weapon of conservation groups during the 1970s. Agencies had to take NEPA seriously or run the risk of the courts finding their EISs inadequate for the decisions made. The threat of lengthy court delays forced many agencies to take seriously the social and environmental consequences of their actions.

A somewhat underrated contribution of NEPA is the requirement that the analysis reflect multiple disciplines, including natural, physical, biological, social, and economic sciences. This multidisciplinary framework represented an important step in infusing new approaches to decision making into many federal agencies dominated by one discipline (e.g., foresters in the Forest Service).

Evolution of Federal Multiple-Use Lands Administered by the Bureau of Land Management (BLM)

Predecessor Agencies to BLM: General Land Office and Grazing Service

The Bureau of Land Management (BLM) is the federal government's other multiple-use land management agency. However, the evolution of this agency into a multiple-use land management agency is quite different from that of the Forest Service. Whereas the Forest Service's evolution started when the federal government wanted to retain lands in federal ownership for their conservation, the BLM's historical antecedent (the General Land Office) was in the business of disposing lands. In 1812 the land disposal functions of the federal government were centralized into the newly formed General Land Office, which was first located in the Department of Treasury and then transferred in 1849 to the newly established Department of Interior (Culhane 1981:75). The General Land Office oversaw the sale of lands, homesteading, and other disposal programs discussed in the beginning of this chapter.

As land settlement moved further west from the fertile Midwest and into the arid intermountain West, where water was the limiting factor, the character of homesteading changed from farming to ranching. Only small tracts with access to water were suitable for homesteading. In addition, economical ranch operations required far greater acreages than could be homesteaded. The unhomesteaded surrounding federal land was treated as a grazing commons. That is, neighboring ranchers and nomadic sheepherders all treated the federal grazing land as a free source of forage. The only way to acquire a "property right to the grass" for one's own livestock was to have one's own livestock out there first and in greater numbers than someone else. The resulting tragedy of the commons resulted in overgrazing, soil erosion, and little incentive to conserve the range by neighboring ranchers (Foss 1958, 1960).

Establishment of the Grazing Service.   The continued degradation of these lands for nearly fifty years eventually led to the passage of the Taylor Grazing Act in 1934. This act and its amendments in 1939 made three important changes. First, grazing districts were set up with Grazing Advisory Boards to define which ranchers had grazing privileges (not rights!) to specific areas of federal land, usually near their privately owned ranch (base) property. This helped create order out of chaos and provided some minimal interest by ranchers in managing livestock on what they perceived as "their" grazing district (even though it still remained federal land). These grazing privileges (i.e., the number of cattle allowed) were to be based on the range's carrying capacity and the rancher's historic use levels. Unfortunately, past use levels often exceeded carrying capacity, and the effort to adjust livestock numbers to range carrying capacity has been a challenge to this agency for decades.

Second, the Department of Interior was authorized to establish a small Grazing Service to administer the use of the grazing districts by ranchers. This small group was understaffed and underfunded for the task, and the ranchers (who were far more knowledgeable about "their" specific grazing districts) maintained effective local control over management of the grazing districts for decades to come (Foss 1960; Dana and Fairfax 1980; Culhane 1981).

The third major provision was for establishing a fee for grazing livestock on federal lands. Although the Forest Service was charging a grazing fee six times the Grazing Service's 5 cents per animal unit month (one cow for one month), there was much western resistance to higher fees. The appropriate level of the grazing fee for public lands is a source of considerable antagonism between ranchers and conservationists even today.

In conjunction with the Taylor Grazing Act, President Franklin Roosevelt withdrew from homesteading all the remaining unclaimed federal land west of the one-hundredth meridian (except Washington and Alaska) and placed that land into the grazing districts. This too is an important historic event that to many signaled the closing of the frontier.

Although ranchers were happy to have the force of law to "close the commons" to migratory sheepherders, they were not interested in active government management of the range and certainly not at their expense. Attempts by the federal government to raise the grazing fee to fund the Grazing Service and determine the appropriate level of grazing animals were met with stiff resistance by Senator

McCarran of Nevada, who headed the Senate Public Lands Committee. McCarran got a pledge from the Department of Interior to freeze the 5-cent grazing fee until his study of grazing fees was completed, a task he stretched out for six years. During this time, the Grazing Service budget and staff fell from 250 to eighty-six, and it had to close eleven of its sixty district grazing offices (Muhn and Stuart 1988:48). At the same time, there was growing evidence of duplication of responsibilities between the General Land Office (which administered mining and other uses of these federal lands) and the Grazing Service.

BLM Emerges from the Grazing Service and General Land Office

From what remained of the Grazing Service and the General Land Office, the Bureau of Land Management was born in the Department of Interior by President Truman's Executive Reorganization Plan in 1946. This was largely a combination of Interior's General Land Office (including all federal mineral leasing programs and all public domain lands in Alaska) with what was left of the Grazing Service.

The new agency was now responsible for implementing by one estimate "3,500 laws passed during the previous 150 years" (Muhn and Stuart 1988:54), including administering the grazing districts under the Taylor Grazing Act, onshore mineral leasing programs, and (in 1953) offshore mineral leasing. The Department of Interior now had one federal natural resource agency that was to manage millions of acres, but under quite conflicting legal mandates as the Taylor Grazing Act of 1934 and the Mineral Leasing Act of 1920.

Organization of BLM.   In keeping with the tradition of the Grazing Service's emphasis on "home rule on the range" (Muhn and Stuart 1988:39), the new organization would be more decentralized than the General Land Office. The organization was originally organized similar to the Forest Service with a Washington, D.C., office, several regional offices, and then district offices. The district offices were organized around the grazing system's district offices and the Oregon and California (O & C) lands forestry offices.

BLM's First Years.   The first director of the BLM was an economist named Marion Clawson, whose task it was to interrelate somehow the many disparate "missions" of the BLM into one agency. One of the carryover missions was to bring livestock use down from its historic levels to the range's carrying capacity. This process of performing soils and vegetation surveys and then using the information to adjudicate grazing privileges down to livestock carrying capacity, much like the grazing fee issue, has continued to occupy every director of BLM since Clawson. The effort to reduce livestock numbers provided Clawson's new agency with its first test of who ultimately controls use of these federal lands: the agency or the rancher who leases the land. After several years of legal appeals, BLM was given the legal support for its position that it could determine livestock carrying capacity and adjust grazing privileges down to this carrying capacity (Culhane 1981:92). Nonetheless, BLM continues to meet strong local resistance from ranchers as it tries to exercise its authority to regulate the conditions of grazing on federal lands.

Further decentralization of BLM took place under the Eisenhower administration with the creation of state offices in each of the eleven western states. With the abandonment of regional offices, the current BLM structure of the Washington, D.C., office; state offices; and district offices was largely in place in 1954.

Much like the close ties that Forest Service employees had with forestry schools and forestry professional societies, BLM personnel reflecting their grazing ancestry began developing close ties with the range management schools and professional range management societies. In the next few decades the BLM would seem to become identified with range management, even though it maintains an active forestry program in many states, particularly on the Oregon and California (O & C) lands.

BLM Moves Toward a Multiple-Use Agency

One of the concepts the agency adopted under Clawson was the Forest Service's ideas related to multiple use (Muhn and Stuart 1988:62). However, the BLM had no explicit authority to use the principles of multiple use and sustained yield in its management. Nonetheless, it began infusing concerns about wildlife habitat into grazing district management concerns and wildlife representatives into grazing advisory boards in the mid-1950s (Muhn and Stuart 1988:86). By the early 1960s the joint effort by Secretary of Interior Stewart Udall and then director of BLM Charles Stoddard brought the BLM its first formal multiple-use management mandate with the Classification and Multiple Use Act of 1964. This act was patterned after the Forest Service's Multiple Use, Sustained Yield Act of 1960 (Muhn and Stuart 1988:104). Now wildlife, recreation, and water resources would be included along with BLM's traditional emphasis on livestock, mining, and land disposal. This act also instructed the BLM to follow multiple use in its land use planning. At this time, however, BLM's planning still followed functional resource plans, the most visible of which were livestock grazing plans called Allotment Management Plans. Although these plans took account of other resources, they focused primarily on one dominant use at a time.

The multiple-use mandate forced several organizational changes within BLM as well. Within the offices of BLM, new divisions of wildlife, watershed, and recreation were added to reflect the multiple-use mandate. As active multiple-use management began, district office staffs became large groups of individual resource specialists. This often required locals to deal with different people for different resource issues. To maintain the close contact to public land users, the Resource

Area Office concept was introduced in 1966 (Muhn and Stuart 1988:117). The resource area manager would be the line officer with day-to-day authority to make site-specific field-level decisions involving all the multiple-use resources.

The force of the Classification and Multiple Use Act expired in 1970, leaving BLM with no permanent authority for multiple-use management. The agency continued to embrace the principles of multiple use but felt nervous about its legal authority to do so.

BLM's Attempts to Comply with NEPA

As BLM entered the environmental decade of the 1970s it sought to comply with the National Environmental Policy Act (NEPA) by preparing Environmental Impact Statements (EIS's) on its various programs. Rather than develop site-specific EIS's, BLM initially issued programmatic EIS's, which attempted to discuss the environmental effects of the BLM's nationwide program for that resource on all its lands. One was issued for the entire livestock grazing program on BLM land and one for the entire mineral leasing program. These programmatic EIS's only described in general terms the environmental impacts associated with these programs. In two successive and successful lawsuits by Natural Resources Defense Council (a nonprofit environmental law organization), the courts ruled that BLM's programmatic EIS's were inadequate to meet the requirements of NEPA. The court-approved settlements required BLM to prepare a series of site-specific EIS's on its grazing and mineral leasing programs. This effort would be a significant focus of BLM personnel until the mid-1980s.

Wild Horse and Burro Act

The Wild and Free Roaming Horse and Burro Act of 1971 has also had a direct effect on BLM. It limited BLM's techniques for removing ''wild'' horses and burros (basically abandoned animals that had become self-sustaining on the BLM lands) and required protection of these animals. This significantly changed the direction of BLM's management of these animals for years to come. Instead of killing or allowing others to kill them, the BLM developed its current ''Adopt a Horse'' program. This program is viewed as a more humane response to the growing population of these nonnative species and the adverse effects they have on native plant and wildlife populations on BLM lands.

Efforts to Obtain a Multiple Use Organic Act.   Not only did the 1970s find BLM without statutory authority to manage for multiple use, but BLM lands were also excluded from wilderness suitability review under the Wilderness Act of 1964. As the agency attempted to respond to the environmental interests of the public it found itself hindered by its patchwork of antiquated laws reflecting frontier values.

BLM's Multiple-Use Organic Act: Federal Land Policy and Management Act

When the federal government's Public Land Law Review Commission presented its recommendations for improving management of all public lands in 1970, the BLM saw this as an opportunity to push for permanent congressional authority for multiple use (Muhn and Stuart 1988:167). Such an organic act would reconcile the hundreds of conflicting public land laws it administered into one internally consistent set of laws. In 1971 a National Resource Lands Management Act was submitted to Congress. Over the next several years various forms of the act would be passed by one house of Congress or the other, but it was not until 1976 that both houses of Congress finally agreed upon the sweeping legislation that would reconcile many public land laws and update the management of BLM lands. The law became known as the Federal Lands Policy and Management Act (FLPMA) and became law in October 1976, the same month the Forest Service's National Forest Management Act passed. As we shall see, there are several similarities between them.

FLPMA has numerous important provisions. The most noteworthy of these are the following:

1. The act requires land and resource management that follows principles of multiple use and sustained yield. This gave legal sanction to BLM's existing approach.
2. The actual multiple-use and sustained-yield practices must be employed "in accordance with the land use plans developed" by BLM. Thus the intent was that comprehensive all-resource planning would direct on the ground resource management.
3. The act requires planning to use a systematic interdisciplinary approach.
4. BLM lands became eligible for consideration under the Wilderness Act of 1964. Therefore BLM was required to inventory, evaluate, and recommend specific roadless areas to Congress as suitable or unsuitable for Wilderness designation.
5. The act revised mineral leasing laws and gave BLM greater control over the exploration and development of locatable minerals, such as gold, silver, and uranium. Thus, although the antiquated 1872 Mining Law was not repealed, minimal controls were put in place so that excessive damage to other multiple-use resources would be minimized.
6. BLM was asked to identify as part of its planning process "Areas of Critical Environmental Concern" (ACECs) that required protective management to preserve their unique features.
7. The act eliminated many old land disposal laws and served notice that nearly all BLM's current lands would continue to be retained in federal ownership. Several old laws were updated in FLPMA to reflect changes in the last several decades.

8. FLPMA limited the role of the Grazing Advisory Boards to advising managers on allotment management plans and use of range improvement funds. To balance the livestock users, Multiple Use Advisory Councils were established to reflect all resource users.
9. FLPMA authorized yet another grazing fee study (to be performed with the Forest Service).
10. The act established the twelve-million-acre California Desert Conservation Area.
11. FLPMA provided BLM with its own law enforcement authority, something the agency particularly needed as it tried to manage the off-road-vehicle user groups in the California Desert Conservation Area.

Multiple-Use Planning Under FLPMA

FLPMAs emphasis on interdisciplinary, multiple-use planning as a basis for land management required BLM to rethink its pre-FLPMA planning process. Although the old Management Framework Plans (MFPs) had served the BLM in a pre-FLPMA and pre-NEPA era, there was much debate about whether to shore up the MFP approach or to start from scratch with a comprehensive planning process similar to what the Forest Service was investigating for its similar charge in NFMA. At the time, BLM was running what some have described as several "de facto planning processes": formal planning through MFPs, court-mandated grazing EIS's that involved limited amounts of planning to deal with forage allocations between livestock, wildlife, soils, and water quality; and a minerals leasing program that also required planning land uses associated with coal suitability.

In 1978 the new director of BLM, Frank Gregg, determined that to develop the "resource allocation decision process in the multiple-use plan as required by FLPMA" a new Resource Management Planning process would be instituted to "substantially upgrade the planning system to meet the needs of the 1980s." The planning regulations to implement planning requirements in FLPMA were issued in 1979. Much like the Forest Service, much effort was devoted over the next several years to developing a comprehensive, interdisciplinary planning process that would carry the agency through the 1980s.

The same change in presidential administrations that forced a revision to the newly released NFMA planning regulations had an even more dramatic effect on BLM's new Resource Management Planning system. In the name of "streamlining" and "focusing" the BLM's planning process on problem solving, the emphasis on fully comprehensive planning of all resources on all lands was reduced. Parallel planning processes for wilderness recommendations and RMPs would take place. The details of BLM's planning process will be discussed in chapter 10. Nonetheless, the promise of the 1979 regulations for directing the agency to balanced multiple use was short-lived. FLPMA held the potential to move BLM from what one Interior Department official called the Bureau of Livestock and Mining to what Director Gregg called it the Best Land Manager (Muhn and Stuart 1988:196). We will reserve judgment on which of these terms best describes the BLM under FLPMA until after we review the results of BLM's new streamlined planning process in chapter 10.

Attempts to Combine Multiple-Use Agencies into One Department

Before we end our discussion of multiple-use agencies in this chapter, it is worth noting that there have been numerous attempts to consolidate the BLM and the Forest Service into the same cabinet-level department. Given the similarity in mission, there would be many economies of scale in management and research. However, attempts at consolidation, including the most recent by President Carter, have failed. The idea of geographically consolidating lands into BLM blocks and Forest Service blocks was proposed by the Reagan administration. This proposed transfer (which also failed) was just one more round in a recurring public land management issue: coordination of land management regulations and practices between federal land-holding agencies in different departments of the federal government. The difficulty of split agencies persists but has been reduced by a series of cooperative agreements whereby the agencies have adopted similar grazing regulations and other land use regulations to reduce the complexity that public land users must deal with on any actions involving both National Forests and BLM land.

National Parks and the National Park Service

The idea of preserving federal land in National Parks, as opposed to multiple uses in National Forests, was rooted in the more protectionist wing of the conservation movement of the late 1800s. The movement was personalized by John Muir, who stressed preservation of nature for reflection, spiritual inspiration, and recreation. To Muir, industrial man could seek renewal and salvation in the wildness that nature offered. And many of these western areas were truly inspirational: Yosemite Valley, the geysers of Yellowstone, the Giant Sequoia trees, Mount Rainier.

To preserve these natural wonders, Congress reserved them from land disposal laws. For Yosemite Valley, this was done by ceding the land to the state of California in 1864, whereby California would manage it as a public recreation resource. Soon, however, the federal government began designating these areas as National Parks. The first of these, Yellowstone, was so designated in 1872. In 1890 Yosemite reverted back to federal ownership and became a National Park. That same year in California, Sequoia and General Grant National parks were established. Within ten years Mount Rainier, in Washington, and Crater Lake, in Oregon, were also set aside as National Parks.

In all this activity on National Parks two key elements were missing. First, what was the exact purpose of these reservations; that is, how were the lands to be managed? Second, who was to protect these areas from unauthorized uses and who was to manage them? Various purposes for reserving the lands had been mentioned, including protecting the scenery. In Yosemite, the land was protected from live-stock grazers by the U.S. Army, which was also protecting Yellowstone. Some of the smaller parks were under the authority of the Department of Interior's General Land Office, and some were under the auspices of the Department of Agriculture's Forest Service. The Forest Service's philosophy under Gifford Pinchot of wise use was seen by some as a potential threat to preservation of the special character of these National Parks.

In addition to the potential concerns about park management by the Forest Service, there was a real threat to Yosemite's Hetch Hetchy Valley from dam builders at the turn of the century. As the preservationists fought to protect National Parks from development projects such as dams, it became even more apparent that the National Parks needed leaders who would act as their defenders and managers. With the flooding of Hetch Hetchy Valley in Yosemite National Park for water supply and hydropower to San Francisco, the impetus for establishing the National Park Service was made even stronger. Many conservationists felt that without a single agency dedicated solely to protecting National Parks, developments such as Hetch Hetchy would continue (Foresta 1984:17).

The Sierra Club and a variety of preservation-minded individuals held National Park conferences in 1911 and 1912 that developed recommendations for a parks bureau (Foresta 1984:17). Their strategy was to distance what was to be a preser-vation-oriented agency from the more utilitarian Forest Service in the Department of Agriculture by placing the new parks bureau in the Department of Interior. Conservationists were leery of letting Pinchot and his agency, who were more interested in harvesting and extracting resources, be in charge of National Parks, which they wanted to preserve. The Department of Interior saw the National Parks as an opportunity to become established in the public lands management arena. After several years, a National Park System was established in 1916, with the National Park Service within the Department of Interior to manage these lands.

Objectives of National Parks and Establishment of the National Park Service

The objectives of the new agency are best summarized in the commonly cited quotation of the bill: ''To conserve the scenery and the natural and historic objects and wildlife therein and to provide for the enjoyment of same in such manner and by such means as will leave them unimpaired for the enjoyment of future genera-tions.'' The secretary of interior, in a letter to the first director of the National Park Service, Stephen Mather, provided three principles that should guide administration of the National Parks:

1. The National Parks should be maintained in absolutely unimpaired form for the use of future generations as well as those of our own time.
2. They should be set apart for the use, observation, health, and pleasure of the people.
3. The national interest must dictate all decisions affecting public or private enterprise in the parks.

The primary duty of the National Park Service (NPS) is to preserve the parks for posterity in essentially their natural condition.

Although it appeared that preservation was the primary mandate for the parks, with visitor use second, Mather felt he had to build public support for the NPS to generate the budget to manage the parks and to get new ones designated by Congress. As such he encouraged visitor use and accommodated that use by providing roads, hotels, and concessions within a portion of the National Park, but leaving the bulk of the park natural.

The legacy of these efforts is still evident in many National Parks, such as Yellowstone (Old Faithful Lodge), Yosemite (Awahnee Hotel), and so on. This also set a precedent for the type of development that would be allowed in the park. Once these types of facilities and precedents were set, many local congressional representatives saw National Parks as more for tourism than for preservation.

This view of National Parks as tourist attractions continues up to the present day. Examples include the Moab, Utah, Chamber of Commerce's request in 1978 that the NPS build a high-standard road and bridge to allow visitors to drive to the confluence of the Green and Colorado rivers in Canyonlands National Park. More recently, promotion of tourism was one of the rationales for the creation of the Great Basin National Park in Nevada in 1986.

Park Service Gains Management of National Monuments. In 1933 President Franklin Roosevelt expanded the National Park Service management responsibilities to include National Monuments that had been designated under the Antiquities Act of 1906 (even if within National Forests), historic sites, battlefields, and even scenic highways (e.g., National Parkways). The specifics of each of these NPS land classifications will be discussed later.

National Outdoor Recreation Planning and the National Park Service

In 1936 NPS began its first of what was to be many years of close involvement in national recreation planning. The NPS was asked to develop a recreation plan for all federal lands outside of the National Forests. This function would continue within the NPS until the establishment of the Bureau of Outdoor Recreation within the U.S. Department of Interior in 1963. The national recreation planning function returned to the NPS about twenty years later, when the Reagan administration, in an attempt to reduce the federal role in recreation planning, abolished the Bureau of Outdoor Recreation (which had been renamed Heritage Conservation and Rec-reation Service, or HCRS) and transferred its reduced functions back to the NPS. Thus the Park Service presently serves as the lead federal agency for recreation planning and provides much assistance to state and local governments in recreation land acquisition and management.

National Parks from the 1950s to the 1960s: Mission 66 Program.   With the rapid growth in National Park visitation that accompanied the post–World War II auto-mobile boom, the Park Service saw a need to upgrade the capacity of the National Parks to accommodate the continued growth in use. Thus in the 1950s, the Park Service's Mission 66 program was to bring all units of the National Park System up to a high standard of preservation and staffing by 1966, the fiftieth anniversary of the NPS. The other goal of the Mission 66 effort was to increase visitor use of the National Parks to bolster public support for them and their budgets. Under this policy paved roads were built into both Olympic National Park and Mt. Rainier to provide public access (Foresta 1984:55). It was under the Mission 66 program that the modern visitor center was born as a centralized facility to inform and educate the visitor about that park and the National Park Service in general.

Mission 66 reflected a period of emphasis on visitor use in the NPS dual man-date of preservation and use of the parks. In 1963 Secretary of Interior Udall brought his views of the ''new conservation ethic'' into the department. As a result, the Park Service began to abandon the tourism promotion and development goals associated with Mission 66. The pendulum began a long swing back to preservation. Although this movement would be momentarily interrupted by Secretary of Interior James Watt from 1981 to 1983, NPS emphasis on preservation continues today.

Management Objectives of the National Parks System

The National Park Service continually updates and refines its mission statement to reflect its changing emphasis on use versus preservation. In 1978 the Park Service's manual named five long-range objectives:

1. To conserve and manage for their highest purpose the natural, historical, and recreational resources of the National Park System.
2. To provide for the highest quality of use and enjoyment of the National Park System by millions of visitors.
3. To develop the National Park System through inclusion of additional areas of scenic, scientific, historical, and recreational value to the nation.
4. To use the Park System to communicate the cultural, natural, inspirational, and recreational significance of the American heritage.
5. To cooperate with others to protect and perpetuate natural, cultural, and recreational resources of local, state, regional, and international importance for the benefit of humankind.

Thus the Park Service mission was not only to manage its own lands but to act as a partner with local and foreign governments to protect natural and recreational resources worldwide.

Current Structure of the National Park System

Although much of the public's attention focuses on the large western parks, which are sometimes referred to as the "crown jewels," Congress has given the National Park Service the responsibility to manage a wide range of natural, historic, and cultural resources throughout the United States. The very name National Park System only begins to convey the wide range of natural, cultural, historic, and recreational resources managed by the NPS. Each land classification has its own special and sometimes legal attributes.

1. *National Parks.* The most obvious units within the National Park System are the National Parks. As of 1987 there were forty-nine National Parks with 47 million acres total. National Parks are normally large areas that contain several nationally outstanding natural and/or cultural features. The areas of land are large enough to ensure the protection of these features from influences outside the park boundaries, although we have begun to recognize the need for ecological boundaries even for National Parks as large as Yellowstone, since wildlife and hydrological functions often require millions of acres. This issue will be discussed in more detail in chapter 13. The establishment of National Parks takes an act of Congress.

2. *National Monuments (NM).* National Monuments are areas of land, normally smaller than a park, that are set up to protect just one nationally significant resource. There were seventy-seven National Monuments in 1987, totaling nearly 5 million acres. National Monuments can be established by the president. Such authority for establishing National Monuments dates back to the Antiquities Act of 1906. President Theodore Roosevelt used this act to establish numerous National Monuments, many of them aimed at protecting archaeological sites. Sometimes the critical features of an area will have been protected as a National Monument while Congress debates National Park status for a larger area containing the National Monument. This has been the case for many National Parks, including Arches National Park in Utah. One of the most recent large-scale uses of the Antiquities Act was by President Carter in 1979. He used it to protect millions of acres in Alaska while debate continued over which areas would be designated National Parks.

3. *National Preserves.* The National Preserves classification is also aimed at protecting selected natural features. Unlike National Parks and Monuments, National Preserves may allow for such uses as hunting or mining if they do

not impair the resources of the preserve. Many National Preserves are found in Alaska because subsistence hunting is a way of life for many rural peoples there. There were twelve National Preserves in 1987, totaling about 22 million acres.

4. *National Seashores and Lakeshores.* This classification protects water-related natural and recreational values of national significance. There are National Seashores on the Atlantic, Gulf, and Pacific coasts. Some of the most intense recreation use of any Park Service units are National Seashores, such as Fire Island, in New York; Cape Cod, north of Boston; and Point Reyes, north of San Francisco. There are ten National Seashores with 600,000 acres. There are four National Lakeshores as well, which are centered around the Great Lakes.

5. *National Historic Sites, Memorials, and Battlefields.* These are small areas that commemorate nationally significant people (e.g., past presidents, such as Washington and Lincoln) or events, such as Revolutionary or Civil War battles. There are a total of twenty-four military/battlefield parks, twenty-four memorials, and sixty-three historic sites, not including the White House and Capitol Parks. Unlike the National Parks, many of these Historic Sites, Memorials, and Battlefields are naturally located in the eastern United States, with several in Washington, D.C.

6. *National Recreation Areas (NRAs).* This classification emphasizes protection of an area for its primary use as a recreational resource. Other uses are allowed (such as mining) only to the extent that it does not impair recreational use of the NRA. NRAs are often established on lands surrounding reservoirs created by other agencies. Such areas include Flaming Gorge NRA, Lake Powell NRA, and Lake Mead NRA all of which are along federal water projects on the Colorado River or its tributaries. The NPS manages fifteen NRAs totaling 3.7 million acres. In some cases, NRAs have been created out of National Forest land, and they continue to be managed by the Forest Service, as in the case of Shasta Lake NRA in California or Sawtooth NRA in Idaho. The Forest Service has thirteen NRAs totaling 1.5 million acres. There are several NRAs to provide urban recreation, such as Golden Gate NRA, near San Francisco, and the Santa Monica Mountains NRA, near Los Angeles.

7. *National Parkways.* This designation protects scenic resources along travel corridors. There are four of these, including the Blue Ridge Parkway in the Appalachian Mountains.

As a careful review of this list indicates, the National Park System is made up of far more than just "parks" and includes special provisions for recreation at NRAs and National Seashores. The planning and management of these units within the National Park System are the subject of chapter 13.

Federal Wildlife Policy and U.S. Fish and Wildlife Service

Evolution of Federal and State Wildlife Law

Wildlife and fisheries management in the United States represents an interesting division of roles between the federal and state governments. A series of Supreme Court cases from the 1870s through the 1890s clarified that the states have the authority to control and regulate wildlife within their boundaries. To this day, resident fish and wildlife are considered the property of the state in which they reside. Nearly all game laws and hunting and fishing regulations are set by the respective states. Enforcement activities are also carried out by the state Fish and Game agencies, regardless of whether the harvesting is on federal lands, such as National Forests and Wildlife Refuges, state, or private lands. However, there were a few very important exceptions.

The federal government could ban all hunting on its land, as was the case in Yellowstone National Park in 1894 (Bean 1983:21). In addition, it could regulate the taking of wildlife on any of its own game refuges. The first broad authority for establishing these game refuges was given to the president under the Antiquities Act of 1906. A National Bison Range was established in 1908, and the National Elk Refuge was established in 1912.

Another important exception relates to migratory birds. Because migratory birds, such as waterfowl, often cross state and international boundaries, federal treaty-making powers give the federal government legitimate power in this area. That is, hunting regulations and conservation provisions of migratory waterfowl are covered by international treaties, such as the Migratory Bird Act of 1913, which is naturally the responsibility of the federal government.

Passage of the Migratory Bird Act of 1913, although helpful for the conservation of migratory birds, was deemed inadequate to maintain migratory bird populations without some provision for habitat acquisition (Bean 1983:120). In particular, the United States was losing wetlands at a very rapid rate because of draining of and conversion to farmland. As such, Congress eventually passed the Migratory Bird Conservation Act in 1929 to allow the U.S. Department of Interior to purchase or rent areas to be protected as refuges or sanctuaries for migratory birds. In 1934 the federal Duck Stamp program was established to fund acquisition of waterfowl habitats. Since 1934, hunters have been required to purchase a federal migratory bird hunting stamp (popularly known today as the Duck Stamp), with the funds used to finance acquisition of migratory bird habitats. This program has been a success story that has been helpful in slowing the decline of migratory birds in the United States. About 2 million acres of National Wildlife Refuge lands have been purchased or leased with revenues from the Duck Stamp.

Evolution of Federal Fish and Wildlife Service

Paralleling the legal process of defining state and federal roles in wildlife management was determination of which federal agencies would implement these federal laws. Given the importance of fish and wildlife as food supplies in the 1800s, wildlife and fish population trends were closely watched. In 1871, Congress authorized the president to appoint a commissioner of Fish and Fisheries to study the decrease in supply of coastal and lake fisheries (Dana and Fairfax 1980:79). A Bureau of Fisheries was later established in the Department of Commerce. In 1905 the Bureau of Biological Survey was established in the Department of Agriculture to study birds and mammals in the United States.

By the mid-1900s, there had been numerous federal wildlife laws, reservation of game refuges under the Antiquities Act, establishment of migratory bird refuges under the Migratory Bird Conservation Act, and so on. However, federal authority and management responsibility was split among several agencies in three or more departments. To remedy this, President Franklin Roosevelt created in 1940 a centralized federal fish and wildlife agency in the Department of Interior by combining the Bureau of Fisheries and the Bureau of Biological Survey (from USDA) into the new U.S. Fish and Wildlife Service (Verburg and Coon 1987:20).

Roles of the U.S. Fish and Wildlife Service (USFWS)

The U.S. Fish and Wildlife Service (USFWS) plays at least six major roles in federal wildlife management. As discussed earlier, it has the primary responsibility for management of migratory birds. This includes development of hunting regulations, habitat acquisition, habitat protection, and enforcement of international treaties on migratory birds. The USFWS effort on habitat acquisition associated with the North American Waterfowl Management Plan will be discussed in chapter 12.

The USFWS is also the primary federal wildlife agency. Its broadest source of authority is contained in the Fish and Wildlife Act of 1956 (Verburg and Coon 1987:20). Under this act, the agency is instructed to take steps required for the protection, conservation, and advancement of fish and wildlife resources (Verburg and Coon 1987:20). Another broad source of federal wildlife authority is the Fish and Wildlife Coordination Act of 1934, under which the USFWS reviews any federal project or one requiring federal permission to evaluate its effect on wildlife or its habitat. As part of its assessment of impacts, USFWS recommends how any adverse impact can be mitigated by modification of the project or through compensatory protection of similar habitats adjacent to or near the project. This role in wildlife mitigation will also be more fully discussed in chapter 12. The USFWS implementation of the Coordination Act has been most visible in protecting wildlife and fisheries in federal development projects such as U.S. Army Corps of Engineers and Bureau of Reclamation dam projects. To improve the evaluation of wildlife and fisheries impacts from these types of development projects, USFWS has produced standardized evaluation techniques, such as the Habitat Evaluation Procedures (HEP) and Instream Flow Incremental Methodology (IFIM). The HEP habitat models will be discussed more in chapters 5 and 12.

The National Environmental Policy Act of 1969 provided a more formal vehicle for USFWS to express its concern about the impact of federal projects or federally permitted projects of all types on fish and wildlife. Since an Environmental Assessment (and often an Environmental Impact Statement [EIS]) is required for nearly all significant federal actions, this provides an avenue for USFWS to achieve recognition of fish and wildlife concerns in project decision making and to mitigate unavoidable fish and wildlife losses. The USFWS, as the lead federal wildlife agency, spends a great deal of time reviewing the EIS's of other agencies. NEPA also significantly changed the way the USFWS performed its day-to-day business. USFWS must also prepare EIS's on its own wildlife projects and refuge plans as well as assist agencies in meeting the mitigation requirements associated with NEPA.

The most obvious role of USFWS is managing 452 federal wildlife refuges encompassing nearly 90 million acres. This role will be discussed in greater detail later and in chapter 12. The USFWS also aids state wildlife management by providing wildlife research and technical leadership via national research laboratories and by maintaining Cooperative Fish and Wildlife Units at selected universities to encourage the training of fish and wildlife biologists. Using funds collected from federal excise taxes on firearms under the Pittman-Robertson Act and for fisheries restoration using excise taxes collected on fishing equipment under the Dingell-Johnson Act, USFWS provides monetary grants to the states for fish and wildlife conservation projects. Since 1955 the USFWS has conducted a national survey on hunting and fishing that tracks trends in visitation and hunter/angler expenditures.

The USFWS also operates about ninety fish hatcheries to augment state hatchery efforts. The fish from federal fish hatcheries are transported to streams and lakes throughout the United States (Reed and Drabelle 1984:13).

Lastly, the USFWS is the lead agency associated with implementing the Endangered Species Act. Given the dominant roles both endangered species and refuges play in the activities of the USFWS, these two issues are discussed in more detail later. Nonetheless, the overview of the six major roles of USFWS illustrates the breadth of responsibilities of the agency. It spans management of its own refuge lands and operating fish hatcheries to consultation with other agencies and research. As such, the USFWS is a relatively decentralized agency with much activity taking place at its seven regional offices and numerous field offices throughout the United States and at the refuges themselves.

National Wildlife Refuge System Administration Act of 1966 and the Modern-day National Wildlife Refuges

From the first federal Wildlife Refuge at Pelican Island in Florida in 1903 to the early 1960s, refuges were added for various purposes: game refuges, elk and bison ranges, waterfowl production areas, and so on. In fact, there are at least five different routes by which a refuge might come into existence (Reed and Drabelle 1984). These include (1) executive withdrawals of otherwise designated federal land and management by USFWS as a refuge. This has sometimes been done under the Antiquities Act of 1906 and has been used by presidents as recently as Carter in 1980; (2) purchase of wetlands or related migratory bird habitat using funds from the sale of Duck Stamps; (3) purchase of lands using funds from the Federal Land and Water Conservation fund (money that originates as a portion of revenue from federal offshore oil and gas leasing); (4) an act of Congress specifically designating an area as a refuge; (5) donations of land from individuals or groups.

Because of the many different origins of refuge lands and numerous management mandates, it was necessary to think about these different wildlife areas as a system. In 1966 the National Wildlife Refuge System Administration Act consolidated wildlife refuges, game refuges, and management areas into one new National Wildlife Refuge System under the U.S. Fish and Wildlife Service (then called the Bureau of Sport Fisheries and Wildlife).

Unfortunately, the National Wildlife Refuge System Act did not provide detailed guidance on the explicit objectives of this system. In fact, the act permitted broadening the uses of Wildlife Refuges in some ways. With the act, not only were compatible recreation uses allowed, but so were other compatible uses, such as agriculture, livestock grazing, and even mining and timber harvesting in some cases. This expansion has resulted in many ''compatible uses'' being established; only later did it become apparent that these uses were really incompatible. By then it was too late to exclude many of them. To make matters worse, the original act did not completely strip away joint agency management of some wildlife refuges, such as the Charles M. Russell Refuge in Montana. Thus a multiple-use agency such as BLM would share authority over management with the Fish and Wildlife Service. In many cases these built-in federal agency conflicts led to inadequate protection of wildlife habitat. It was not until 1976 that the act was amended to specify that all units of the Refuge System would be managed by the USFWS (Bean 1983:128). Nonetheless, even on a National Wildlife Refuge, wildlife can be at best considered the primary use, but in many cases it is not the sole use.

Expansion of Alaskan Refuges

The last major imprint on the National Wildlife Refuge System was the Alaska National Interest Lands Conservation Act of 1980 (ANILCA). Although a large amount of what was formerly BLM land was designated as National Parks, by far the largest land designation was National Wildlife Refuges. The USFWS received nine new wildlife refuges and six of the seven existing refuges were enlarged for a total of 77 million acres. An important feature of this act was the establishment of purposes and priorities for these refuges. The highest-priority purposes related to (1) conserving fish and wildlife populations and habitats in their natural diversity; (2) fulfilling international treaty obligations with respect to fish and wildlife; (3) ensuring that water quality and necessary water quantity were available to the refuge (a major recognition of the importance of water in wildlife and fisheries management, a lesson learned the hard way in the lower forty-eight states, where refuges were established in Nevada and California without adequate water supplies to fulfill their mandates). Subservient to these three provisions was the additional purpose that each refuge, other than the Kenai, was to provide the opportunity for continued subsistence use by local residents (Bean 1983:134). A significant feature of the law is that it mandated that all wildlife refuges in Alaska conduct a comprehensive planning process so as to determine how it would achieve these purposes (Bangs et al. 1986). The procedures used by the U.S. Fish and Wildlife Service to plan and manage its refuges will be studied in chapter 12.

Wilderness and Wild/Scenic Rivers as Wildlife Refuge Land Uses

As mentioned earlier in this chapter, refuge lands are eligible for classification as Wilderness. Rivers on refuges are eligible for designations under the Wild and Scenic Rivers Act. The resources are still managed by the USFWS but with the added requirements associated with the specific land classification acts. As of 1984, Congress had designated Wilderness on fifty-seven refuges with acreage totaling nearly 20 million acres (Reed and Drabelle 1984:28). Proposals for additional Wilderness designations will continue, especially in Alaska, where ANILCA required Refuges to consider potential suitability of Refuge lands for Wilderness when developing their Comprehensive Conservation Plans.

Endangered Species Act of 1973

The Endangered Species Act of 1973 and amendments in 1978 are the most visible federal wildlife laws and have created special responsibilities for all federal agencies, but especially for the USFWS. This legislation contained four key provisions:

1. It shifted responsibilities for endangered animals from state control to the federal level and included threatened or endangered plants.
2. It set up a process whereby species would be studied for recommendation for either threatened or endangered status. An endangered species is defined as any species or subspecies of fish, wildlife, or plant that is in danger of

extinction throughout all or a significant portion of its range (Fish and Wild-life Service 1983:55100). A threatened species is any species or subspecies of fish, wildlife, or plants that is likely to become endangered within the foreseeable future throughout all or a significant portion of its range. Five factors are considered in whether to list a species as endangered or threatened (Fish and Wildlife Service 1983:55100). These include (1) present or threatened destruction or modification of its habitat; (2) overharvest of species; (3) disease or predation; (4) inadequacy of existing regulatory mechanisms; (5) other natural or man-made factors affecting its continued existence.

3. It required that all federal agencies seek to conserve endangered species and take all measures necessary to restore threatened and endangered species and their habitats to the population levels that would place them out of danger of extinction. The act generally prohibited any entity (private, state, or federal) from killing, harming, or harassing any threatened or endangered species.

4. It emphasized protecting habitat, including designation of critical habitats whereby management of endangered species would be the dominant concern.

The USFWS interpreted the prohibition on harming endangered species as meaning that an agency could not engage in any action on public (or private lands) that would jeopardize the continued existence of an endangered species. A section in the act (Section 7) requires federal agencies to obtain a "biological opinion" from USFWS regarding the potential for the agency's action to jeopardize the continued existence of any listed threatened or endangered species. USFWS has many specialists whose job is to render such jeopardy opinions. The Section 7 consultation was intended as an early-warning system to agencies to modify their projects to eliminate effects on endangered species. However, the most famous first applications of this section of the act applied to projects that were to a large extent already under way. Perhaps the most famous and the one that led to amendment of the Endangered Species Act was the Tellico Dam incident in 1978. We will now turn to that episode.

During a Section 7 review of the Tennessee Valley Authority's construction of the Tellico Dam, it was discovered that the dam would destroy what was believed at that time to be the only habitat of a species of fish called the snail darter. In June of 1978, the Supreme Court ruled that construction on the dam was to be halted. In some sense, the Court ruled that Congress was clear in the act and that nothing should take precedence over an endangered species. Although the ruling was a strictly correct interpretation of what Congress had specified in the 1973 Endangered Species Act, the fact that a 2-inch fish could stop a dam was held up to ridicule by developers as clear indication that environmental regulations had gone too far. Later that same year, Congress amended the Endangered Species Act. The amendments allow a balancing of other interests with biological ones to be made by a Cabinet-level committee (sometimes referred to as the ''God Squad'') in cases where conflicts between development and endangered species were involved.

The next major test of the revised Endangered Species Act and the extent of balancing that would be allowed under the amendments came in 1991, with the controversy over protecting the northern spotted owl in old-growth forests in the Pacific Northwest. Continued clear-cutting of the remaining old growth was determined by biologists to be a major threat to the survival of the spotted owl. An interagency panel of scientists (led by Forest Service research biologist Jack Ward Thomas) determined that preventing extinction of the owl would require preservation of several million acres of low-elevation old growth as habitat protected from logging.

Preservation of such large habitat areas could reduce the amount of old growth available for timber harvesting in National Forests in Oregon, Washington, and northern California. These states and their associated private timber industries were concerned about the potential for loss of several thousand jobs if such large areas were protected as owl habitat. Even further, listing of the owl as a threatened species could delay timber sales and make logging of new sales more difficult. Under pressure from Oregon's congressional delegation, Secretary of Interior Lujan and President Bush had indicated that some balancing would be necessary between preservation of owls and the timber industry.

In June of 1990, the USFWS released its decision that the northern spotted owl would be listed as a threatened species throughout its range from northern California to Oregon and Washington. However, the extent of critical habitat to be designated as part of the northern spotted owl recovery effort has still not been fully resolved. The USFWS initial estimates of over 10 million acres of critical habitat were scaled back to about 7 million acres in the final critical habitat determinations. As of 1992 the Bush administration appears reluctant to support even this reduced acreage as critical habitat. The high costs of preserving the northern spotted owl along with several ''races'' of threatened salmon in the Pacific Northwest have resulted in numerous legislators calling for major revisions in the Endangered Species Act when it comes up for reauthorization in 1993

Summary of Interrelationships Between Land-Managing Agency and Land Use Classifications

In this chapter we have reviewed the origin and current laws governing the four main federal land-managing agencies. Together the Bureau of Land Management, Forest Service, Fish and Wildlife Service, and National Park Service manage over 90% of all federal lands and over 25% of all lands in the United States. However, other federal agencies, such as the Bureau of Reclamation, U.S. Army Corps of

Table 2.4
Relationship Between Federal Agencies and Land Use Classifications

Land Classification	Agencies			
	NPS	USFS	BLM	FWS
National Park	X			
National Monument	X	X		
National Seashore and Lakeshore	X			
National Recreation Area	X	X	X	
National Forest		X		
National Wildlife Refuge				X
Wild and Scenic River	X	X	X	X
National Trail	X	X	X	
Wilderness Area	X	X	X	X

Engineers, and the Tennessee Valley Authority, manage federal lands associated with federal water projects. These three agencies often interact with the other four federal agencies (particularly the Fish and Wildlife Service). As such, these three agencies use many of the same resource decision aids such as matrices for impact evaluation, optimization models, and benefit-cost analysis that will be discussed in the following chapters. Many of the principles presented in chapters 3–8 are applicable to and used by these other three federal agencies. Most of the principles and techniques discussed in chapters 3–8 are also applicable to management of state lands, such as state forests or state parks.

However, to set the stage for chapters 9–13, which are specific to the four federal land management agencies discussed here, table 2.4 provides an overview of the interrelationships between these four federal land agencies and the common land classifications discussed in chapter 2. This illustrates that certain land classifications are managed by the agency that managed the lands prior to its reclassification. This is the case for NRAs, Wilderness Areas, and Wild and Scenic rivers.

3

Economic Rationales for Continued Government Ownership of Land

Market Failures as One Reason for Government Ownership

As the review of the public land laws in chapter 2 demonstrated, Congress and the president often had very specific reasons for reserving various types of lands in federal ownership. This included protection of unique scenic and historical resources of the United States that would be lost if the lands were developed by the private sector. A common theme for reserving lands for National Parks, Wildlife Refuges, and Wilderness Areas was to protect and provide services of natural resources that would not be adequately supplied by private development. To make sense of the economic reasons for continued government ownership of land we must understand the deficiencies of private markets in dealing with some facets of natural resources. We must also understand how federal landownership relates to the general functions of the federal government. In this way we can evaluate the reasonableness of the current dual public and private ownership of land and resources in the United States.

Externalities as One Source of Market Failure

The original reservation of the National Forests was justified in part by the "cut-out-and-get-out" timber-harvesting practices of private firms. This left cut-over lands with no replanting of trees or cover to protect the watershed from soil erosion and downstream flash flooding. This harvesting was privately profitable in the 1800s for two reasons. First, there was so much virgin timberland that it was cheaper to move on to more of it than to replant. Second, logging firms were able to avoid bearing the downstream costs of soil erosion and flooding.

Implicit in the concerns about the downstream consequences of clear-cutting is one of the fundamental rationales for government ownership: market failure and externalities. The latter represent the effects of a firm's action that spill over to other people outside of the firm's employees or customers and that therefore are external to the profit-and-loss decisions of the firm. For example, timber harvesting high in a watershed increases peak runoff downstream, which can overflow the natural riverbank, spilling onto adjacent land. This land may be used for agriculture or urban uses. Timber harvesting can also result in substantial quantities of soil erosion, as vegetation cover is removed by cutting and access road construction. The increased turbidity of the downstream river flows negatively affects fisheries and many beneficial uses of water in downstream cities and towns. Figure 3.1a illustrates this interaction.

Figure 3.1a illustrates the separate ownership of timberlands and downstream city lands that take water from the river. When the timber owner calculates the costs and benefits of harvesting timber on his or her private forestland, the owner considers just the necessary costs. These include the private costs of constructing timber access roads and cutting the trees. Since the owner does not own the downstream water supply used by the city, he or she bears no costs associated with the secondary effects of timber harvesting that are imposed on downstream landowners and water users. These downstream city residents are third parties that are ''external'' to the private timber landowner decisions on profitability of timber harvesting. Because people who live downstream do not own the forestland, they have no legal right to ask the timber operator to constrain his or her timber-harvesting practices to minimize peak flood flows or downstream soil erosion. Nonetheless there is a very real economic cost of upstream timber practices on downstream residents in the form of flood damages and water quality degradation.

If the land area is small enough and the damages to the town or city large enough, there may be incentives to ''internalize'' the external costs by having the residents negotiate with the upstream forest landowner over his or her timber-harvesting practices. The city could pay the timber owner some of the money it would save on reduced water treatment and flood cleanup if the owner cut fewer trees so that less soil was displaced..

If the upstream area is a major water source for the town, it may simply be in the town's best long-run interest to purchase the forested land in its watershed. Now that the town owns the forestland, the external costs of timber harvesting are internal to the city's own profit/loss calculations. This is illustrated in figure 3.1b. For example, harvesting some of the timber would provide revenue to the citizens of the city. But now with a single owner of both the forestland and the water supply system, this must be balanced with additional water treatment costs resulting from increased sediment from timber harvesting. Since both these costs are faced by the same single owner, these costs imposed by timber harvesting will show up as higher water treatment costs to the same entity making the timber-harvesting decision.

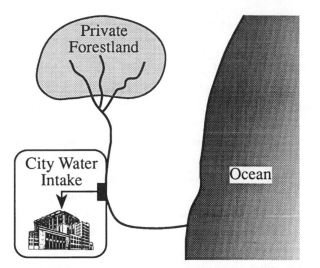

Figure 3.1a.   Separate Ownership and Externalities

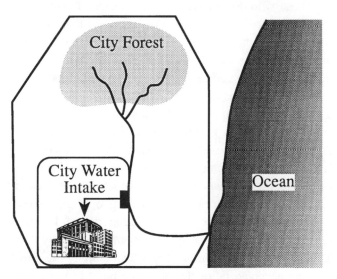

Figure 3.1b.   Single Ownership and Internalization of the Externality

Thus a balancing of timber-harvesting revenues and water supply treatment costs will occur until the gain from additional timber harvesting would be just offset by the increased cost of water treatment. Consequently, the first economic reason for government ownership of land relates to the logic of ''internalizing externalities.'' Government ownership helps ensure that commodity production decisions on lands that also provide downstream wildlife, fisheries, and water supply reflect a balancing of those commodity benefits with the costs imposed on natural resource services from those lands.

Public Goods as Source of Market Failure

Another economic rationale for public ownership relates to another source of market failure, the inability to supply the optimal amount of *public goods*. These are
goods or services that, once a given quantity has been provided to one person, can
also be costlessly consumed by another. For example, if a clean atmosphere free
of air pollutants was provided to you where you live, that same air quality would
necessarily be provided to all who lived in the same area. An atmosphere is not
technically divisible; that is, one person riding a bike to work cannot consume clean
air while another in front of him receives dirty air. This is of course in contrast to
private goods, such as hamburgers. For example, if you consume an entire hamburger, that same hamburger is no longer available for your friend's consumption.

As will be discussed in more detail in chapter 6, the competitive, free-enterprise
system does a good job of supplying the optimum amount of private goods when
there are no externalities in the production or consumption of the private good.
Private goods are allocated to those who are willing to pay the price necessary to
produce them. Those who don't pay, don't get the good. Thus nonpayers can be
economically excluded from consuming private goods. This is of course efficient,
since if you do not value the good at least equal to the cost to produce a unit for
you, the resources should be used to produce other goods.

However, this market logic breaks down with public goods. How can a firm
stay in business when it supplies a public good to one person, and all others in that
area can enjoy the good without paying? It often cannot stay in business just selling
to a few paying customers. In general, private for-profit firms will not produce
public goods. Even when it is possible to produce some limited amount of a public
good, the fact that many people can get benefits without purchasing the good results
in the firm just producing the quantity justified for its paying customers. But there
is an "external benefit" to additional nonpaying people that implies the level of
the public good (e.g., size of the park) should be increased. However, private for-
profit firms ignore the benefit to nonpaying customers, since by definition they are
nonpaying. The net result is that private firms undersupply the public good.

There is a long-accepted rationale for government provision of public goods
such as national defense. One house cannot be protected from air attack while that
of the nonpaying neighbor next door is not. The same rationale underlies government provision of clean air. The protection of the atmosphere from ozone-damaging
chlorofluorocarbons (CFCs) is an international public good.

Natural and environmental resources often have many of the characteristics of
public goods. Protecting the habitat for the northern spotted owl in the National
Forests of Oregon for one person preserves the habitat for the northern spotted owl
for millions of other people. Inventorying, studying, and stabilizing archaeological
ruins result in new knowledge about our past that is available to all people, not just
to the person or group who paid for the research. Preserving the geysers in Yel-

lowstone for one person's enjoyment provides those same geysers to millions of other people. Protection of the nesting habitat for migratory waterfowl in South Dakota provides duck hunting opportunities to millions of people in the numerous states all along the Central Flyway. Preservation of the Snake River through Hell's Canyon in Idaho as a free-flowing river for one's own grandchildren and great-grandchildren also preserves it for future generations of other people's children. Protection of a unique species or natural environment is therefore a public good that occurs on many types of federal and state lands.

Since all people can benefit from consuming public goods whether they pay or not, there would be an incentive for people in a market to try to avoid payment for a public good, since one consumes the good anyway and can free-ride off the payments of others. If all people free-ride, little or none of the public good will be supplied, because without payments to cover the costs of initially producing the public good, a firm would lose money supplying it. In a few cases where the private benefits of supplying a small amount of a public good exceeds the supply costs, some small quantity of a public good might be supplied by the private market. However, the privately profitable quantity to supply will be less than the social optimum. This occurs because the private firm ignores the additional benefits that nonpayers receive. Thus at the privately supplied quantity, the benefit of an additional unit of the public good to society exceeds the costs of supplying an additional unit.

Since all members of society may receive benefits from supply of a public good, government is a key element (as a direct producer or financier) in the optimal provision of public goods. Society cooperates through its government by agreeing to share the costs of providing this public good. We agree either through our elected representatives or directly through an initiative process to force everyone to pay for the good through tax assessments of some form (e.g., income taxes, sales taxes, or property taxes). Through payment by taxes, there is little or no free riding, as all consumers pay for the provision of the public good.

Thus the economic rationale for government ownership of land relates to its ability to internalize externalities and supply the optimum amount of public goods from these lands. In some sense, this rationale relates to government's ability to overcome the deficiencies that would result if many of the natural resources found on public lands were allocated exclusively through private markets.

Government Must Still Determine Optimal Quantity of Public Goods to Supply

Of course government financing and provision of public goods do not automatically guarantee that the socially desirable amount of the public good is provided. Provision of increasing supplies of public goods (e.g., more migratory waterfowl and larger national parks) still involves *opportunity costs,* which are the costs that result from forgoing other valuable goods or services. Opportunity costs of public goods can take the form of diverting available tax monies from supplying other government services that people value (e.g., public housing for the homeless, food stamps) or forgoing other valuable uses of the natural resources involved in producing the public good. For example, more land devoted to production of waterfowl may mean less land for agricultural or mineral production. Thus government must determine the optimal amount of public goods to supply, given the trade-offs between benefits and costs to society. Because market prices do not exist (and even if they did, they would be very misleading for the reasons discussed earlier), the government must turn to other expressions of public value for alternative quantities of public goods. (The details of these techniques will be discussed in chapter 6.)

Determining the optimal mix of public goods and private goods (e.g., timber and minerals) to supply from public lands and the optimal quantities of these public goods is one of the fundamental purposes of land and resource planning by federal agencies. Because this requires balancing benefits of additional public goods against competing uses of those resources, the logic of economics may often be appropriate. However, as will be discussed more later, attempts to transplant a textbook market model blindly to the provision of these natural resources will be disastrous. As described earlier, the conditions for operation of a private, for-profit, perfectly competitive market do not exist for most of these natural resources. The presence of market failures, externalities, and public goods justifies government ownership and thus provides feedback that the narrow view of a private, for-profit firm is a misleading basis for public land management.

Alternatives to Public Ownership of Land

The reason for stressing this rationale for public ownership is partly the recurring themes of privatization of public lands. Privatization proposals often take one of three forms: (1) sale to individuals (the latest manifestation being the 1980s Sagebrush Rebellion; see ''Another View'' at the end of this chapter); (2) operating the public lands on a ''pay-as-you-go basis'' (O'Toole 1988); (3) managing public lands as quasi-independent corporations (Nelson 1981). Although each of these proposals has some element of truth, all ignore the fundamental rationale for continued government ownership and management: most of the natural resources on public lands do not fit the market model of perfectly divisible resources, the production of which does not impinge upon any third parties. That is, forests are more than timber; they are watersheds, wildlife habitat, and recreation areas; cutting timber will affect the value these other resources provide to people outside the timber transaction. Rivers are more than potential hydropower and agricultural water supply; they are fish habitat and in many cases outstanding recreational resources. Thus although many of the resources on public lands have some private marketable component, what makes them special is that they contain a large nonmarketed public goods component that would be largely undervalued by private market transactions

Improving Current Public Land Management: Which Way Do We Go?

All three of the competing approaches to privatizing public lands—disposal (Stroup and Baden 1983), ''marketizing'' (O'Toole 1988), or ''corporatizing'' (Nelson 1981) attempt to address the less than optimal management of public lands under current government ownership. There is little question that the public lands could be managed for greater benefits to society. There are numerous examples of misguided management on public lands, some of which reflect unintentional mistakes by agencies that had imperfect information at the time about the ecological costs of their actions (e.g., fire suppression policies of the early 1900s). But such unintentional mistakes are not the sole province of public agencies. For example, a review of companies with sites listed on the Superfund list of hazardous waste sites reads like a listing of the Fortune 500 firms in the United States. Much of that pollution resulted from lack of information by private firms about the long-term consequences of their disposal actions. As society develops better information on environmental consequences, both public agencies and private firms adopt this better information in their management either voluntarily or by regulation. (Of course some pollution resulted from failure to impose waste disposal costs on agencies, firms, and their customers rather than on society as a whole.)

Another alleged mismanagement of public lands reflects failure to manage solely on the basis of maximizing profits from the lands (Walker 1985). In many cases this is hardly mismanagement, as the discussion of public land agency legislation in chapter 2 demonstrates. For example, Congress had many more motivations in mind than financial return when it set the objectives of the National Forests, National Parks, and National Wildlife Refuges (e.g., protecting watersheds and protecting resources for future generations). As discussed earlier, public land management driven by a pursuit of maximum profits often leads to market failure in the form of negative externalities and undersupply of public goods. Thus some alleged mismanagement reflecting failure to maximize profits is often due to pursuit of other objectives that Congress requires government agencies to pursue. It is to these other objectives or roles of government that we now turn.

Three Roles of Government in the Economy

It is well accepted that government plays three major roles in most economies (Musgrave 1959):

1. Promoting economic efficiency in using resources to obtain the maximum benefit for society.

2. Stabilizing the overall economy to prevent severe downturns (e.g., avoiding depressions) and stabilizing the overall level of prices (e.g., avoiding hyperinflation).
3. Promoting equity by ensuring everyone of minimum levels of material goods.

Economic Efficiency

The objective of promoting economically efficient use of society's resources helps to ensure that society receives the most benefit available from its scarce resources. This task involves efforts to keep markets competitive rather than monopolistic (e.g., the Antitrust Division in the Department of Justice), since competitive markets usually result in greater output and lower prices. When large economies of scale make a single producer more efficient, various levels of government will grant a monopolistic right in exchange for the firm agreeing to government regulation of its prices and profits. Thus a single government-regulated electrical utility often provides a given city with electricity. Within this objective of promoting economic efficiency is the government's role to correct for market failures and supply public goods. A market failure such as a negative externality represents an *inefficient* use of resources. When a negative externality is uncorrected, society uses too many resources to produce too much of that good. For the last unit of a negative externality generating activity, the benefits are often much less than the opportunity costs, where opportunity costs include the private input costs and the costs imposed on third parties from the externality. Although environmental regulation of industry is justified by internalizing the negative externalities, control of other externalities occurs through government ownership of land. Thus one justification for government ownership of land is that public ownership is consistent with the government's role of promoting economic efficiency in the use of natural resources.

Economic Stabilization

The stabilization function of government includes controlling unemployment and inflation as well as promoting economic growth. Much cyclical unemployment (above the temporary frictional unemployment associated with job switchers) is related to either lack of aggregate demand for all products in the economy or the ups and downs of the overall business cycle. The national government can deal with this cyclical unemployment using fiscal and monetary policies. Fiscal policy involves changing the overall level of government taxing and spending in the economy to stimulate or slow down the overall demand for goods and services. Monetary policy relies on directly influencing interest rates or changing the nation's money supply (which can also influence interest rates). Changing interest rates changes a key determinant of business investment spending and some consumer spending. However, some temporary unemployment is due to structural changes in the economy, such as movement from an industrial economy to a service economy or automation of factories, which reduces the amount and nature of labor required in the production process. This type of unemployment is best dealt with via worker retraining programs to facilitate the flow of labor from industries needing less labor to those needing more.

This stabilization function provides little in the way of justification for government ownership or management of public lands (Krutilla and Haigh 1978:383). Government ownership of land will, by itself, neither reduce nor increase national employment levels significantly. In addition, changing the management of public lands will not effectively deal with a national recession. For example, if the national economy goes into a recession there will be reduced demand for large single-family homes nationwide. Thus there will be a decrease in demand for timber on National Forests and privately owned commercial forestland. This will translate to reduced logging employment in these areas. As we saw from our review of public land laws in chapter 2, creating jobs for those who are temporarily unemployed because of the reduction in housing demand is not the role of the public land agencies. The solution is for the stabilization function of the federal government to implement the appropriate fiscal and monetary policies to move the economy out of the recession, not for public land agencies to change otherwise sound land management practices.

Distributional Equity

The distributional equity function of government is concerned with reducing the inequality in well-being that results from reliance on a purely market-driven distribution of income. The federal and state governments utilize a "tax and transfer system" to transfer a portion of income from middle- and upper-income households to the lowest-income households. For example, various cash payments (welfare), in-kind payments (food stamps, public housing), and specially targeted programs (e.g., Aid to Families with Dependent Children, Headstart) attempt to provide a minimum standard of living to those unable to contribute directly to the economic system or who do so but are still not self-sufficient.

The tools available to public land managers to aid in improving the distributional equity are quite limited and not targeted directly at those most in need. Providing free firewood (if you have a pickup truck to drive there to cut it) or free public land recreation does little to aid the neediest members of society. As sympathetic as land managers might be to the plight of the poor or homeless, the resources at their disposal (often on land areas many hundreds of miles from the urban poor) would make an ineffective contribution to improving the well-being of such people.

Efficiency As a Primary Economic Rationale for Government Ownership of Land

As Krutilla and Haigh (1978) have argued, the justification for continued government ownership and management of public lands rests on the promotion of economic efficiency. A large part of this involves internalizing negative externalities and providing public goods from public lands. Krutilla and Haigh (1978) found much congressional direction to the Forest Service to focus on economic efficiency in both the 1974 Resources Planning Act and the 1976 National Forest Management Act.

Researchers at the Conservation Foundation (Shands et al. 1988) have extended this argument by suggesting that public lands emphasize the supply of public goods rather than commodity production, such as timber, livestock, and minerals. Their argument is that private lands have a reward system built in via the price system to produce these private commodities. That is, consumers pay firms directly for supplying additional quantities of private goods. In the absence of significant externalities in the production of private goods, the price mechanism, operating through the laws of supply and demand, will ensure that the market will meet changes in consumer demand for private goods. Since the price system does not operate accurately to reward the production of public goods, these will rarely be supplied on private lands, and what little supply there is will be just what is produced jointly with the commodity. Thus these researchers recommend that public lands fill a special place in natural resource management: emphasize the supply of those goods the market will not supply or provides less than the optimal amount, such as wilderness or migratory wildlife habitat.

In many cases this prescription fits the comparative advantages of much public land in two ways. Much of the land owned by such agencies as BLM was not wanted by farmers, miners, and developers; consequently, they were still in federal ownership in the mid-1900s. Thus they were the less productive lands in terms of supplying marketable commodities, such as beef, timber, or minerals. Second, much of the land reserved by the federal government as National Parks, Monuments, Wildlife Refuges, Wilderness Areas, and so on, was designated as such to preserve unique scenic or ecological values. That is, much of what made many areas worthy of reservation was the outstanding public good attributes of these lands to current and (if protected) future generations.

The logic of this specialization of public lands to produce nonmarketed natural resources rather than commodities (which can be produced elsewhere on private lands) has become visible as one of the alternatives in the 1990 Resource Planning Act national program for the Forest Service (1990a).

Improving Management of Public Lands: The Role of Planning and Evaluation

Having suggested that one of the primary criteria for managing public lands is attainment of efficiency in their use (i.e., maximizing benefits in excess of costs), one must ask how society is to ensure that this objective is met. In part, that is the subject of this book: improving the conceptual foundation and techniques available to identify and guide public land managers to efficient choices.

However, the critics of public ownership correctly point out that even when economically and environmentally optimal resource allocations are demonstrated to managers by their staffs (or others outside the agency), many managers continue to pick inefficient allocations. One reason for this may be that the managers do not receive rewards for making efficient choices or penalties for making inefficient ones. That is, public land managers do not have sufficient incentives to pick efficient choices, and as O'Toole and others would argue (Stroup and Baden 1983), these managers may actually have incentives to pick inefficient resource allocations. Numerous examples illustrate the plausibility of these statements (and they will be explored in more detail in "Another View" at the end of this chapter). Certainly, anyone who has ever studied timber sales on the Tongass National Forest in Alaska in the 1980s can attest to the irrationality of continuing these sales in the face of overwhelming evidence of persistent and substantial economic and environmental losses. The perpetuation of this misallocation is not solely the fault of the managing agency, however. As one Forest Service economist remarked repeatedly throughout his career, much of what happens on the Forest is because "Congress told us to do it that way!" In the case of the Tongass, it was the Alaskan senators who originally convinced their fellow members in Congress to put into the Alaska National Interest Lands Conservation Act of 1980 explicit timber targets of 400 million board feet of timber annually (regardless of whether it was economically efficient to supply this amount) and $40 million annually to meet this target (with the money not subject to the normal appropriations process).

Three Roles of Analysis in Public Land Management

The ultimate role of analysis is to improve the management of public lands and avoid economically and environmentally unsound practices. Analysis can help to avoid or eliminate unsound management practices by identifying and quantifying the magnitude of the inefficiencies and environmental impacts associated with situations arising because "Congress told us to do it that way." For as surely as Congress gives bad prescriptions, it can, when repeatedly shown their high costs (both economic and environmental), take those prescriptions away. This has recently been the case on the Tongass National Forest, where Congress passed the

Tongass Timber Reform legislation in late 1990. In the face of pressure by conservationists and economists from the Wilderness Society, it eliminated the two inefficient legal requirements. The mandated level of timber harvest has been eliminated, as has the preset level of funding for timber sales.

Information on the benefits and costs of alternatives is also useful to local line managers within a public agency, who often exercise a significant amount of discretion over the types of management practices that are carried out on a piece of land. A careful integrated analysis of both marketed and nonmarketed resources will help the manager to balance competing demands for different mixes of multiple-use outputs. The manager and his or her staff can work interactively to identify the multiple-use trade-offs associated with modifying land management practices to meet concerns of one interest group or another. Analysis can assist the manager in determining which interest group's statements of ''dire consequences'' are credible and which are designed mostly to sway public opinion to the interest group's cause. Thus objective analysis can be a manager's best defense in avoiding major misallocations of resources and in reconciling contradictory demands of competing user groups.

In addition, a federal land management agency's analysis is useful to the other public agencies, including the Office of Management and Budget (in the Executive Branch). The analysis can be an asset to an agency in dealing with Congress, just as lack of analysis can be a distinct liability in dealing with Congress and its various oversight committees. Analysis performed by the agency itself or the General Accounting Office or Congressional Budget Office often evaluates how well an agency is implementing legislation. Similar studies evaluate whether laws must be amended to give an agency more specific direction on a particular issue.

In all three of these cases the analyst's job is to produce the facts and let the analysis speak for itself. The analyst cannot guarantee that his or her findings will be the final determinant of policy or land use allocations. Sometimes public land management decisions more closely reflect what is politically acceptable to key members of the administration or Congress than what is economically or environmentally sound. However, analysis can ensure that society is aware of the economic and environmental benefits and costs of adopting something less than the most efficient alternative. If there are not good reasons for choosing the most efficient alternative, the agency may be ordered by a federal judge to go back and rethink its approach. However, there are often good reasons for not selecting the alternative with the highest economic benefits when it unreasonably sacrifices other objectives left out of the benefit-cost comparison. Drawing on Clawson (the public lands scholar and former director of BLM), we provide in chapter 4 a listing of factors that might guide public land management and suggest how these factors might be integrated into a comprehensive analysis.

Another View: Privatization, Marketization, and the Sagebrush Rebellion

Almost since the beginning of efforts by the federal government to reserve land for public use, arguments have been raised against such reservations. At times it does seem incredible that in a country boasting of its free enterprise system, one-third of the land is owned by the federal government. However, private mining, logging, and ranching operations have access to most of this government land, so a somewhat regulated version of free enterprise use of these lands still continues.

Nonetheless, within a couple of years of the enactment of the Federal Land Policy and Management Act of 1976 (FLPMA), concerns by a few western counties about loss of local control over federal lands coalesced into the "Sagebrush Rebellion." FLPMA with its requirements for review of BLM lands for Wilderness, designation of Areas of Critical Environmental Concern, multiple-use planning process, and so on hinted at the potential of transforming the BLM from a custodial agency mainly responsive to local initiatives over land management into an active management agency responding to national direction for multiple-use management. Locals were concerned that the BLM might actually transform itself from the "Bureau of Livestock and Mining" into (in Director Frank Gregg's words) the "Best Land Manager," in the manner of the Forest Service.

The solution of rural counties in Nevada and Utah to this threat of losing local control was to call for the federal government to give these lands to private individuals or to the states (who would at least exercise relatively local control). This Sagebrush Rebellion remained a fairly extremist position until the election of Ronald Reagan as president and his selection of James Watt as secretary of the interior in 1981. These individuals added political support to the cause of the Sagebrush Rebellion. The Reagan administration implicitly used the threat of the Sagebrush Rebellion to urge land management agencies to return their management to a "good-neighbor" policy that gave more weight to the preferences of people living adjacent to these federal lands. Such a policy was implemented by the Reagan-appointed BLM director, Robert Burford, in a variety of ways during the early 1980s, some of which will be discussed in more detail in chapter 10. The most dramatic change may have been largely to abandon efforts to identify and implement reductions in livestock grazing on BLM lands.

Although the Sagebrush Rebellion did not reach its goal of transferring lands back to private ownership, it did broaden the debate about new institutions and rules for managing them. Several individuals began talking about the importance of marketlike incentives to elicit efficient behavior in government agencies (Stroup and Baden 1983). Examples of economically and environmentally unsound land management practices, such as money-losing pinyon-juniper land-clearing projects for livestock (called chainings, because two large caterpillar tractors knock down the trees by dragging a large anchor chain between the tractors) and money-losing timber sales (identified by Thomas Barlowe and colleagues [1980] of the Natural Resources Defense Council) suggested public agencies were answering not the call of conservation but that of their own self-interest in the form of higher agency budgets.

The argument for private ownership of the public lands appeared simple: private owners would not subsidize money-losing timber sales or range improvement projects. The private owner would bear all the costs of such money-losing actions, and self-interest would preclude unprofitable projects that also happened to be environmentally destructive.

But why would a public land manager engage in money-losing timber sales? There was little penalty not to engage in such sales, and some authors allege that the manager and agency can actually gain (O'Toole 1988). The problem is that public ownership drives a wedge between those that make decisions and those that must pay for the consequences of those decisions. In essence, the public land manager loses the taxpayers' money, not his or her own! If the public land manager is efficient and restricts his or her management to only economically efficient actions, the money saved is returned to the taxpayer, and his agency budget and personnel would shrink. The tendency to be efficient and accept a smaller budget and staff ran counter to what some believed was the inherent inclination of all public agencies (not just public land agencies): when in doubt, maximize your budget! The budget maximization hypothesis has been applied as an explanation of public land management agency behavior by several authors, including Stroup and Baden (1983) and O'Toole (1988).

Given the assumed validity of the budget maximization hypothesis and the beliefs that separation of authority to make decisions from responsibility to bear their consequences has reduced efficiency of public land management, several solutions have been offered: (1) put the land under private ownership, so that authority and responsibility would be merged together, as they are in a private company (where the owner must "eat" his or her mistakes but is also rewarded for efficient decisions); or (2) retain public ownership but make the agency self-financing by putting all resources on a pay-as-you-go basis. O'Toole (1988) believed that the alleged Forest Service bias in favor of timber production relates to the fact that the service could increase its budget by selling timber but not by selling recreation, wildlife, wilderness, and other amenities. His solution is to provide the same kind of monetary incentives for production of these amenity resources as the agency has for timber production. Coupling this with the requirement that the agency be self-financing from its sales receipts (no more taxpayer funding via Congress) would, O'Toole believes, result in a balanced treatment of all resources. No more would the agency have an incentive to engage in money-losing timber sales or range improvement projects. No longer would recreation and wilderness be slighted for timber.

These discussions have brought new thinking to the management of our public lands. They highlight important issues related to inappropriate incentives to managers unintentionally present in many laws that Congress passes for other purposes. This reminder of the importance of incentives helps to improve the making of new laws. It also aids in designing ways to ensure that new public land laws, such as NFMA and FLPMA, are implemented as intended (in particular, designing regulations and performance evaluations that harness employees' self-interest to move the agency to economically and environmentally sound management).

But although there are many elements of truth in the positions supporting privatization or marketization of public lands, there are several major weaknesses as well. The first argument worth questioning is whether the budget maximization hypothesis is in fact a good predictor of how agencies behave. Although much has been written about the appeal of this as a hypothesis, there has been little formal statistical testing of this hypothesis. What limited systematic statistical testing that has been done on the Forest Service shows either no support whatsoever (Loomis 1987) or a limited influence but one that is small compared to other factors (Sabatier et al. 1990). Thus there is little immediate reason to abandon public land management agencies if the primary reason for doing so is that they operate as budget maximizers.

The second and perhaps the key weakness of the privatization position is that many of the resources on public lands cannot be allocated efficiently through the market. Thus whether lands are transferred to private ownership or the public agency attempts to market these amenity resources, both actions are doomed to fail. The private landowner will find it much more profitable to concentrate on producing private goods, the full benefit of which can be captured from the buyer and transferred to the landowner. Since very little, if any, of the benefits of providing wildlife habitat for migratory or wide-ranging species (e.g., waterfowl, elk, whooping cranes, wolves) or preserving water flows for downstream recreation or preserving wilderness areas can be captured by the landowner (whether private or public), these resources would naturally be slighted under a pay-as-you-go policy. As will be discussed more fully later, the failure of these resources to pay their own way in a market setting is an indication not of their lack of value to society but of the inability of the market to reflect the benefits these resources provide. Because of the public goods attributes and beneficial externalities (beneficial spillovers to third parties) associated with these amenity resources, private ownership or allocation on a financial basis will result in market failure. Thus private ownership and marketization are institutions that, although they work very well for private goods, are far from perfect solutions for providing public goods.

However, one of the useful insights of the privatization and marketization discussion is that in many ways current public land management institutions are also imperfect solutions to the provision of public goods. Distorted incentives are provided to agencies when they are allowed to keep a substantial portion of timber receipts but little of recreation receipts and are precluded from earning anything on Wilderness recreation (since Congress precludes charging for Wilderness access). Certainly these distorting incentives have some influence on agency decisions. Incentives provided to western counties to encourage more timber harvesting arise from the fact that these counties receive 25% of the *gross* timber receipts for their county schools and roads. Thus what is needed is neither wholesale revisions of public land institutions nor perverse incentives to provide amenity resources. Rather, two responses are called for: removal of misdirected incentives that exist for commodity production (and if necessary simply making a fixed "payment in lieu of taxes" to the counties) and a planning process that communicates the benefits of marketed and nonmarketed resources to public land managers and other decision makers. We now turn to a discussion of the specific criteria for public land planning and management.

4

Criteria and Decision Techniques for Public Land Management

Determining the desirability of a particular management action requires some standards to guide judgments. Analysts often call these evaluation standards or measures of desirability *criteria*. For example, an increase in the number of wild horses on BLM lands may be judged by some as desirable, whereas others see it as deleterious. Often difference of opinion over natural resource management stems from differing frames of reference or criteria by which each group judges the action. For ecologists and many wildlife biologists, wild horses are nonnative species that further disrupt the ecological balance of vegetation and soils; hence increases in wild horse populations are undesirable. However, for devotees of wild horses, an increase in their number brings a sense of nostalgia for freedom of the open range. Thus, interpretation of management actions on the ground (e.g., clear-cutting, reseeding) and their results (e.g., cleared ground, more livestock) as desirable or undesirable depends on the individual's evaluation criteria.

The goals and objectives set by Congress or the president in establishing a particular area of federal land often provide some of the criteria by which to judge the appropriateness of different management actions. The public land laws (reviewed in chapter 2) and the logic for retention of public lands (discussed in chapter 3) identified several criteria by which to judge the desirability of public land management actions. Clawson (1975) formally grouped many of the legal and other rationales for public land management into five criteria, which are discussed in more detail later.

However, before we introduce particular criteria for resource management it is important to understand that the particular application of these criteria must be tailored for the legal mandate under which each type of agency operates. Congress has given the National Park Service a much-more-well-defined task, with a much smaller range of management options (e.g., no mining, timber harvesting, water development projects) than that assigned to the BLM, with its broad multiple-use management. How each of these agencies must tailor and integrate the following criteria will be discussed after all five are presented.

Five Potential Criteria for Evaluation of Public Land Management Alternatives

The following are the five criteria for evaluating public land alternatives:

1. *Physical and biological feasibility.* This criterion is essential to the success of natural resource management. All management practices and alternatives must be within the capability of an ecosystem. With sustained-yield management of any renewable resource, the harvest rate must not exceed the growth rate. Although humans can increase that growth rate some, ecosystem variables such as precipitation, soil type, length of growing season, and so on impose definite limits. Because wetlands require substantial amounts of water, it would not be biologically feasible to develop wetlands in many arid areas, no matter how much public demand there is for waterfowl hunting. Thus knowledge of physical, chemical, and biological factors is critical for successful resource management. In particular, these data must take two forms: knowledge of the current stock of renewable and nonrenewable resources present in the area to be managed and the rate of change in the stock (or resource flow). The first type of data can come from such inventories as soil surveys, vegetation mapping, stream records, wildlife censuses, and so on. The second type of data on current and potential growth rates of renewable natural resources is more difficult to obtain. Specifically, data are needed on the natural rates of change of the resources (natural growth rates of renewable resources, erosion rates of the soil, and so on), and on how these rates of change would be altered by different resource management activities that might negatively impact or enhance the resource.

Without some basic agreement on present resource stocks and their natural rates of change there is little foundation for planning or management. Knowledge of what is physically and biologically possible is critical to narrowing down the range of possible management actions. Many a resource management plan has failed because of poor understanding of the ecological limits of a particular area.

2. *Economic efficiency.* As indicated in chapter 3, economic efficiency is one of the legitimate concerns of government and one of the rationales for continued public ownership of land. By *economic efficiency,* we mean that a management action generates benefits to society (e.g., consumers and producers) that are greater than the costs of that action. The economic efficiency criterion is often the next logical one to apply after biological feasibility. If a

management action is physically and biologically possible, does it create more social benefits than it costs? Economic benefits are broadly defined as increases in well-being (not limited to just financial or material well-being) to both consumers and producers. Costs include both financial (i.e., cash) and opportunity costs (in terms of other resource uses forgone). These definitions of benefits and costs will be made more operational in chapter 6.

Using the economic efficiency criterion, we see that just because an area has trees does not mean that timber harvesting makes economic sense. If the sum of the cost of building the roads to harvest the trees (including the forgone primitive recreation and reduced water quality, fisheries, and wildlife costs), of logging, and of transporting the logs to market exceeds the price of the trees (plus the value of the roaded recreation created by the forest roads), it makes no economic sense to harvest them. Just because a particular mineral is present does not mean the land should be leased for its development. Often the losses to the other resources exceed the *net* value of the minerals (the price of the mineral minus the extraction, transport, and refining costs). That is, there may be $100 million worth of coal in the hills, but if it would cost $90 million to get it out, move it to market, and process it, there is only a $10 million gain to society over and above the direct costs of producing the coal. If this $10 million gain is less than the external costs coal mining imposes on other resources, then mining is not economically efficient, since society as a whole (the sum of consumers and producers, collectively) is worse off by leasing the public land for coal mining.

Implementing economic efficiency as an objective is not different from implementing biological feasibility: inventory information is required. Instead of inventorying the plants and animals one must inventory people to determine their benefits as consumers and producers from these plants and animals. The way this is done will be thoroughly discussed in chapter 6, but for the time being we may just note that such valuations are made through the market for most commodities (e.g., timber, beef, minerals). The techniques for revealing the values people have for resources and public goods that do not pass through markets are a little more difficult, but they offer no insurmountable problems. It is no more difficult to predict the habitat value of some newly created marsh to a heron than to predict the value people will get from viewing the heron at this marsh. Both require models (to be discussed in chapter 5), data, and some quantitative analysis.

Although economic efficiency provides information on whether there is a net gain to society beyond what was sacrificed to obtain this gain, it does not by itself (not at least without some additional effort) tell us about how those benefits and costs are distributed between various segments of society, such as the rich and the poor or different geographic areas of the country. Since society has some concern about who receives the gain and who bears the costs of public land management actions, this is criterion 3.

3. *Distributional equity.* As discussed in chapter 3, one of the functions of government is to ensure a minimum level of well-being to the less fortunate members of society. This is accomplished using a tax-transfer system to tax gains and transfer a portion of that gain to less-well-off members of society. As was also discussed in chapter 3, public land management is not, in and of itself, an effective tool for equalizing the distribution of income or well being in society. Public lands supply too limited a range of goods and services to equalize material well-being adequately.

However, when public land agencies do take management actions, how those actions redistribute benefits and costs geographically or among income groups is a factor that should be given some consideration (Clawson 1975). Thus the total or aggregate benefits and costs calculated as part of the economic efficiency analysis in criterion 2 can be disaggregated to show the distribution of those benefits and costs. The distribution of benefits and costs can be shown by income groups (low, middle, and high), by age categories (e.g., under nineteen years, twenty to thirty, thirty-one to sixty, sixty-one and over), by geographic distribution (e.g., rural versus urban), by race (e.g., Hispanics, Asians, and so on) or by occupational category (e.g., loggers, hunting guides, restaurant workers, miners, and so on). This information on employment effects is often calculated using a technique called input-output models and will be discussed more fully in chapter 7.

It is important to display the distributional consequences of public land management actions for at least two reasons. First, in a democracy, people should be informed about the consequences of different publicly funded actions. In fact, the National Environmental Policy Act (discussed in chapter 2) requires agencies to display and consider the effect of their actions on the human environment. Second, if all other consequences of two actions are similar, society often would prefer the resource management action that does not put a disproportionate amount of costs on segments of society (either races, age groups, or occupational categories) that are least able to bear them.

4. *Social and cultural acceptability.* People, through their elected officials and courts, have shown that they want decisions of public agencies to be made with awareness of their impact on the various cultures in our society. Sensitivity toward particular species of wildlife, such as eagles or salmon, or special tracts of land that have religious significance for Native Americans are important examples. There are several examples of the desires of Congress, as reflected in legislation, or of particular court opinions that maintain Native Americans' historic rights to utilize fish and wildlife resources that are not granted to ordinary citizens. In Alaska, subsistence uses are the priority uses of fish and wildlife on all federal lands (Bishop 1981). Fish and wildlife are culturally important to Eskimo cultures and are a substantial part of their food supply.

Public land managers must recognize that interpretations about what is socially or culturally acceptable change over time. It is important for managers to keep abreast of changing social norms. The environmental decade of the 1970s seemed to be replaced by the conspicuous consumption of the 1980s only to be seemingly replaced by an even more fervent and broader environmental concern in the early years of the 1990s. To some, smoke-belching power plants represent a symbol of a country's industrial might; to others, the same sight represents a fouling of the air we breathe. In many cases public land managers must learn to accept the variety of perceptions different segments of society will have of the same phenomenon. All these different perceptions of reality must be considered when determining which land management action is optimal.

This consideration of social and cultural norms of others is not limited to the present generation. Protection of historical or archaeological sites and artifacts of earlier cultures must often be a key consideration in public land management. One of the most obvious examples is in the Southwest, where pictographs, petroglyphs, and cliff dwellings of Anasazi Indians are found in many areas of BLM and NPS land.

Social and cultural values are expressed through people's actions or choices, some of which include voting, writing letters to agencies, participating in public meetings, and choosing to live in a particular area and do certain work. In other cases, sociologists, informal group discussions, observational studies, and surveys may be required to generate information on social values.

5. *Operational practicality or administrative feasibility.* In addition to all the preceding, a manager must always ask the hard question, "Do we have sufficient personnel and funding to perform this management action successfully? For example, a particular timber-harvesting system may have proven physically and economically feasible in an experimental or research forest, but it may not work on a large scale without constant supervision by forest engineers. A new livestock grazing system may enhance vegetation for a particular wildlife species when graduate students are constantly monitoring pastures and adjusting stocking rates, but it may not work with the part-time ranchers who have jobs in town. These are important considerations. The fact that a system ought to work does not mean it will work in a different administrative environment.

Budget and personnel are always real-world constraints on what is possible. In the short term, these are often rigid limits that must be adhered to. In the long term, cost-sharing arrangements with other agencies or groups, use of volunteers and justifications for additional funding (based on analysis of the benefits) can modify these limits.

The other major factor that I have chosen to group under Clawson's (1975) heading of administrative feasibility relates to adherence to existing federal laws. These would include not only the act that governs all aspects of the agency (e.g., NFMA for the Forest Service or FLPMA for BLM) but also such acts as the Clean Water Act, Endangered Species Act, Surface Mine Reclamation Act, Clean Air Act, and so on. For example, a permit could not be issued by BLM for a strip mine if it would result in violation of the strict Class I air-quality standards at a nearby Wilderness Area or National Park. A timber sale could not be approved if it would violate state water quality standards. An off-road vehicle race could not be permitted by BLM if it threatened the continued existence of an officially listed endangered species.

This criterion of administrative practicality must balance reality with the costs of change. If analysis identifies opportunities for more beneficial land management, such actions should not be discarded simply because they run counter to existing procedures and policies (many of which are often leftovers from a previous era). Variances to agency procedures and policies can be granted when a clear case is made for an alternative approach. Laws can and are often changed. In the Pacific Northwest, Congress passed a law in the early 1980s allowing timber companies to sell back to the Forest Service public timber-harvesting rights they had bought a few years earlier from the Forest Service. With the drop in timber prices, the firms would lose money on the sales and thus would have defaulted on them.

Techniques for Integrating the Five Criteria in Decision Making

Taken individually, each one of the preceding criteria provides a different standard or measuring scale to determine what is a desirable action or alternative. In some cases the different scales (wildlife populations, dollars, employment, and budget) may all rank one particular management action as the best. Often these different criteria will yield quite different ratings for the desirability of two actions. In some cases, the more inefficient a project is in terms of both the environment and economics, the more likely it will be to enrich local merchants and increase employment in the local area. For example, conversion of oil shale into oil would generate hundreds of jobs in western Colorado, but it would also result in substantial degradation in air quality, water quality, and deer habitat. In addition, the cost of producing the oil from oil shale would likely exceed the price for which oil can be purchased on the world market. Thus only with generous subsidies would this economically inefficient project be financially viable to private industry.

The problem is that an action that may look desirable on one criterion may not be desirable on others. What is a manager to do with such conflicting information? There are several approaches for integrating the information from these five criteria so as to identify the "best" choice available (given the criteria). The common trait in all these approaches is a formal decision structure to relate the five criteria. Six alternative decision structures or frameworks that will allow integration of these five factors follow.

1. *A matrix approach to integrate the five criteria.* The most straightforward approach is to develop a matrix that lists the five factors down the left-hand side and the alternative actions along the top columns. Table 4.1 illustrates such a matrix.

There are several ways to use this framework to help identify which alternative is preferred. One is to look for an alternative that does just as well as some other alternative on several of the criteria but better on the remaining criteria. Such an alternative is said to *dominate* the other ones. By this we mean that an alternative is better or the most desirable and hence is preferred to the others.

A numerical example may help to clarify how this search for dominance is carried out. Table 4.2 illustrates three alternative locations for acquisition of a

Table 4.1
Simple Decision Matrix

Criteria	Alternatives				
	A	B	C	D	E ... Z
Biological					
Economic					
Distributional					
Social/cultural					
Administrative					

Table 4.2
Dominated Alternative Example

Criteria	Alternatives		
	A	B	C
Physical/biological			
Water use (acre-feet)	1,000	700	1,500
Salmon populations	500	500	750
Species/acre	35	45	65
Economic			
Net benefits ($)	$5,000	$6,500	$4,000
Distributional Equity			
Progressivity Index	0	0	0
Social/cultural			
Native Americans	+ 1.2	+ 1.2	+ 1.5
Chamber of Commerce	+ .5	+ .7	+ .9
Audubon Society	+ .2	+ .5	+ .7
Administrative No.			
Personnel Hours to Required	50	50	65

wildlife refuge. Alternative A requires large purchases of water and the habitat at this location supports only minimal species diversity. Alternative location B requires that less water be purchased yet produces the same number of salmon and has higher wildlife diversity.

One of the first differences to note between table 4.1 and table 4.2 is that the latter lists one or more *performance indicators* or measures for each of the five criteria. This is known as operationalizing the criteria. Performance indicators provide a measurable indicator that will be used as a proxy to reflect the concepts expressed in the criteria. In this example, we have three performance indicators for physical/biological feasibility. They deal with water required (the less water required, the better), salmon populations (the higher, the better), and species diversity (the higher, the better). The economic criterion is represented by the difference between benefits to people minus the costs (both financial and opportunity costs). Distributional equity is quantified using an index that measures whether the alternative locations will result in a distribution of net benefits that favors higher-income households or lower-income ones. If one location would result in benefits accruing primarily to upper-income households, the index would take on a negative value; if the benefits accrue primarily to lower-income households, a value of greater than zero would result (because the inequality in well-being would be lessened); and if benefits are distributed in the same proportion as income is distributed, then the index would take on a value of zero (i.e., distribution of project net benefits would neither worsen nor improve the distribution of well-being in society).

Social and cultural factors are represented by acceptability of alternative locations to various groups that have taken an active interest in the creation of the new wildlife refuge. To develop an index of their views, a survey might be administered whereby each alternative location is described and the members of the group are asked to state whether they would "strongly support (scored $+2$), mildly support ($+1$), neither support nor oppose (0), mildly oppose (-1) or strongly oppose (-2)" selection of a particular location. This is just one example of incorporating social and cultural factors. Other approaches include the manager's perception of the different group's support for each alternative (as groups have expressed them in newspaper articles, TV interviews, and so on).

Lastly, administrative factors are represented by the number of personnel hours that will be required to manage a new wildlife refuge in this particular location, to monitor its success, and to make any adjustments the monitoring suggests. Since personnel hours are scarce within the agency, the fewer of these required on a particular project, the better.

Having identified the performance indicators, we can now compare Alternatives A, B, and C. By performing a pair-wise comparison on each factor, it is clear that Alternative B dominates Alternative A. That is, Alternative B provides at least as much salmon and more species diversity but requires less water. Alternative B also has a higher level of net benefits (in part because it may use less water). Alternatives A and B are equivalent in terms of effect on distributional equity and on agency personnel hours. In terms of social/cultural acceptability, Alternative B rates just as well, if not better, with the three interest groups.

Since this pair-wise comparison shows that Alternative A is inferior to B, Alternative A can be dropped from further comparisons with other alternatives, such as C. Thus looking for dominant alternatives and dropping inferior alternatives helps to simplify a complex problem so that it is dealt with more easily.

2. *A criteria-ranking approach to integrate the five criteria.* If we look at table 4.2, we see that one cannot reject either Alternative B or Alternative C as inferior. These alternative locations for the wildlife refuge provide a different mix of contributions to each of the five criteria. In particular, by purchasing additional water rights, the agency can maintain higher in-stream flow for salmon and extend the duration of seasonal wetlands (hence greater species diversity). Thus on the biological criterion, Alternative C is preferred to B. But on the economic criterion, the cost of the additional water seems to exceed the additional benefits of having more salmon (recreational and commercial fishing benefits) and greater species diversity (greater wildlife-viewing benefits); thus net benefits have fallen. But the greater salmon production in Alternative C is favored by the Native American Council on cultural grounds and by the Audubon Society because of greater species diversity. Unfortunately, Alternative C takes more agency personnel hours to monitor the timing of water deliveries.

Unlike the comparison of Alternatives A and B, a comparison of B and C reveals no clear winner. But some decision must certainly be made about which location for the new wildlife refuge is to be selected. The available land and water will not be for sale forever! There are several ways to determine which is the most desirable alternative in this case.

The first approach is to have the managers and/or public officials *rank* the five criteria from most important to least important. That is, using pair-wise comparisons between each of the criteria, determine which is the most important, the next most important, and so forth, until the least important is reached. For example, since this is a wildlife refuge, the public officials might appeal to the legislative history authorizing the new refuge to determine that biological and social/cultural factors were more important than economic ones. Discussions with higher levels in the organization might reveal that additional seasonal personnel hours might be forthcoming if adequate justification for the additional personnel were provided. Therefore administrative feasibility is not critical for this decision and would be ranked low. Since distributional equity is similar across alternatives, it need not be explicitly ranked.

Given this ranking (from most to least important) of biological, social/cultural, economic, and administrative feasibility, Alternative C would be the most desirable. This is of course due to the fact that it performs better than Alternative B on the most important criteria (biological and social/cultural), which outweighs its poorer performance on less important criteria (economics and administrative).

A slight variation of this approach of ranking criteria would be to rank the performance indicators associated with each criterion. In cases where there are more than one performance indicator per criterion the performance indicators would be ranked within each of their overall criteria, so that their number would not influence the weight given to any particular one.

3. *Weighting and multiattribute utility approaches.* Some decision specialists have advocated more elaborate methods than 1 and 2 to make comparisons with multiple criteria. These approaches involve the decision maker weighting each criterion by some number that reflects its relative importance or usefulness (utility). The weights may be developed from an averaging of different decision-maker rankings (Fish and Wildlife Service 1980; Cherfas 1990:367). Or they may be developed directly by managers. For example, biological factors might be weighted ten points (out of ten); economics, five points; distributional equity, two points; and so on.

The primary advantage of the weighting approach is that once the weights are established, one overall value (incorporating all five criteria) can be computed for each alternative. That is, a normalized (e.g., in percentage terms, such as percent of maximum) performance score of each alternative on a given criterion is multiplied by the weight of that criterion. Then the criterion-by-criterion scores are summed for each alternative (i.e., one score for each column). The decision maker simply selects the alternative with the greatest "weighted" score or in some sense the "highest utility." (See Merkhofer and Keeney [1987] or Keeney and Raiffa [1976] for a detailed discussion of this method.) This approach of determining relative values is one of the techniques used by the U.S. Fish and Wildlife Service (1980) to weight different species when evaluating the overall impacts to wildlife habitat from development projects. That is, some land management actions (e.g., wetlands creation) that enhance habitat for some species (e.g., waterfowl) reduce carrying capacity for other species (e.g., upland birds). The agency's assessment of the overall desirability of the management actions depends on the weight given to different wildlife species.

Even with computational aids, one challenge of this method is agreeing on the relative weights of each of the criteria. The decision maker must go beyond a mere pair-wise ranking of the criteria. It is no longer enough to say biology is more important than distributional equity; weighting requires the decision maker to reveal (directly or indirectly) how much more important biology is than distributional equity. If biology ends up with a weight of 10 and distributional equity has a score of 2, this implies that biological performance is five times more important than distributional equity.

Determination of the weights is a critical step, for a different weighting may often dramatically change how a given alternative scores relative to other alternatives. This change in scores arises not because an alternative's performance has changed on a particular criterion, but because the criterion now has been given more importance (i.e., a greater weight).

Given the importance of the weights in determining which alternative is best, it is of great concern whose opinions or preferences go into setting these weights. Are the weights determined by the public, by interest group negotiation, or by a select few managers in the agency? All of these could technically be used. Since several studies have shown that technically trained managers often do not have views that match the public on natural resource management or attitudes toward risk, letting agency personnel dictate the weights may not lead to socially optimal decisions. Involving the public is worthwhile, but care must be taken to obtain a representative sample of the public, not just the most vocal interest groups.

For many resources the public has already revealed weights for different mixes of outputs; these are called prices. Prices reflect the sacrifices people are willing to make to obtain more of some good or service. Although discussed more fully in chapter 6, we note here that they tell us the "importance" or worth of additional lumber, beef, water supply, hydropower, and so on. Implicit markets for unpriced wildlife and general recreation will be discussed in chapter 6. From these implicit markets we can infer a price based on the dollar sacrifices people make to use different natural resources (i.e., view trees and wildlife, go fishing, and so on). In other cases we can develop simulated markets for people to state the degree of sacrifice (in dollars) they would make to know more land would be protected as Wilderness (also to be discussed in chapter 6).

Rather than have each land management agency invent its own weighting system for reconciling biology, social values, and economic efficiency, people have already implicitly developed weights for many of the different performance indicators of these criteria in deciding how to allocate their scarce dollars (and for public goods, their scarce votes) among competing concerns. To a large degree factors such as biological or administrative feasibility can be put into common units, dollars. Of course, some dimensions of various criteria, such as minimum biologically viable populations and cultural factors, cannot be readily expressed in dollar terms that are meaningful to everyone in society. Other factors, such as distributional equity, although not strongly influenced by public land management, must be given some weight in the decision process. Thus using dollars as weights may be useful primarily when alternatives vary in terms of economic contributions.

One solution to the problem of incommensurable criteria is to maximize attainment of one of the criteria (the one that most closely ties to the legislative mandate of the agency and reflects the most comprehensive performance measure available) subject to attainment of at least minimum levels of the other criteria. In a National Park this might be maximizing naturalness, with the requirement that the alternative selected be administratively practical, socially acceptable to a majority of current visitors, and not worsen distributional equity significantly. Just because these other criteria are required to have at least the minimum does not mean that only the minimum will always be provided. Some alternatives may provide far more than the minimum on one or more of these other criteria.

There are three decision techniques to implement such a process of maximizing one criterion, subject to minimum levels of the others: screening, benefit-cost analysis, and mathematical optimization. We will start with the simplest of the three, *screening*.

4. *Screening as an integrating approach.* In the screening approach, the incommensurable criteria are used as filters that "screen out" alternatives that do not meet at least the minimum requirements of the criteria. For example in figure 4.1, Alternative F drops out at the first screen, since it contains some management actions that are not biologically feasible (e.g., the alternative may propose to produce both more oil and waterfowl on a given area of land at the same time). Alternative E drops out as culturally unacceptable (e.g., E may fail both to provide minimum protection to archaeological and historic sites and to provide adequate amounts of wildlife for subsistence use). All remaining alternatives meet the minimum requirements for being administratively feasible because none requires more employees or more budget than is likely to be available. Alternative D might drop out because it imposes substantial costs on one segment of society (e.g., D results in a doubling of unemployment rates in one rural town). All remaining alternatives may pass the economic efficiency test of having benefits at least equal to costs (not shown in figure 4.1). Therefore the public land manager would pick from among the remaining alternatives, based on the one that maximizes the key criterion. If this were a Wildlife Refuge purchased with Duck Stamp funds to serve as a waterfowl breeding area, the preferred alternative might be the one with the highest sustainable waterfowl production. In this example, waterfowl production would be maximized, given that other criteria are not violated.

5. *A benefit-cost analysis approach.* As discussed in chapter 3, continued public ownership of lands is largely justified on the grounds of minimizing market failures and supplying public goods, factors associated with economic efficiency. Some public lands legislation (especially for the Forest Service and BLM) stresses the importance of economic efficiency as a criterion for managing lands (Krutilla and Haigh 1978; Bowes and Krutilla 1989:7). The Forest

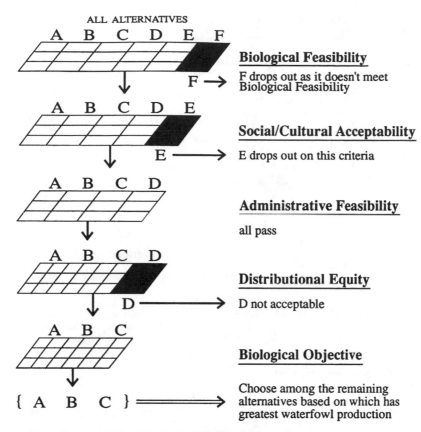

Figure 4.1.   Screening for Wetlands Wildlife Refuge

Service planning training manual (Jameson et al. 1982:4) states, "NFMA now indicates a shift toward increased emphasis on efficiency." As will be discussed in chapter 6, economics provides a framework to express most (but not all) resource outputs and costs in commensurable units (dollars) that reflect the value of these outputs to people in society. This suggests that economic efficiency would be a relevant criterion to maximize subject to varying levels (all of which would be above the minimum) of the other criteria.

Measuring the contribution of each alternative to economic efficiency can be done using *benefit-cost analysis* (BCA). This approach could be implemented in one of two ways: (1) perform BCA on just the alternatives that meet the screening process, as illustrated in figure 4.1, or (2) perform BCA combined with increasing levels of attainment of the other factors.

The screening approach is performed as previously described; then BCA is performed on just the alternatives that pass through all screening factors (e.g., which demonstrates that the remaining alternatives provide minimum attainment of the other factors). The preferred alternative is that with the greatest

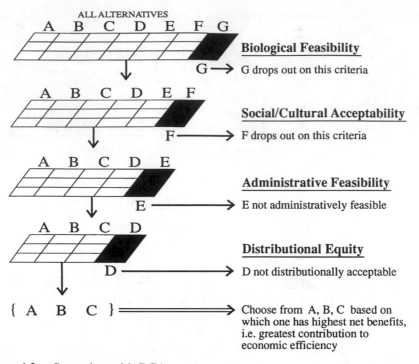

Figure 4.2. Screening with BCA

contribution to economic efficiency (i.e., the one with the largest amount of benefits in excess of costs). This process is illustrated in figure 4.2.

The second BCA approach involves designing a range of alternatives. All alternatives would still have to meet the test of biological feasibility. But different alternatives would provide more than the minimum on one or more criteria. For example, one alternative might just meet the minimum levels for administrative feasibility but go well beyond the minimum on social/cultural acceptability or distributional equity. The net benefits of these alternatives with more than the minimum of one or more criteria would be compared to alternatives that just met the minimum on all criteria. The manager would then be able to assess what the costs (i.e., forgone benefits) were of attaining more than the minimum level of these other criteria. This would help the manager determine which alternative represented the best balance in attaining the five criteria. For example, a more culturally acceptable alternative might give a great deal of protection to all fish and wildlife species of symbolic value to Native Americans. A more distributional-equity-oriented alternative might provide more free firewood permits. Once again, net benefits would be calculated for each of these alternatives. Thus the decision maker is given an array of alternatives, some of which surpass the minimum levels for the other criteria. But presented at the same time would be the opportunity cost in terms of reduced net benefits (and hence reduced attainment of economic efficiency) associated with greater attainment of other objectives. We demonstrate graphically in the next chapter the reason that economic efficiency associated with attaining more of other factors may be reduced. Needless to say, knowing the opportunity costs of gaining more of the other factors is helpful in making informed trade-offs between the five factors. This second type of benefit-cost analysis is greatly facilitated using the sixth approach, mathematical optimization.

6. *Mathematical optimization.* An optimization model selects one of the five criteria and maximizes it subject to meeting prespecified levels of the other four criteria. As long as all criteria (or their performance indicators) can be represented by an equation, it can be maximized in a mathematical optimization model such as linear programming, quadratic programming, nonlinear programming, and so on. As will be discussed more in chapters 8 and 9, these optimization models are very useful for many types of natural resource allocation problems and are widely used by government agencies and businesses to aid in decision making.

At its simplest level, the basic structure of mathematical optimization models is an objective function representing the criterion to be maximized and all other criteria represented as constraint equations. To develop an objective function, the analyst must select one criterion and then express it as an equation. For example, choose to maximize wetland acres *or* board feet of timber harvest *or* acre-feet of water yield.

The selected physical unit would then be maximized, given what is described mathematically in a series of equations as being physically possible (given equations for available land, labor, and money), as being socially acceptable (also represented by one or more equations), and as meeting distributional equity (also represented by one or more equations). The computer model would then search all the possible combinations that are technically feasible to find the one that has the highest attainment of the criterion being maximized. Although similar in spirit to the screening approach in figure 4.2, this approach is far more sophisticated. But this sophistication comes at a price of being far more demanding in terms of informational requirements and technical expertise of staff and managers to structure, solve, and, most important, interpret the resulting allocation of resources.

Optimization is often used to maximize net economic benefits, since this measure is actually a summary of numerous outputs and costs into one number. That is, the drawback to maximizing elk numbers is that everything else must be represented as a constraint. If one is also interested in fish, water yield, and recreation visitation, these can only be put in as constraints, since one cannot simultaneously maximize elk numbers, fish numbers, acre-feet of water, or visitor days. That is, one cannot maximize more than one criterion at a time, unless there is some way to collapse all these considerations into a meaningful aggregate. One cannot add fish and acre-feet of water to visitor days, as the resulting number is meaningless because of different units of measurement. But if the benefits of having these different resource flows can be converted into a meaningful common unit, then one could maximize the aggregate in an optimization model.

Frequently, what is done in practice is to measure the economic benefit of the various outputs in dollar units. This allows the various outputs to be scaled in commensurate units, so that they can be added together meaningfully. Now maximizing this aggregate allows us to give the various resources a potentially broader influence on the solution than just being a constraint. For example, if the number of acre-feet of water is treated as a constraint that is required to be at least 1,000, there is little indication that having more than 1,000 is also beneficial. By including the value of acre-feet of water in the function that is being maximized, the aggregate benefits continue to rise as more water is produced. Thus it is likely that we may get closer to the optimum beneficial water flow by including it in the aggregate to be maximized rather than simply as a constraint. When we also subtract the costs from the measure of aggregate benefits, we obtain a measure of the contribution to economic efficiency (i.e., net benefits). The optimization approach can also be structured like the straight BCA approach to evaluate the sacrifice in attainment of the primary criterion to attain higher and higher levels of the other four criteria.

As will be discussed in more detail in chapters 8 and 9, optimization models such as linear programming are used extensively by the Forest Service in performing multiple-use resource allocations. This model has many conceptual and operational advantages, but it may not be appropriate for every public land decision.

Choosing the Appropriate Decision Aid

Which one of these broad decision framework techniques is appropriate depends on the degree of equality of the five criteria and on the structure of the resource allocation problem. As table 4.3 illustrates, if one treats all the criteria equally, then a matrix approach may be most appropriate. Using the matrix approach, the analyst attempts to identify the dominant alternative (if one exists). Next, if one or more criteria can be ranked as more important than others, this may greatly simplify the selection of the most preferred alternative among nondominated ones. If the decision maker can go beyond pair-wise rankings and state or infer the degree or absolute amount of importance between criteria, then each alternative can be reduced to one weighted average using a multiattribute utility (MAU) approach. This weighted average would reflect the contributions to all the criteria. The alternative with the highest weighted average or score is then selected. In a weighting approach, one is, in a sense, maximizing, as the goal becomes maximization of the

Table 4.3
Decision Aid Continuum

Matrix				→ Optimization
All objectives weighted equally	Rank objectives	Weight objectives (MAU)	LP	BCA

sum of the weighted criteria. Toward the optimization end of the spectrum in table 4.3, one selects one criterion to maximize subject to attainment of minimum levels of the others. On public lands, where the specific law has given more emphasis to one of the five factors, an optimization approach on that one factor might be useful. For example, on a National Wildlife Refuge, maximization of habitat for a particular indicator species might be appropriate. However, when dealing with true multiple-use agencies, the discussion in chapter 3 and in Krutilla and Haigh (1978) on the rationale for continued government ownership of these lands suggests that emphasis should be given to economic efficiency, with the other criteria being at least at some acceptable levels and exceeding these levels wherever possible. At the extreme right-hand side of table 4.3 a BCA decision approach selects economic efficiency as the criteria to maximize subject to acceptable levels of attainment for the other criteria.

Another factor that might determine which decision framework is most appropriate relates to lumpiness or indivisibilities in resources. By indivisibilities we mean the resource is naturally available in one big unit that cannot be subdivided into smaller units without losing its functionality. Thus manageable roadless areas often come in 5,000-acre blocks, or 9,000-acre blocks, and even 1-million-acre continuous areas. These are not functionally equivalent to five hundred units of 10 acres each, or nine thousand 1-acre parcels, and so on. The same may be true of certain wildlife habitats: a grizzly bear may require 1,000 unroaded continuous acres, not twenty parcels of 50 acres. If the decision structure is such that (1) resource allocations are very lumpy or exhibit indivisibilities (i.e., must treat an entire roadless or habitat area as one integral unit) and (2) the other four factors cannot be represented mathematically, then a BCA (with or without a screening approach) is most useful for determining which alternative maximizes attainment of economic efficiency.

Optimization models, such as linear programming (LP), are less well suited to lumpy or indivisible management actions than BCA. As will be explained in more detail in chapters 8 and 9, the typical structure of an LP model is such that every time a particular management activity (e.g., harvest timber, build a campground, improve a stream) is performed, the size or scale of that activity must be the same and must have the same average output response, the same average cost, the same average effect on other resources, and so on. But nature is heterogeneous; not every area suitable for a campground comes in 10-acre blocks. Now, of course, building a 10-acre campground could be represented by one management activity; building a 20-acre campground by a different management activity; and so forth. However, this adds to the model complexity. Also, trying to maintain geographic detail in LP models is somewhat difficult because of computer memory limitations. Although computer memory limitations are rapidly becoming less and less of a problem, there is still a great deal of complexity in setting up such a detailed model.

However, there are decision structures where, despite the preceding drawbacks, optimization models such as LPs provide an excellent decision framework. When resource allocations involve selection among several different competing management activities on many relatively homogeneous tracts of land, subject to land, labor, budget, and resource constraints, many of which may change over a fifty-year planning horizon, LP is a very valuable decision aid. Part of its contribution lies in the LP accounting system, which is an excellent way to organize all the required information. Of course, its strongest asset is the program's ability to compare rapidly the many hundreds of possible combinations of management actions and associated land allocations and identify the best combination, given the constraints representing the four other factors. This type of decision structure is more common in multiple-use agency decision making than in National Park management, but the basic modeling structure would still be useful for many issues faced in park management as well.

Chapters 6, 8, and 9 will explore in detail models useful for evaluating multiple-use decision making with emphasis on benefit-cost analysis and linear programming. However, both of these decision aids and any of the matrix approaches share one thing in common: they all rely on formal or informal models of physical and human systems. All these decision aids require models (ranging from intuitive word models and pattern recognition models to mathematical and statistical models) to determine what is physically, hydrologically, and biologically feasible. In addition, all the decision aids require some *predictive model* to determine how the quantitative performance indicators for each criterion will change under different management actions (e.g., timber harvesting, livestock grazing, mining, and habitat improvements) associated with each alternative plan. It is to this discussion of the pervasive and increasing role of models in public lands management that we now turn.

5

Roles and Appropriate Uses of Models in Natural Resource Management

Usefulness of Models

Literally dozens of times each week a field biologist, botanist, hydrologist, or recreation planner must make a natural resource decision that might be improved by using a formal or informal model. A field biologist might be required to determine whether the minimum in-stream flow for trout should be 100 cubic feet per second (cfs) or 150 cfs. A botanist or range conservationist may need to decide what type of plant species would be best for surface mine reclamation.

Historically, these decisions were made using the employee's best judgment from years of experience. However, this approach has two limitations. First, a new biologist or botanist assigned to an area may not have years of experience with that geographic area (e.g., climate and soils) and may not be an expert on the wildlife species in that area. Thus some guide to what plants survive well in different soil types and climates would be very valuable in helping the new botanist avoid poor decisions on the plant species to require for revegetation. Even a simple table indicating the critical variables to collect data on (soil characteristics, such as texture, percent clay, and records of precipitation and temperature) would help. In some ways this table would be a simple model.

The second reason for models is that the general public, the courts, and public officials are requiring a well-documented rationale for many public land management actions. A fisheries biologist can no longer say, "It is my professional opinion that 150 cfs would be the minimum in-stream flow" without having some systematic, objective basis for supporting that 150 cfs and not 100 or 200 cfs is the minimum in-stream flow. Once again, models are relied on as commonly accepted procedures to make complex determinations about the allocation of natural resources.

What Is a Model?

A model is an abstraction of reality. It is a simplification of reality that provides useful guidance in attaining a particular management or policy goal. For example, a map is a type of model. Of course, maps are simplifications of the real world, but because they are *useful* simplifications, they work! However, choosing the right map for the particular type of trip is as important as choosing the right model to solve a particular problem. If you want to take a cross-country skiing trip in a remote wilderness area, a topographic map would be appropriate. If you want to find out how to get to Kennedy Airport from downtown New York City, a topographical map would be useless, but a city street map would minimize your chances of getting lost. (Notice I said *minimize* your chances of getting lost, not eliminate your chances of getting lost.) Any map or model is just a guide to help you reach your intended objective, but it cannot guarantee you won't make a mistake. If you use the right model for the job, you will reduce your chance of making an erroneous decision, but you must still use the model with your own judgment.

As this analogy to maps illustrates, there are many types of models, and each is more or less appropriate for some problems. Models can range in complexity from simple rules of thumb through single-equation relationships with two or three explanatory variables, to complex mathematical models with thousands of equations that must be solved simultaneously using a supercomputer. An old model of a grazing system was a "take half, leave half" approach. That is, livestock or wildlife could, on a sustained-yield basis, consume about half the vegetation and leave half for maintenance of plant vigor or regeneration, retention of soils, and so on. This rule represents a simple model of how plants respond to grazing. A more complex model may not necessarily result in an improvement over this simple model.

We often evaluate models in terms of their ability to predict and explain the phenomenon of interest. If the simple model's guidance is followed for a few years, we can field-test or verify the model. If half the vegetation is consumed each year, we can later evaluate whether adequate vegetation has been retained to provide a sustainable grazing system and keep soil erosion to a minimum. If this simple model can also explain how a grazed ecosystem will respond to deviations from "take half, leave half," then such a simple model is preferred to a more complex one. Greater complexity does not always result in greater accuracy. Often it causes greater chance for error.

Models are helpful in problem solving and determining which alternative best meets some specified goal or objective. Without a formal model to compare alternatives it is often quite subjective as to which management action or policy best meets a given objective. In this sense, models need not always be quantitative, but they must provide a systematic framework that illuminates how each alternative contributes to attainment of the objective. This can be done with a simple matrix, a single equation, and often more complex optimization models (although this is certainly not required).

Types of Models

There are many types of models, including equations, graphs, and physical models, which are small-scale replicas. A model can be as simple as a single equation. A simple model of the relationship between trout production (TP) in a stream and stream flow and stream temperatures is given by equation (5.1):

$$TP = B_0 + B_1 FLOW - B_2 TEMP \tag{5.1}$$

where TP is measured as the number of adult trout per mile, FLOW is measured in cfs, and TEMP is daily maximum water temperature.

This equation, or model, states that trout production (TP) is positively related to stream flow (FLOW) and negatively or inversely related to stream temperature (TEMP). In this model, B_1 is the rate of change in TP as FLOW rises. The exact magnitude of B_0, B_1, and B_2 would be calculated by performing a statistical regression on trout production levels, FLOW, and TEMP data. The resulting model can be applied or "run" to predict TP levels with a large variety of FLOW and TEMP (not necessarily just the combinations observed in the data) by inserting expected values of FLOW and TEMP under different water management regimes and calculating trout production.

A more complex model might be a model of the market conditions for a particular natural resource. This model involves linking three equations: an equation describing the demand for the natural resource, another describing the supply of the resource, and a third describing the conditions for equilibrium in the market for this natural resource.

For example, consider the market for coal. A simple demand for coal given by equation (5.2):

$$Q_d = A_0 - A_1 P_c + A_2 GDP + A_3 P_{SUBS} \tag{5.2}$$

where Q_d = quantity demanded
P_c = price of coal
GDP = Gross Domestic Product reflecting derived demand for energy
P_{SUBS} = Price of substitute energy sources such as oil

The quantity supplied would be a function of

$$Q_s = B_0 + B_1 P_c + B_2 RES - B_3 P_{INPUTS} \tag{5.3}$$

where Q_s = quantity supplied
RES = reserves of coal available
P_{INPUTS} = price of inputs used to mine coal such as coal miner wages

Equilibrium in the market for coal occurs where quantity demanded equals the quantity supplied. Thus the last equation in this model is

$$Q_s = Q_d \tag{5.4}$$

This, then, is a simple model of the market for coal.

Role of Assumptions in Models

Assumptions play several key roles in allowing us actually to build and apply models. First, assumptions tell us what the key or critical variables are that influence the system we are studying. The assumptions are in some sense a working hypothesis about what really influences the system we are interested in modeling. In our introductory example on revegetation, we assumed that precipitation, temperature, and soils were the key variables. Certainly hundreds of things might influence plant survival, but if we want a workable model capable of evaluating alternative vegetation options, only a few of those hundred variables are probably critical. Although our predictions will never be completely accurate without all two hundred variables, they will likely be much better than a "best guess" of the effect. The existing scientific literature can guide our assumptions about what the critical variables are. Thus one role of assumptions is to limit intelligently the number of variables that an agency must collect data on. An agency rarely has the time or money to collect large volumes of inventory data on every conceivable environmental parameter. The model helps to focus the inventory and data collection efforts and avoids collecting extraneous data.

Second, assumptions are used not only to simplify the number of variables considered in a model but also to simplify the relationships between the variables (i.e., reduce the complexity of the relationships). This is important in quantitative models, since complex nonlinear relationships are often difficult to deal with in many standard statistical programs or optimization techniques (to be discussed in detail later). A review of the literature in the particular area of research might indicate that an assumption of linearity might be plausible. If not, the literature might give guidance as to which one of the many nonlinear relationships is most appropriate. For example, is the relationship between stream flow and fish habitat quadratic or logarithmic?

Third, assumptions inform the agency as to when the model is applicable. That is, if the assumptions of the model are met, the model should provide a reasonably accurate prediction and explanation of the phenomenon under study. If the assumptions are closely approximated, the model still may be useful. If, in fact, the assumptions of the model depart dramatically from the case at hand, the model should not be used for that case. In this case, if the model is used, it would not be surprising that its predictions were far from accurate.

Rationale for Use of Models in Analysis, Management, and Policy

There are several advantages of models to both the planner-analyst and the decision maker. For the planner, models help make a very complex system (natural or man-made) more understandable. By increasing his or her understanding of the system, better policies can be designed to bring about desired changes, or more accurate predictions can be made. Models aid the planner in developing an understanding of a complex system by forcing him or her to identify the key variables and relationships that are most relevant in making up the basic structure of the natural or policy system under study. This abstracting from reality helps prevent the planner from losing sight of what the crucial parameters are that must be analyzed if the plan is to be successful.

Models also provide a framework or logical system by which to determine the data needed and how to relate these data to the problem one is trying to solve. Models provide a common point on which the many resource specialists can focus their thinking. Having the various adversarial parties build a simple cause-effect model about management of the resource will identify the areas of agreement and disagreement and will highlight what information is needed to resolve any differences.

Models provide decision makers with at least three benefits.

1. They furnish an expedient means for the decision maker to arrive at conclusions to many "what-if" questions. These types of questions include the following: What if 20% of the land on our forest is designated Wilderness? What if we increased livestock grazing? If in-stream flow is reduced, what happens to water quality?
2. They provide decision makers with an objective way to evaluate the consequences of alternative policies or programs on the goals or objectives desired. For example, will Policy X bring about the desired biological diversity, or is Policy Y better to attain this goal? Models also provide a means to communicate among various parties. That is, a model is a common framework so that each group can express its concerns about a policy. A group's concerns can be represented by adding variables or constraints, for example.
3. They provide evidence to the decision maker's supervisor and to his or her constituents that policies and plans were arrived at in an objective, scientific way. Unfortunately, this aura of scientific management is provided by the model even when the decision maker's bias may be in one of the constraints. Nonetheless any subjective elements are discoverable if one carefully reviews the model structure and assumptions.

Phases in Model Building

The steps in building a model generally involve five phases. The first and most important phase is *problem definition*. Without a carefully defined yet manageable problem statement, the remaining phases of model building will likely be doomed to failure or at least not produce a usable tool. Critical steps in this first phase is determination of

1. What questions must the model be able to answer (as distinct from what questions it would be nice to know answers to)?
2. What are the values at risk in the decisions to be made with the model? (Is loss of life possible, or just aesthetics, for example?) The greater the values at risk in the decision, the more effort on careful model building is justified.
3. What are the budgetary and time limitations for building the model, collecting the data, and analyzing the data?
4. Will this model be used only for this decision, or is it likely to be used over and over again in the future for recurring similar decisions? The more repeated use is expected, the more attention to generalizability of the model is warranted.

Once answers to these questions have been agreed upon, *initial formulation* of the model becomes the next phase. In this phase one should draw from the relevant theory(ies) of the system(s) one is trying to model. In some cases, theories of human behavior (economic, psychological, sociological), engineering, and earth and biological sciences may have to be integrated as well as possible to provide a workable model. This integration often provides the necessary linkages between variables under the decision maker's control (policy or management variables) and the variables reflecting accomplishment of the desired outcome (changes in human behavior or improvement in the physical environment).

Drawing from these theories, the critical variables for the model are identified. Next, the appropriate level of aggregation is determined. By *aggregation,* we mean the degree of detail necessary to measure. If theory tells us that household income or soil type is the key variable, there can still be many alternative measures, ranging from broad averages to point estimates taken at very small intervals across the possible range of values. Of course, the budget and time available to build the model will also influence the precision by which the variables are measured.

The next step within the initial formulation phase is to hypothesize about the specific relationships among the variables and among the sets of relationships (e.g., equations) themselves. Questions about whether the relationships are linear, multiplicative, and so on, must be determined. Often the functional relationships implied by a well-developed theory may have to be simplified to keep the model computationally simple. Alternatively, less-well-developed theories may provide little guidance on functional form. Here several forms may be initially proposed and later tested against the data or in simulation exercises to assess the plausibility of each functional form.

Once the separate parts of the model have been specified, these components must be related to one another. Questions to be answered here include whether one set of relationships may dominate others either legally (water laws, property rights), biologically (threshold), or economically.

Once the initial model is formulated, the relevant data indicated as necessary by the model are collected in the *data collection* phase. Here primary data (e.g., field inventories, surveys, and so on) are collected and assembled with existing data in a format required for the next step.

This next phase is to *calibrate or statistically estimate* the quantitative values of the model. In our trout production model we would need to estimate the size of B_0, B_1, and B_2 to know the exact magnitude of the change in trout production with changes in stream flow and temperature. To do this normally requires a statistical technique called *multiple regression,* although other statistical techniques are frequently used in modeling as well. When available time or funding makes collection of sufficient data for statistical analysis impossible, quantitative estimates for the model can be developed from a consensus of field specialists or from the scientific literature in the field.

The final phase is to attempt to *validate the model.* Validation could take the form of testing the model's performance against historical data or running the model through a series of simulations to see if the results are plausible. In some cases such "testing" is impossible, because of the type of model or because of insufficient historical data. In these cases, peer review of the model often must suffice as a method to validate the model or suggest areas of the model in need of reformulation to improve its accuracy.

Once a reasonable model has been formulated it can be applied to the planning problem at hand. Here application may be broadly defined as use of the model to predict the consequences of alternative management actions. This is where a model would be used to predict the performance indicators associated with each of the management criteria discussed in chapter 4. Based on the results of the model's evaluation of alternatives, an alternative is selected and documented with regard to the model.

Problems in Model Application

Problems in applying models to real planning questions involve data limitations that make the desired model impossible or too costly to implement, limitations in the current theory that make developing the structure of the model difficult, and finally, computational limitations that may require reducing the degree of detail in the model. However, these deficiencies or limitations are present in almost any approach to resource decision making or policy analysis. These problems only become more apparent when the modeling process is made more explicit and involves persons beyond the decision maker. In addition, some resource specialist or the decision makers may feel that the effect or ''optimal'' solution is obvious (at least to him or her) and that a rigorous model to evaluate alternatives is unnecessary. But everyone implicitly uses some model of how the world works to make most decisions, even if they are unaware of it. This model may be a rule of thumb or based on what has worked in the past. The more formal the model and the more persons participating in the decision, the more useful it is to write down just what assumptions these rules of thumb are based on. So although decision makers or analysts may criticize a particular model or the entire modeling process, the defects of commonly used ad hoc models are largely hidden. (For example, setting output this year at the same level as last year's level implies several very specific models of the world.)

One commonly encountered problem in the application of models to real planning or management problems is the incorrect application of a model or the failure to follow a crucial step in the model-building process. Another problem in real-world application of models is that they become ends in themselves (often to the model builder) rather than simply analytical tools to aid in decision making. When the problem definition phase is skipped or not followed, models end up answering the wrong questions or become overly large and complex as they try and answer every possible question. Other pitfalls include models built to answer questions of interest to the researcher rather than to the decision maker. Problems in application of existing models are also encountered. Use of models beyond the limits they were designed for often lead models to generate counterintuitive results. The reaction to such results is often to scrap the entire modeling process rather than to recognize that incorrect application of any tool (a hammer, for instance) can lead to disastrous results.

Key Features Needed in Ecological Models for Natural Resource Management

To determine both what is biologically feasible and how natural resources respond to and can be enhanced by land management actions, models must have certain features. The key feature is that the model must relate, either directly or indirectly, the output variable (e.g., number of animals, pounds of forage per acre) to variables influenced by resource management. Another important feature is to relate the output variable as closely as possible to performance indicators people care about (e.g., animals, plants) or ones identified by legislation as important (e.g., the old Kenai Moose Range in Alaska).

A good example of a well-known physical science model whose variables can be related to land management is the universal soil loss equation (USLE). Here soil erosion (measured as tons per acre) is a variable of interest for several reasons. First, it is sometimes the policy variable. More often, predicting changes in soil erosion in response to alternative land management practices is of interest because soil erosion is highly correlated with other variables of interest, such as water quality or soil productivity. The other reason is actually an extension of this last reason: soil erosion may serve as an input variable into another model. For example, erosion rates are often a variable in models of fish populations or water treatment costs.

In the USLE, soil erosion is related to five main factors (U.S. Department of Agriculture 1978; U.S. Agricultural Research Service 1982). These five factors are rainfall (intensity and duration), inherent soil erodibility, slope (slope length and steepness), vegetative cover, and land use (e.g., tillage practices in agriculture). In this model rainfall, inherent soil erodibility, and slope are ''state-of-nature'' variables that are taken as given and not directly subject to changes by land managers. However, they must be considered in a complete model because they influence soil erosion. Vegetation cover and land use are policy variables because people can directly influence them. For example, strip mining, timber harvesting, road construction, frequent use of off-road vehicles, and intensity of livestock grazing will all affect the vegetative cover variable. Although the USLE was developed primarily to predict erosion from farmlands, it is often used in public land management as well. The U.S. Department of Agriculture has recently developed more sophisticated soil loss equations that can be generalized to a wider variety of land types.

In the next section we provide some examples of fish and wildlife habitat models that can be linked directly to land management practices. The first example is for anadromous fish, such as salmon and steelhead.

Description of Fish Habitat Index Models

As is discussed in more detail in the case study of National Forest planning in chapter 9, the Siuslaw National Forest (located on the coastal range of western Oregon) has both a high density of fish and productive forestland. Timber harvesting has the potential to impact anadromous fish habitat significantly. To be able to determine proactively how different timber-harvesting layouts and harvesting practices would affect fisheries, a Fish Habitat Index (FHI) model was developed by the Siuslaw National Forest with the assistance of Oregon Department of Fish and Game (Heller et al. 1983).

The FHI model predicts the carrying capacity for chinook salmon, coho salmon, and steelhead under both natural (e.g., undisturbed) and logged watershed conditions. A separate FHI is estimated for each species. The FHI for natural conditions was developed from an inventory and analysis of thirty-eight undisturbed watersheds in Oregon (Heller et al. 1983:11).

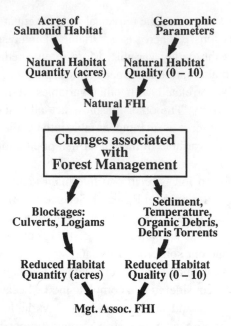

Figure 5.1. Fish Habitat Index (FHI) Model

Figure 5.1 illustrates the basic factors influencing the FHI model. In this model natural or pristine habitat quality is influenced primarily by five key geomorphic parameters: watershed basin relief (measured in change in elevation), watershed or basin area (measured in square miles), length of streams, basin length-to-width ratio, and drainage density. A multiple regression was estimated that relates natural or pristine habitat quality to the value of these geomorphic variables in the thirty-eight pristine watersheds. The output of this model is a predicted pristine, or Natural Habitat Quality (NHQ), rating. Multiplying this natural habitat quality rating times the amount of habitat in a particular land type (e.g., heavily forested, lightly forested, and so on) yields the "natural condition" FHI for a given land type. Equation 5.5 presents this as a simple equation:

$$\text{Natural FHI} = \text{NHQ} \times \text{area of available habitat} \qquad (5.5)$$

The natural, or pristine, FHI provides a measure of the habitat carrying capacity under pristine watershed conditions, such as those that would exist if the area was undisturbed by roads, logging, mining, grazing, off-road vehicles, and other surface disturbances.

As shown in the lower half of figure 5.1, the Natural FHI or quantity of habitat is reduced by five timber-harvest-related impacts on anadromous fish habitat quality. These five factors are (1) stream blockages from logjams or road culverts; (2) sediment increases (measured as cubic yards per acre); (3) water temperature increase; (4) reduction in regulated supply of small organic debris; (5) increase in debris torrents that scour upstream and bury downstream habitat. These five factors are individually well established to be critical variables in determination of habitat quality for several life stages in both salmon and steelhead (Heller et al. 1983).

An equation is used to combine the influence of these factors into one number, called the Watershed Condition Index (WCI). The WCI has a value between 0 and 1 (i.e., WCI is a fraction). Thus increases in sediment, stream temperature, or debris torrents will result in a WCI of less than 1. Since WCI is a fraction, when it is multiplied by the Natural FHI, the result is a decrease in carrying capacity for that fish species. The new carrying capacity resulting from the multiplication of Natural FHI by WCI is called the Management-Associated FHI, given in equation (5.6).

Management-associated FHI
$$= \text{natural FHI} \times \text{reduced habitat area} \times \text{WCI} \quad (5.6)$$

Converting Habitat Indices into Fish Production

To convert habitat carrying capacity (represented by FHI) into production of young salmonids (called smolts), a regression equation was estimated. This equation related smolt production to FHI for a sample of streams in Oregon. A separate equation is estimated for coho and chinook salmon as well as steelhead.

To calculate catchable adult salmon and steelhead, smolt production is multiplied by coefficients that measure percentage of smolts surviving through various life stages. For example, about 5% of salmon smolts survive to adults. The production of adult salmon is utilized, or "consumed," in one of four ways: (1) ocean commercial harvest, (2) ocean sport harvest, (3) freshwater sport harvest, (4) escapement of salmon to spawn. The first three uses of salmon generate an immediate economic return that can be calculated as described in chapter 6. The implicit value of salmon that escape to reproduce is reflected in the catch of the next generation. From fish-tagging studies biologists can provide the geographic distribution of adult salmon and steelhead catch according to these three categories for most major river systems.

Linking the FHI Model to Forest Plan Alternatives

The basic management or policy variables (i.e., the variables influenced by land management decisions) are the five variables in the lower half of figure 5.1 that affect WCI or the amount of habitat area in equation (5.6). By translating different amounts of logging in different land types into the five variables, timber harvests and habitat quality are linked. For example, increasing the width of stream-side buffers that will not have timber harvesting will reduce sediment, debris torrents, and stream temperature. At the same time the retention of trees will ensure a continual supply of small organic debris. No timber harvesting at all in the watershed will mean that WCI will equal 1, and therefore management-associated FHI will equal natural FHI. Scheduling a large percentage of the watershed for timber harvesting in a particular decade will cause substantial increases in sediment, debris torrents, and stream temperature, thereby reducing salmon and steelhead catch. On the Siuslaw National Forest, reducing timber harvest from 15 million cubic feet to 9 million cubic feet results in increases of two thousand salmon caught and five hundred steelhead caught.

The quantitative relationships used in these models are based on the best available information obtained from a combination of site-specific studies, from response coefficients obtained from the literature and modified by professional judgment to fit the Siuslaw National Forest, and finally, from a Delphi technique to obtain best estimates from knowledgeable fisheries biologists. Although these approaches are not meant to be substitutes for comprehensive site-specific studies, the Forest Service relied upon these approaches in preparing its Forest Plan for the Siuslaw National Forest. For each Forest Plan alternative the number of smolts as well as sport and commercial fish catch is calculated. Although the case study of the Siuslaw National Forest in chapter 9 will illustrate how these FHI model results fit into the overall planning process, the reader interested in more details on application of the FHI model to timber management on the Siuslaw National Forest should see Loomis (1988). At present, this FHI model represents one of the best-integrated efforts linking anadromous fish habitat to timber-harvesting activities and suggesting how anadromous fish habitat can be improved (i.e., by addressing those limiting variables in the WCI).

Example of U.S. Fish and Wildlife Service Habitat Evaluation Procedure (HEP) Habitat Suitability Index Model

The U.S. Fish and Wildlife Service (USFWS) as the lead federal fish and wildlife agency has developed a habitat accounting system that is based on Habitat Suitability Index (HSI) models. The HSI models describe habitat preferences of a particular species with respect to food, cover, and other physical features of habitat that are often responsible for observed differences in animal abundance (Farmer et al. 1982). The design of these HSI models can be very specific, measuring a single habitat component viewed to be a factor limiting populations (such as food during the winter season) or very broad, measuring the interaction of all habitat components. In these broad models, not only must the suitability of each individual habitat component be modeled but some measure of spatial "interspersion" or the physical proximity of the various habitat components to each other must be modeled as well. For example, are the various habitat components close enough together from an energy balance standpoint that they can be used effectively by a particular species?

The USFWS has developed over a hundred HSI models that cover individual species of fish, birds, and mammals. Since each of these models identify key habitat variables, they serve as a linkage to management actions that can be designed to enhance fish and wildlife populations. Recently, the USFWS has merged its HSI models with a simple cost accounting system called Habitat Management Evaluation Model, or HMEM (Farmer et al. 1989). HMEM provides the linkages between the habitat variables and management actions that might be carried out on the ground. For example, if a species such as antelope responds positively to amount of grass cover, the model may offer several management actions, such as prescribed burning, seeding, and controlling livestock grazing to increase habitat suitability. Each action has a different effectiveness for increasing grass cover and has a different cost. The model displays how much of an improvement in habitat carrying capacity results per dollar of each of the three different land treatments. Other approaches to the same task involve linking an optimization model (to be discussed in chapter 9) to these habitat models to find the least-cost way to attain a given increase in carrying capacity for a particular species (Matulich et al. 1982). The next section illustrates the linkage of an HSI model for mallard ducks to management activities.

A Habitat Suitability Index Model for Mallards

A schematic overview of mallard duck breeding habitat is presented in figure 5.2. (For the details of this model see Matulich et al. 1982.) Note that the model just focuses on breeding habitat and not on other annual habitat requirements, such as wintering habitat.

In the middle of figure 5.2 this schematic HSI model displays two major components of mallard breeding as nesting and wetlands. To the left of these two components is the measurement of the spatial interspersion between these two components. To the right of the middle are more-detailed habitat components of nesting and wetland habitat. For wetland habitat, this is submersed food and brood-rearing habitat (both of which are further related to the amount of moving and standing water and several other variables). Nesting habitat is related, in part, to emergent vegetation density, which is defined as the percent coverage of vegetation emerging from the water. This is the variable we will examine to see how it is influenced by land management.

Figure 5.3 illustrates the three land management actions that can influence density of emergent vegetation (VD_T). These actions are water-level manipulation; percent of the area protected from human, vehicle, and livestock disturbances; and number of livestock animal unit months (AUMs) of forage per acre. All these variables are subject to either being directly changed by the land manager to improve mallard habitat or being influenced by land management practices designed to accommodate other multiple uses, such as livestock grazing or off-road vehicles.

To focus in on a management action's relationship to density of vegetation, we shall consider the livestock AUMs per acre. Research reported in Matulich et

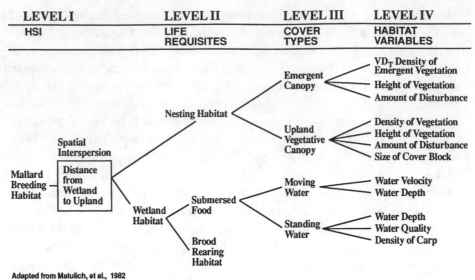

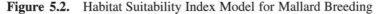

Adapted from Matulich, et al., 1982

Figure 5.2.   Habitat Suitability Index Model for Mallard Breeding

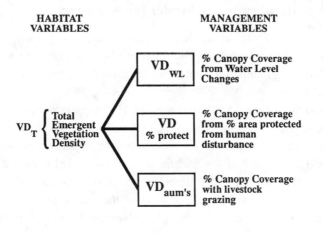

Adapted from Matulich, et al., 1982

Figure 5.3.   Linkage of Management Variables to Habitat

al. (1982) uses the simple linear relationship given in equation (5.7) to quantify the relationship between percent canopy coverage per acre (i.e., vegetation density, or VD) and the number of AUMs per acre in eastern Washington:

$$VD = 70\% - 35 \text{ (AUMS)} \qquad (5.7)$$

This equation states that the maximum attainable percentage emergent vegetation canopy closure, even with no grazing, is 70%. Every additional AUM per acre cuts this percentage of canopy closure in half. Thus with a stocking rate of two AUMs per acre, the percent canopy closure is zero. This livestock-vegetation relationship is based on two effects of cattle grazing: livestock trampling of vegetation and grazing of the vegetation.

The land manager interested in balancing livestock grazing and mallard production (to support wildlife and wildlife-related recreation) could use this part of the model to evaluate the changes in mallard habitat brought about by changes in livestock grazing. Of course, figure 5.3 tells the manager there are other avenues by which to increase emergent vegetation density rather than by just changing the level of livestock grazing. For example, the model could be used to see what compensatory water-level increases would be needed each year to offset the loss of emergent vegetation from additional livestock grazing. Finally, the three management variables influencing emergent vegetation density are combined to yield the net effect using a simple equation that reflects the relative contribution of each variable.

Many of these HSI models have been computerized, so that land managers and biologists can quickly calculate the effect of changing any of the land or water management variables and observe the resulting effect on HSI. Once again these predictive models take on the form of simulation models in providing biologists and managers with feedback on what the most effective combinations of land and water management practices are for enhancing wildlife populations. The use of these types of HSI models will be mentioned in the discussion in chapter 12 of management planning on the Charles M. Russell National Wildlife Refuge.

6

Applying Economic Efficiency Analysis in Practice: Principles of Benefit-Cost Analysis

Throughout the earlier chapters of this book, we have often alluded to the fact that dollar values could be inferred for a variety of the social benefits arising from natural resources. In general, these values can be determined from human actions, such as market transactions, number of recreation trips taken, housing locations, and so on. The time has come to describe the concepts and techniques used to value the many types of social benefits that arise from natural resources, particularly those found on public lands.

What Is Formal Benefit-Cost Analysis?

Benefit-cost analysis (BCA) is a process of comparing in common units all the gains and losses resulting from some action. A complete BCA compares alternative actions to determine which one provides society with the most economically efficient use of its resources. BCA accomplishes this by calculating a single index of the overall effect of a given alternative on both present and future consumers and producers.

One view of BCA states that it is an informational input on the economic efficiency effects of a project or policy. This information is combined by public officials with information on equity, administrative feasibility, and other criteria to determine the alternative that increases social well-being.

An alternative view is that the results of a BCA should be used as a decision rule. Specifically, BCA determines the alternative with the greatest economic efficiency and recommends that alternative for adoption over all others. In many respects the BCA ''makes the decision.'' Although the mechanics of performing BCA are the same in both cases, the use of the BCA in determining the final decision is quite different. Since there will always be a few social concerns that will not be quantified into dollars, expecting the BCA to make the decision can be somewhat heroic (and perhaps dangerous). The limitations of BCA (described later) as a methodology become critical when it is the sole criterion in making natural resource decisions.

However, a government agency may decide that in ranking similar competing public land management alternatives, the one to be selected is that which is most economically efficient. For example, in evaluating alternative wildlife projects that all contribute similar unquantifiable benefits for the same nongame species, the agency will choose the one with the greatest dollar benefits.

Regardless of how the BCA information is used, the information can allow the public, agency managers, and elected public officials to separate the truly inferior alternatives from those that merit serious discussion. A BCA done following the techniques presented in this chapter will shed considerable light on three central features of a public management action, project, or policy: (1) the comparative benefits of alternative management emphases, including the particular mix of multiple uses offered in the alternative; (2) the optimal size or scale of a public land management action (e.g., acres of campground, range improvement, wilderness area, and so on); (3) the optimal timing of implementing the components of the management action or policy.

Even when BCA is used as a decision rule it is not limited to simply determining the acceptability or unacceptability of alternative resource actions. In many cases a thoughtfully done BCA can help to isolate the economically inefficient features of a management action that should be dropped, so that the economically efficient features can be implemented. Thus BCA can be used as a guide to designing the most beneficial projects or resource management actions by subjecting each separable feature to BCA. Along the same lines, BCA can provide information on approximately when a project or resource management action should be undertaken to maximize benefits.

Use of the "With and Without" Principle

One of the fundamental principles of making sound decisions is to separate the effects that are due specifically to the management action from those that will occur without it. In essence we wish to know changes in types of resource uses and amounts of the resources that will be produced by a project over and above the amounts that would have occurred even if we do nothing. For example, if we are trying to estimate the benefits of a new fish hatchery, only the added fish catch is a benefit, not the total number of fish caught associated with the current fishery.

Figure 6.1 illustrates this point. Under the current conditions in the first year, anglers catch one thousand fish. In ten years, that catch is expected to go up to twelve hundred fish, even if no hatchery is built, because of habitat improvements and changes in fishing regulations that were put into effect a few years earlier. However, if the hatchery is built, fish catch will rise to fifteen hundred fish in ten years. Analysts often become confused over whether the benefits of the hatchery should be the difference of the before project minus the after project (five hundred fish) or the difference between the with and without project (three hundred).

Clearly, it would be erroneous to compare before and after, since part of the gain in fish catch (two hundred fish) would have occurred even if the hatchery project had not been implemented. No hatchery is needed to produce the first increase (two hundred) in fish catch in the future, and it would be incorrect to attribute that change to the hatchery. The change that results from the hatchery is just the increment over what would occur if nothing is done.

But just because nothing is done does not the mean the future without-project level equals the current level. The future without-project trend could be increasing, as in figure 6.1, or even decreasing, as in figure 6.2. As figure 6.2 shows, if the future without is a decreasing trend, simply maintaining the current levels results in a net gain. Although the before-and-after levels are identical, certainly the project or policy has made a difference. For if nothing was done, anglers would catch only eight hundred fish in ten years rather than the current one thousand. Thus avoiding a future loss is a project benefit.

The "with and without" principle is also very helpful in zeroing in on exactly which resources need to be fully evaluated in a BCA and which can be ignored. If the level of a particular resource output would be the same in the with and without alternatives, then it need not be analyzed as part of the BCA. The bottom line in a BCA is driven only by changes in types of resource outputs and changes in the quantities of resource outputs. Although the resource may seem important, if the project or policy does not affect it, there is no need to include it in the analysis. Of

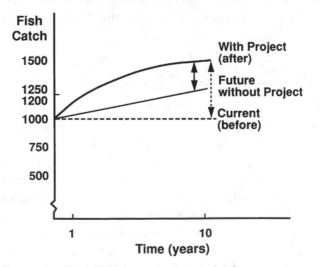

Figure 6.1.   Comparing With-Without to Before and After

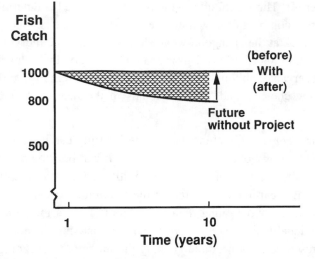

Figure 6.2.   With and Without in a Declining Future

course, some preliminary analysis may be necessary in the early planning or "scoping" phase of an analysis to determine if the resource would be significantly affected.

Conceptual Foundation of Benefit-Cost Analysis

One overall objective of a BCA is to determine which combinations of resource uses produce the greatest net gain to society. These gains and losses are measured from the viewpoint of individual members in society. Since people prefer more to less, reallocations of resources that increase the benefits to one individual without reducing the benefits to anyone else would be preferred.

Defining Benefits

To determine if a resource action actually represents an improvement in social well-being, one must be able to measure the net gain in benefits. Much like beauty, the benefits of a resource action are in the eye of the beholder. Are the jobs created by the project a benefit? Are the additional minerals produced an economic benefit? What about cost savings to society from producing electricity with hydropower instead of more expensive fossil fuels? The answer to these questions depends on the analyst's valuation philosophy and accounting stance.

Valuation Philosophy

As discussed in chapter 1, the anthropocentric view states that people are the ultimate source of value. Goods and services provide benefits only if individual people are made better off. The magnitude of the benefits received is determined by each individual's own judgment of how much better off he or she is. One way of measuring this is to look at the change in what economists call the individual's level of utility (e.g., level of satisfaction). Unfortunately, an individual's level of utility is not directly observable, although we can indirectly determine the utility provided by observing the choices made by individuals in the face of different prices or incomes.

From Utility to Income.   Graphing a consumer's utility using indifference curves (see Varian 1990 or any other price theory text), it can be shown that the change in income is a valid measure of the change in utility. Thus we measure an individual's benefits of a particular resource as the maximum amount of income they would pay for the resource rather than go without. This is the sum of money at which the individual is indifferent between paying that sum of money and having access to the resource or project and keeping the money and forgoing the project.

Income as a Proxy for Sacrifice of Other Goods.   We will illustrate in more detail this concept of willingness to pay later in this chapter. At this point we wish to stress that willingness to pay in the form of income is simply a proxy for willingness to give up other goods and services to have the resource or project under study. In essence, maximum willingness to pay is often driven as much by a willingness to substitute the new good for the old one as it is by level of income. For example, a person who really likes bird watching might be willing to give up three trips to the beach to have one additional bird-watching trip. In this sense, income influences willingness to pay only because we use dollars as the measuring unit, not because the individual is directly forgoing income. Rather, the individual is forgoing three beach trips to gain one bird-watching trip. But if we wish to have a dollar measure of the three beach trips forgone to gain one bird-viewing trips, we must convert the beach trips forgone into their income equivalent in dollars. In this way we can compare the value of the bird-watching trip gained to the cost society must incur to provide the bird-watching opportunities.

From Individual Benefits to Social Benefits

Now that we have an initial definition of what benefits are from an individual's point of view, we need to see how this information would be used to determine if some resource management action is desirable. To do this, we assume that what is best for society is what is best for each individual in society. That is, society's benefits are the simple sum of the benefits individuals derive. For example, assume society is made up of just two people Mr. Smith and Ms. Barney. There are two possible bundles of goods that cost the same to produce. Society is attempting to decide whether the highest-valued use of resources is to produce Alternative B rather than the current Alternative A. Table 6.1 shows the benefits to each of them with Alternative A and Alternative B.

Table 6.1
Consumption Benefits with Different
Bundles of Goods and Services

Alternative	Ms. Barney	Mr. Smith
A	$100	$150
B	$125	$150
C	$ 40	$200

Review of table 6.1 shows that Alternative B is preferred, since Ms. Barney receives a bundle of goods she values more highly (i.e, is willing to pay more for) than in Alternative A. At the same time, Mr. Smith is neither better off nor worse off. Thus moving from Alternative A to B results in what is called an actual Pareto improvement in society's welfare: Ms. Barney gains ($25) and Mr. Smith stays the same. There is an actual clear gainer and no losers. Thus the "economic pie" gets larger, with Ms. Barney's slice growing and no change in Mr. Smith's slice. Now once we have adopted Alternative B, we have obtained a Pareto optimum, or optimum economic efficiency. That is, there are no other alternative reallocations of resources that make one of them better off without making the other worse off. Clearly, Alternative C cannot do it, as Mr. Smith gains $50 but at the expense of Ms. Barney losing $85. At Alternative B the economic pie is at its maximum size, given the current technology and tastes and preferences of Mr. Smith and Ms. Barney.

Potential Pareto* Improvements

Unfortunately, many resource actions, such as damming a river or draining a wetland, that can make the economic pie larger result in some groups in society losing and others gaining. For example, is Alternative D in table 6.2 better than Alternative B? Alternative D does not represent an actual Pareto improvement, because Mr. Smith is made $25 worse off (compared to B), whereas Ms. Barney is made $75 better off. What can we say about an alternative policy that makes one person better off but another worse off? If we can add individual valuations of Ms. Barney and Mr. Smith together, we can determine if the sum or total of the valuations is the highest with Alternative D. Using the criteria of total benefits, Alternative D is preferred, since it yields the greatest value of total output. In other words, the economic pie is larger than with any other alternative. Alternative D represents what we call a potential Pareto improvement: the gainers (Ms. Barney) could compensate the losers (Mr. Smith) and still have a gain remaining. With regard to Alternative D in table 6.2, Ms. Barney could afford to give Mr. Smith $25, so he is back to his original level of benefits ($150 in Alternative B), and Ms. Barney

---

*A Pareto Improvement is named after Vilfredo Pareto, a nineteenth century economist who first investigated the ideas of social well being.

Table 6.2
*Identifying a Potential Pareto Improvement Using the
Total Benefits Rule*

Alternative	Ms. Barney	Mr. Smith	Total Benefits
A	$100	$150	$250
B	$125	$150	$275
C	$ 40	$200	$240
D	$200	$125	$300

would still be better off by $50. On this basis, Alternative D is the most economically efficient. In no other alternative could the beneficiaries compensate the losers and still have any gain left over. This is sometimes referred to as a compensation test. Such tests can be quite elaborate, including not only whether the gainers can fully compensate the losers and still have a gain left over but whether the losers cannot bribe the gainers into forgoing the program (see Mishan 1981).

For the total or sum of benefits to reflect a meaningful measure of social well-being at least two key assumptions must be made: (1) that we can meaningfully compare different people's valuation of goods; (2) that a dollar's worth of benefits provides the same change in individual utility or well-being to every person.

Economic Efficiency and Equity in Computing Total Social Benefits

The preceding assumptions may be a bit troubling. There are certainly cases where a hamburger to a homeless person who gets one meal a day provides more satisfaction to that person than the same hamburger provides to a rich person who eats three meals a day. But our measure of value is the amount of sacrifice a person would make to obtain the good in question. The poor person has fewer other goods (or income) to sacrifice, so his or her revealed value in the form of willingness to pay would likely be lower. As cruel as this extreme example sounds, we must remember what BCA is all about: it is economic efficiency, not equity. This is the logic of the marketplace. All dollars have equal power in the market.

As in evaluating market outcomes, society certainly is concerned about equity when ranking alternatives in BCA. As chapter 4 explains, how the total benefits are distributed may be as important to social well-being as their amount. The distributional implications of economically efficient resource allocations are best handled by displaying not only the total net benefits, but also the way the benefits are distributed across different income, age, or ethnic groups. In this way the political system can choose to weight benefits to different groups differently. Since any differential weighting system is a value judgment, it should be reserved for the political system and not be carried out by the analyst unless the analyst is explicitly given weights for this project by elected officials.

In the absence of explicit, project-specific weights, we will follow standard practice used in federal benefit-cost procedures (U.S. Water Resources Council 1979, 1983) in this book and weight all individuals' dollars equally. Although this is just one of the many possible weighting systems, it is certainly consistent with much of the public land legislation reviewed in chapter 2 (which rarely mentions preferential treatment for distributional reasons).

Would We Always Want to Pay Compensation?

Another concern often expressed about reliance on potential Pareto improvements rather than an actual Pareto improvement is the failure to pay the compensation to the losers. There are several reasons for not requiring actual payment of compensation, even if failure to pay compensation leaves some people worse off. These reasons may be grouped into ones relating to economic efficiency and equity. The economic efficiency reasons include the following:

1. The high information costs associated with identifying the specific individuals that gain and the specific individuals that lose.
2. The large transactions costs of facilitating actual payments from the gainers to the losers.

The point here is that much of the net gain in benefits to society might be consumed in attempting to carry out the steps necessary to pay the compensation. The equity reasons for not paying the compensation include the following:

1. Payment of compensation could worsen equity if gainers are poor and losers are relatively well off. Certainly, having the poor compensate the rich would run counter to most accepted norms of equity.
2. A progressive tax system (i.e., one in which tax rates rise with the level of income), coupled with an effective system of transfer payments (e.g., welfare, food stamps, rent subsidies, and so on), will result in partial compensation of the losers as gainers must pay an increasing portion of their gain as taxes and losers' taxes will be reduced (Sassone and Schaffer 1978:11).
3. If BCA is applied to most government programs, everyone will eventually be a gainer in some evaluations. We simply recognize that although our piece of the expanding economic pie might get smaller from one project, if the economic pie is continually expanding as resources are moved from low-valued uses to higher-valued uses, eventually our slice will increase (Sassone and Schaffer 1978:11).

Role of Accounting Stance in Determining Transfers

Although benefits are defined from the viewpoint of individuals, often one person's gain is exactly offset by another person's loss. That is, some resource reallocations do not represent net gains in economic efficiency, but rather simply a transfer of economic activity from one person to another or one location to another. For example, if Ms. Barney receives $100 worth of benefits from a new reservoir but Mr. Smith loses $100 of river recreation downstream when the reservoir is in operation, then there is no net gain to this society.

Although this seems very straightforward, if we add one element of realism to this example, we can illustrate a frequent confusion over what constitutes benefits. Assume that the Ms. Barney (who gets the new reservoir) lives upstream in the state of Kentucky but Mr. Smith lives downstream in Tennessee. If the analyst measuring the benefits of constructing the new reservoir worked for the state of Kentucky and took a state view, then he or she would find that the reservoir resulted in a gain of $100 of benefits to residents living in Kentucky. This state viewpoint is one accounting stance. By accounting stance we mean the geographic viewpoint of which benefits and costs matter and which do not. Thus with the state of Kentucky or local accounting stance, only benefits received in Kentucky and only costs incurred by Kentucky residents would count. The state analyst would ignore both benefits and costs outside our region of interest.

From the standpoint of national accounting, such a state accounting position is too narrow to reflect all the benefits and costs to all people. That is, the lost river recreation in Tennessee is a real decrease in well-being and should be counted. As discussed in chapter 2, a national accounting stance is the appropriate one for most actions. This is especially true when dealing with management actions on National Forests, National Parks, and National Wildlife Refuges or when dealing with the expenditure of federal funds. In this case, the national interest is clear: we are all owners of these natural resources or payors of the funds to manage them. As discussed in chapter 3, a state accounting stance in this two-state example would result in failure to incorporate a negative externality of the reservoir into benefit-cost calculations. Since one of the reasons for public ownership of resources is to internalize such externalities into public decision making, clearly at least a national accounting stance is required to ensure that the complete benefits and costs of a resource management action are reflected.

But there are times when even a national accounting stance is too narrow. Consider the case of controlling acid rain in the northeastern United States and southeastern Canada. The costs of controlling air pollution are incurred largely in the midwestern states. The benefits are received downwind, in upstate New York, New England, and Canada. The benefits the Canadians would receive are real increases in their social well-being. They should count when we add up the benefits to compare them to the pollution control costs. Once again, if analysts take an accounting stance that is too narrow, real benefits to individuals (who just happen to reside in a different political jurisdiction) will be ignored. The United States might erroneously reject acid rain controls because the benefits (measured as the gain to just U.S. citizens) are less than the costs. But the true social benefits (including those to the Canadians) might well exceed the costs.

The general guidance to the analyst is to adopt the spirit of the first legislation requiring BCA and measure benefits and costs ''to whomsoever they may accrue.'' Much like dealing with equity in BCA, concerns about which states or nations gain and which ones lose are best dealt with by displaying the distribution of benefits and costs to each political jurisdiction. This is much better than adopting a narrow accounting stance that results in complete omission of certain state's or nation's benefits and costs. Such an omission is often a recipe for economic and ecological disaster.

Gross Willingness to Pay, Cost, and Net Benefits

It is now time to begin to answer one of the first questions posed by this chapter: is the additional employment created by a resource management action a benefit or a cost? One often reads in the newspaper that a new power plant or timber sale will ''create'' a certain number of jobs. The same newspaper article might quote a timber industry official who states that preserving an area as Wilderness results in society forgoing trees that are worth $1 million at the lumber yard. Is this an accurate statement of the opportunity cost of designating the area as Wilderness? We now turn to an example that will allow us to answer these questions.

Table 6.3 presents data on three resource uses that could take place on a specific parcel of land. To keep the example simple, we will assume these are three mutually exclusive uses, although a more complicated example would allow for joint production between the uses. (Joint production will be discussed in chapter 9.) The first use is a sustained-yield timber operation (with the trees already naturally occurring on the site); the second is livestock grazing; and the third is to operate a fee elk hunting operation. The column headings relate to gross benefits, cost of production (or expenditures) and net benefits. Net benefits is equal to gross benefits minus the cost of production. We can use table 6.3 to answer the question about what the losses to society are from maintaining the land area for elk hunting as compared to timber harvesting. The gross or market value of the annual sustained yield of timber at the lumber mill might be $400,000. For example, the mill owner will pay $200 per thousand board feet (mbf) and we estimate an annual timber yield of 2,000 mbf. If we keep the area as an elk hunting operation, does society

Table 6.3
Example of Annual Gross and Net Benefits

Resource	Gross Benefits (Price * Qty)	−	Costs or Expenditures	=	Net Benefits
Timber	$400,000		$350,000		$50,000
Cattle grazing	$ 75,000		$ 35,000		$40,000
Elk hunting	$ 80,000		$ 20,000		$60,000

forgo the $400,000 worth of timber? The answer is certainly no. For it takes $350,000 of labor and machinery to harvest the trees and transport them to the mill. Specifically, workers must build logging roads, cut the timber, yard the timber to the trucks, and then haul the logs to the mill. Employment of these workers, machines, and fuel on this forest means society forgoes the value of the goods and services these inputs would have produced elsewhere. Thus the $350,000 is a cost that must be subtracted from the $400,000 to obtain the net gain society receives from a sustained-yield timber operation on this land. The $50,000 net gain is called the stumpage value.

What would society forgo if the elk hunting operation is maintained? Only the net benefits of timber are forgone. The reason is that although we give up producing timber with a market value of $400,000, we also save $350,000 of cost of production that can be employed elsewhere in the economy. If there is a firm demand for the 2,000 mbf of timber that would be produced from this tract of land, the timber will be cut from some forest, somewhere. In that forest, labor, machinery, and fuel will be employed to produce the 2,000 mbf of timber.

The same logic applies to estimating the net value of the livestock grazing operation. The rancher would receive $1 per pound for the 75,000 pounds of beef raised on the land. But this is not the net value of the beef to the rancher or to society. The rancher must spend $35,000 purchasing his or her initial cattle stock, feed supplements, veterinarian services, and labor to tend the cows and drive them to market. The net gain to the rancher and the nation is only the difference between what society gets ($75,000) and what it gives up ($35,000). The net benefits of $40,000 represent the gain to society beyond what was given up.

Now let's examine the elk hunting operation. For the moment, assume two hundred elk hunters would bid $400 each to go elk hunting in this area. That is, they would pay $400 each year to have the area available for elk hunting rather than go elk hunting at the next best area. This yields a gross benefit of $80,000 in table 6.3. However, the hunters incur harvesting and transportation costs associated with elk hunting of $100 per hunter. In our example these costs total $20,000. The net benefits of the elk hunting operation are just $60,000: the gains minus the costs.

Although some state Fish and Game agencies would like to count the amount hunters themselves spend as a benefit of hunting, it is clearly a cost. Hunters and society give up producing some other goods to devote the resources to producing elk hunting. In addition, if in fact the landowner decided to harvest trees instead of producing elk, what would be lost to society? Certainly, the current expenditures by elk hunters on ammunition and transportation would not be lost to the economy as a whole. The only way such expenditures could be lost to the economy as a whole is if the hunters went home and set fire to the money! Since they are unlikely to do this, what actually occurs is that hunters spend that money on their next-most-preferred activity. This might be elk hunting at a less desirable area, deer hunting, or bowling. In any case, the $20,000 expenditure will still occur in the economy. What is lost is the additional enjoyment and satisfaction that spending that $20,000 provided from hunting elk in that particular area. That is the net loss to society.

As the reader can infer, the $20,000 of hunter expenditures, the $35,000 of rancher costs, and the $350,000 of logging expenses reflect costs, not benefits. In addition, these factors of production and spending will simply be transferred elsewhere in the economy if these resource uses are precluded at this area.

To see the nature of this transfer consider the decision on location of the Intermountain Power Project (IPP) coal-fired power plant. In 1978 this large coal-fired power plant was proposed at Hanksville, Utah, near Capitol Reef, Canyonlands, and Arches National Parks. Because of concern about air pollution reaching these parks, it was decided to build the power plant on the west side of Utah, near the town of Lyndal, and far away from any National Park. Although the town of Hanksville lost all the influx of construction workers and plant operations personnel, this was perfectly offset by the town of Lyndal's gain of the same level of construction workers and plant operations personnel. Viewed from an international, national, or state accounting stance, there was no change in employment of resources. But viewed from a narrow accounting stance of the city of Hanksville, there was a net loss. Viewed from the narrow accounting stance of the city of Lyndal, there was a net gain. As can be seen, both of these narrow accounting stances seriously misrepresent the change in economic social well-being.

Assumptions About Employment of Resources

As discussed earlier, employment of labor, materials, diesel fuel, machines, and other factors of production represents opportunity costs to society. That is, use of a given amount of our scarce resources in one project or management action necessarily means giving up using these resources to produce output from some other project. All of this assumes reasonably full employment of resources. If the labor employed in one project would have been otherwise unemployed for the duration of its employment in the project, then there would be no opportunity cost of using that labor in this project. That is, use of unemployed resources does not require society to give up anything else to gain the new level of output.

How likely is it that the resources employed by a project would otherwise be unemployed? If the project is a long-lived one requiring labor and materials over several years (or even decades), it is very unlikely that one could rely on unemployed resources. Excess unemployment (above the background frictional levels) tends to be transitory or lasting for short periods during recessions. Most resources, whether labor or machinery, are relatively mobile and will seek out employment rather than remain unemployed for years at a time. However, the possibility exists that during some phase of a public land management project, otherwise unemployed resources would be put to work utilizing natural resources provided by those public lands. If this is the case, the analyst can either count the economic costs of such resources at zero (or near zero, depending on the value of leisure time to unemployed workers) or, if labor costs are included on the cost side, count the wages paid as a project benefit. Either treatment will be reflected in the appropriate change in a project's net benefits.

Displaying Changes in Employment and Economic Activity

Often such laws as the National Environmental Policy Act, which governs preparation of Environmental Impact Statements, or local politicians require that the employment related to a public land management action be displayed. Thus it may be appropriate to translate direct changes in project employment into total employment by including spin-off employment in support sectors of the economy induced by the project. In the same way, it may be important to calculate the change in personal incomes in a given county or region that results from an increase in economic activity induced by the project. Although both the induced employment and personal income generated by a project that uses fully employed resources are a transfer, there is often interest in knowing the magnitude of those transfers.

This economic impact analysis often serves at least two useful purposes. First and foremost, it can help local government planners plan the "infrastructure" of roads, schools, hospitals, housing, and parks that will be needed to accommodate the additional workers and their families. Second, politicians may desire to stimulate selected regional economies at the expense of other areas. For example, expanding job opportunities in rural areas and increasing the tax base in rural economies may be a distributional goal. Even though this rural gain may be offset by a loss of the same economic activity in an urban area, politicians may wish to stimulate rural economies to maintain their viability. The techniques for computing income and employment effects of projects via multiplier analysis (using input-output models) will be discussed in chapter 7.

Measurement of Economic Efficiency Benefits

Now that we know what we wish to measure, the next step is to determine how we are going to measure it. The methodology chosen to measure the benefits of resources must allow us to compare those that are marketed and those that are not. For consistency in valuing both marketed and nonmarketed resources, economists rely on values measured from consumers' demand curves and businesses' supply or cost curves.

Benefits to the Consumer

It is easiest to illustrate valuation of a resource with an example. Consider figure 6.3, which presents a consumer's demand curve for hamburgers. At a market price of $2 per hamburger, the consumer wishes to purchase six hamburgers each week

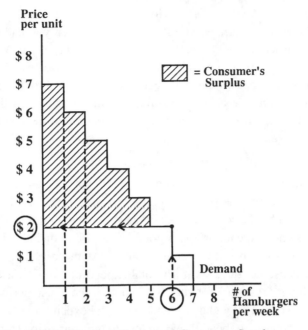

Figure 6.3. Consumer's Demand Curve and Consumer Surplus

(being a real carnivore). What are the benefits of being able to buy six hamburgers at $2? One's first reaction might be that if the consumer paid $2 for each hamburger, they must be worth $2. But if we read up from the quantity axis to the demand curve and over to the vertical price axis, we see that only the sixth hamburger purchased is worth just $2. The fifth hamburger is worth $3, the fourth is worth $4, the third is worth $5, and so forth, to the first hamburger, which is worth $7. The first hamburger one bites into that week provides $7 worth of benefits. That is, if the consumer could only have one hamburger in a week, he or she would pay up to $7 for it (this person is a serious carnivore). Since the market clearing price is just $2, the consumer receives a "personal profit," or what economists call a consumer surplus of about $5 on this first hamburger.

Consumer surplus is defined as the area beneath a consumer's demand curve but above the actual price paid. Thus the net benefits to the consumer from being able to buy six hamburgers at $2 each is the striped triangular area under the demand curve. In figure 6.3 this area is equal to $15. (If the demand curve were continuous rather than discrete consumer surplus would be represented as a triangle the area under the demand curve in excess of price actually paid.) Thus for most goods, consumers receive a surplus or gain in excess of what they pay. For goods and services that are consumable in small units, such as hamburgers, cans of soda, and so on, only the last unit purchased is worth just what the consumer paid. Since the last unit has a value to the consumer that is exactly equal to what he or she paid, there is usually no consumer surplus on the last unit bought.

Consumer Surplus as Real Income.   How real is this consumer surplus? We all have received consumer surplus. Think about a time when you had made up your mind to buy a particular book or recording at its full retail price. You went up to the cash register prepared to pay this price and the sale price was lower than you expected. That difference between what you would have paid for the good and what you did pay was a tangible gain in your real income. You bought the good at the sale price and retained the consumer surplus as added real income (e.g., cash) in your wallet or purse.

Why Actual Expenditures Are Not a Measure of Net Benefits.   Although use of consumer surplus as a measure of the economic efficiency benefits to the consumer seems straightforward, some people become confused over how to deal with the actual money spent by the consumer. A simple example will illustrate why the total amount of consumer expenditure on a good is often a poor indicator of the total benefits the person receives from consuming it. At the present price of $2 per hamburger the consumer buys six, for a total expenditure of $12. However, if the price per hamburger rose to $6, figure 6.3 indicates the consumer would buy two hamburgers per week. This yields the same $12 total expenditure. Using the actual expenditure as a measure of benefits would lead to the conclusion that the consumer would receive the same benefits from consuming two hamburgers that he or she would from eating six hamburgers per week.

Common sense tells us the consumer would prefer to pay a total of $12 to get six hamburgers per week rather than two. If we calculate the consumer surplus of these alternative combinations (six hamburgers at $2 versus two hamburgers at $6) on figure 6.3, we see that consumer surplus is a good indicator of the preferred combination. As shown earlier, the consumer surplus for six hamburgers purchased at $2 is $15. The consumer surplus for two hamburgers purchased at $6 is only $1 ($7 − $6 price). Thus the consumer surplus makes it clear: six hamburgers provides more benefits to the consumer than two.

Although this may seem like an incredible amount of effort to prove the obvious (and it is!), reliance on expenditures as a measure of benefits is, unfortunately, a common mistake in resource decisions. It is a particularly easy mistake to make when actual expenditures are readily observable in the marketplace and consumer surplus can only be inferred after statistically estimating the product's demand curve. The temptation is great simply to use expenditures, rather than to go to the trouble of statistically estimating demand curves. It is hoped that the error of using consumer expenditures as a measure of benefits has been made clear.

Another pitfall in calculating consumer benefits can be illustrated by the question, ''Is not the $2 per hamburger or the $12 spent on all six hamburgers part of the benefit of the hamburgers?'' The key here is to realize that if cows became an endangered species and there were no more hamburgers, consumers would only lose their consumer surplus, not their expenditures. If people could no longer purchase hamburgers at $2 each, the $12 they are currently spending on hamburgers would be spent on some other good. The additional utility or benefits that spending the $12 on hamburgers provided and that are not received from buying the next-most-preferred good (say, hot dogs, soy burgers, and so on) would be lost to the consumer and society.

Remember, demand curves are adjusted for the price and availability of substitutes, so that consumer surplus is the value over and above the next-best substitute. Thus from the consumer's perspective the $12 spent is a cost of hamburgers, not a net benefit. However, the $12 is received by beef producers and hamburger makers. Part of this money pays for costs of production, but part of it reflects benefits to the business producer. We now turn to measuring these benefits.

Benefits to the Producer

To begin with, we will simplify the analysis and measure benefits to a firm or business that is a price taker. A firm is a price taker when its output is a small part of a national or international market for that product. Continuing our hamburger example, one individual cattle rancher's supply would be so small as to have no effect on the price of beef. Figure 6.4 illustrates the relationship between the market for beef (measured in millions of pounds) and the individual cattle rancher (whose output is measured simply in pounds of beef). The equilibrium price is set in a national or international market based on total consumer demand and total industry supply. The demand curve facing the individual rancher is horizontal at the current market clearing price for beef. The reason for this is that the cattle rancher's production makes up such a small fraction of the total supply of beef that doubling his or her beef production or cutting production in half would have no effect on

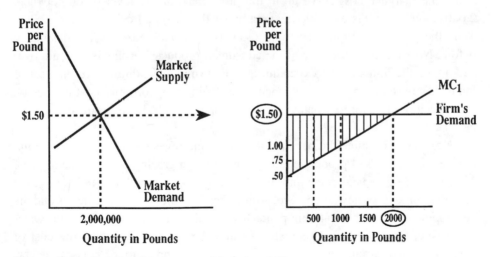

Figure 6.4. Relationship between Market and Firm

price. In addition, the rancher can sell all he or she wants at the existing market price.

Besides the horizontal demand curve, figure 6.4 shows the rancher has an upward-sloping marginal cost curve. That is, as the rancher attempts to produce more beef on a given area of land, his costs per unit of output rise, as he must intensify management and add more capital inputs (supplemental feed, more water tanks, and so on). The marginal cost curve is the rancher's supply curve. It reflects the minimum dollar amount for which the rancher would supply each additional unit of output.

To calculate the net economic efficiency benefits to the rancher we note that the first 500 pounds of beef cost only 75 cents per pound to produce, but the rancher receives the market clearing price of $1.50. Thus the rancher receives a producer surplus or economic profit of a little more than 75 cents per pound on these first 500 pounds of beef. The next 500 pounds of beef cost $1 per pound to produce but are again sold for $1.50 per pound. The producer surplus on these units is 50 cents per pound. It is only the two-thousandth pound in which the price received by the rancher equals the minimum supply cost, and therefore no producer surplus is received on the last unit produced. Thus the total producer surplus for 2,000 pounds of beef is $1,000 ($1.50 − .50 × 2,000 pounds, with the product divided by 2, since producer surplus is a triangle). This is the striped area in figure 6.4. Notice the similarity in logic between derivation of producer and consumer surplus. In both cases, economic efficiency benefits are the net gain over and above the costs.

Another means by which to calculate producer surplus and to see its relationship to a firm's profit is to recognize that the producer surplus triangle is the difference between a firm's total revenue and total variable cost. The total revenue to the rancher is $3,000 ($1.50 × 2,000 pounds). The total variable cost is the area under the marginal cost curve up to the profit-maximizing level of output (here 2,000 pounds). In this example the area under the marginal cost curve is $2,000. Thus the producer surplus is $1,000 ($3,000 − 2,000), just as was calculated before. Note that as in the case of the consumer, producer surplus is the gain over and above the firm's actual expenditures. The firm's expenditures are the cost of inputs. The use of inputs by this rancher results in an opportunity cost to society, since society must forgo whatever else the inputs would have produced in their next-best use.

Next let us examine the benefits of a rangeland investment project that would reduce the cost of beef production to just one rancher grazing on a particular piece of public land. A typical range improvement project performed by a federal agency might be removal of unwanted vegetation to improve production of grasses and the development of a stock-watering pond. If the rancher currently had to haul water to the site, the replacement of a stock-watering pond would reduce the cost of maintaining his current herd. In addition, with extra production of grasses the carrying capacity of the range would increase, allowing higher cattle stocking. These

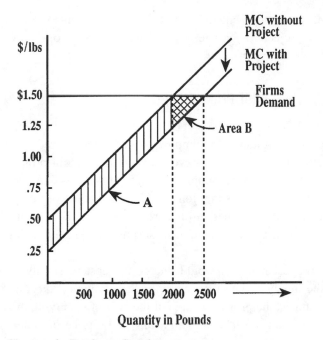

Figure 6.5.   Changes in Producer Surplus

effects would be translated into lower production costs and hence a downward shift in the rancher's marginal cost curve.

As figure 6.5 illustrates, this downward shift in the rancher's marginal cost curve has two effects. First, the rancher receives a resource cost savings, because the original level of output (2,000 pounds) can be produced at a lower cost. Thus these resources can be freed up to produce other goods and services that society values. The resource cost savings on the original level of output is equal to $500 (25 cents × 2,000 pounds). This is denoted as area *A* in the stripped portion of figure 6.5.

Second, with the lower cost of production, it is profitable for the rancher to expand production to 2,500 pounds of beef. With the new marginal cost curve, the cost of producing the two-thousandth pound of beef is now less than the price. As such, a profit-maximizing rancher will expand production until the new marginal cost equals the market price. Thus there is a gain in benefits equal to $62.50 (25 cents × 500 additional pounds ÷ 2, since producer surplus is a triangle). This is the shaded area *B* in figure 6.5. Another way to view this second gain in economic benefits is to realize that society gets $750 worth of beef ($1.50 × 500 pounds) but it only costs $687.50. Thus the net gain is $62.50.

Because this change in output by this rancher is so small relative to the market, there is no change in the price of beef to consumers. If there is no change in price, there is no change in consumer surplus and hence no net economic efficiency benefits to the consumer. In essence this small additional amount of beef is added at the margin, where price equals gross willingness to pay. We now turn to analyzing resource projects or policies that are so large in scope they do affect the market supply and hence price of the final good.

Benefits to Society with Price Changes

Consider a major water supply project or watershed-yield-enhancement project that would reduce a city's reliance on pumping groundwater and replace it with surface water supplies from a proposed reservoir. Given the without-project supply curve (marginal cost curve) for water supply, the price of water is $15 per 1,000 gallons. Given this price, average consumption is 5,000 gallons per person per month. The existing producer and consumer surplus are areas A and B in figure 6.6. As table 6.4 indicates, the consumer and producer surplus are $12.50 each in the without project case.

Now the proposed project would reduce the cost of supplying water to the point where the supply curve shifts out to the Supply with Project in figure 6.6. As this figure illustrates, the project would cut the cost of supplying the current 5,000 gallons of water to city residents. With a fall in the price of water from $15 per 1,000 gallons to $10, this cost savings is shared between consumers and producers. The exact division will depend on the relative slopes of the demand and supply curves. In our specific example the consumers gain $12.50 from the cost savings (area C_1 in figure 6.6) and $12.50 from the transfer of producer surplus to consumer surplus when the price of water falls from $15 to $10 (area B in figure 6.6). The gain to the producer from the cost savings on the existing quantity of water supplied

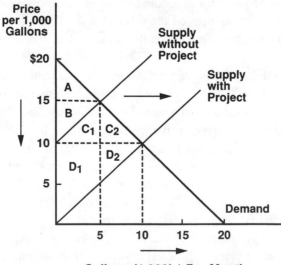

Figure 6.6. Benefits to Society with Price Changes

Table 6.4
Social Benefits from a Large Change in Supply

	Without Project	With Project	Gain
Producer Surplus			
Area(s)	B	(D1 + D2)	
Dollars	$12.50	$50	$37.50
Consumer Surplus			
Area(s)	A	(C1 + C2 + B + A)	$37.50
Dollars	$12.50	$50	
Total Benefits	$25	$100	$75

is area D_1, or $37.50. However, the producer loses the producer surplus in area B ($12.50) when the price of water falls. Thus the net gain to the producer from the cost savings on the original 5,000 gallons of water is $25 ($37.50 − $12.50). Notice that area B is not a net gain, but a transfer of producer surplus realized at the $15 price to consumer surplus when the price of water falls to $10.

In addition to the cost savings on the current quantity of water, the lowered price of water results in the quantity of water demanded increasing to 10,000 gallons. There is a consumer surplus gain from being able to consume the additional water (via greener lawns, cleaner patios and cars, and so on) of area C_2, or $12.50. The producer also gains a surplus from supplying the 5,000 additional gallons of water. This gain is area D_2, or $12.50.

Thus the total benefits of the project to city water users and producer is $C_1 + C_2 + D_1 + D_2$, or $75 per month. This net gain can be calculated by subtracting the original producer and consumer surplus (B and A) from the new total consumer ($A + B + C_1 + C_2$) and producer surplus ($D_1 + D_2$). This is illustrated in table 6.4. Specifically, the total consumer and producer surplus with the project are $50 each or $100 together. Given the without-project benefits of $25, the net gain is $75. Either way one measures the net gain, the project benefits are $75. This gain would be compared to the costs of constructing the project or performing the watershed management action to determine if the overall benefits ($75) outweigh the costs.

Benefits of Environmental Improvement to Recreation Visitors

Demand curves reflect an inverse relationship between the quantity demanded and price, holding all other factors constant. One of the factors held constant is the quality of the good. It makes little sense to estimate a demand curve for cars that lumps together Audis and BMWs with Chevy Novas or Volkswagens. Much like demand shifts induced by changes in income, the demand curve will shift when the quality of the good changes.

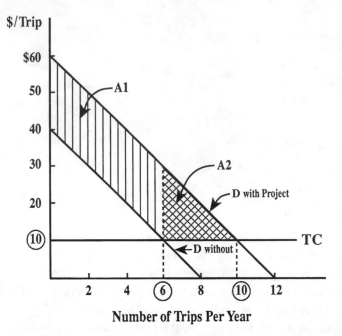

Figure 6.7. How Environmental Quality Affects Visitors Demand Curve

In the case of recreation, the quality of the recreation experience depends on the quality of the natural and environmental resources at the recreation site. If water quality or quantity change, the demand for recreation will shift accordingly. As Maler (1974) demonstrated, we can use the area between these shifted demand curves as a measure of the benefits that visitors to the site derive from the change in resource quality. A simple example will illustrate the point.

Figure 6.7 illustrates how the demand curve for sport fishing would shift if a hatchery or habitat improvement project increased the typical angler's catch rate. At present the angler's demand curve is given by $D_{without}$. At the angler's current travel cost to the site of $10 (i.e., the price paid for recreation if entrance fees are zero), he or she takes six trips. The without-project benefits are $90.

If angler fish-catch rates rise as a result of the project, this increases the enjoyment from fishing at this particular site and shifts the demand curve outward to $D_{with\ project}$. This demand shift has two main effects. First, the angler receives more satisfaction on the current number of trips taken. The gain in consumer surplus on the existing six trips is shown in figure 6.7 as area A_1, or $120, (($60 − $40) × 6 trips). In addition, the higher enjoyment and satisfaction when coupled with the original trip price of $10 implies that taking more fishing trips is optimal. As illustrated in figure 6.7, the angler increases the number of trips taken over the season from six to ten. There is $40 of additional consumer surplus (area A_2 in figure 6.7) on these additional four trips. Thus the total benefits to the typical angler from the fishery enhancement is areas $A_1 + A_2$, or $160.

We can calculate exactly the same answer for project benefits if we subtract the with-project total consumer surplus from the without-project consumer surplus. The with-project consumer surplus in figure 6.7 is \$250, ((\$60 − \$10) × 10 trips ÷ 2). The without-project consumer surplus is \$90. The gain in angler benefits is \$160, just as before.

Comparing the Value of Small Changes in Market Goods with Large Changes in Quantity or Quality of Nonmarketed Goods

A typical public land allocation issue often involves comparing the value of a change in quantity in some marketed output (timber, minerals, beef) with a change in quantity or quality of a nonmarketed good (recreation, fishing quality, and so on). The marketed commodity is often traded in national or international markets. Typically, the public land area under consideration contributes only a very small amount of output to this national market. (Recall the case of livestock grazing on public lands that in total contributes only 7% of the U.S. supply.) Whether this particular public land area supplies zero or its maximum potential output of the marketed output, it will not affect the price of the marketed commodity. In essence, the demand curve for marketed outputs from this tract of public land is horizontal at the market price, just as the demand curve facing an individual rancher would be. (Remember the right side of figure 6.4.) Therefore price can be used as a measure of gross willingness to pay for one more unit commodity output. The net benefits of producing the marketed output from this tract of public land would be determined by subtracting the costs of production of that marketed good from this gross willingness to pay or total revenue. Again this is analogous to the right side of figure 6.4 for an individual rancher. The benefits of commodity production are simply the producer surplus or change in net income realized. No consumer surplus is to be added to the commodity production in this case, even though the first few units of the good traded in the market do provide a consumer surplus. There is no consumer surplus in this case because the change in supply is so small that it does not change the price. If there is no change in price of the good to consumers, there can be no change in consumer surplus. In essence, consumers continue to receive the same amount of consumer surplus they had without the project. If the with- and without-project consumer surplus is the same, no change in consumer surplus is attributable to this land management action.

This same logic would hold if the area of public land provided an identical mix of recreational activities (in the same environmental setting) and of the same quality that could be found at other public lands located exactly the same distance from all visitors as this area. Thus if numerous uncongested perfect substitute sites were available at the same price (distance) and with the same quality (setting, aesthetics, and so on), then the addition or deletion of one such recreation area would result in little change in consumer surplus (i.e., people would not be willing to pay anything more for this site as compared to existing sites).

However, it is rare that there are uncongested, perfectly identical substitute recreation areas located at the same distance to users. In general, if commodity production at one area results in a loss of a recreation area, there will be a price increase for consumers who lived closer to that site than any other site providing the same mix of recreation activities and in the same setting. That is, these consumers will now have to travel further to obtain that same type of recreation. Consequently, the price increase translates into a loss in consumers surplus, just as any other price increase would. In cases where the substitute sites are so congested that they are rationed by advance reservation or permits, no new visitors can be accommodated and the entire existing consumer surplus would be lost. This situation is typical in East Coast states and a few western states such as California. If the other site does not restrict entry of visitors, the additional recreationists may often impose congestion costs on existing visitors to these sites and hence indirectly reduce the quality at the existing sites.

Because loss of a recreation site or changes in recreation quality result in nonmarginal changes that do affect prices or quality, there is a change in consumer surplus. This arises in part because recreation is not a homogeneous product traded in national markets. Because of the high travel costs associated with recreation, 80 to 90% of visitors travel from within a few hours to most recreation sites (i.e, they have localized markets).

There is no inconsistency here in terms of using market price to derive stumpage values (i.e., producer surplus) for timber and then using consumer surplus for recreation. Both producer and consumer surplus are measures of willingness to pay. Price is a measure of gross willingness to pay at the margin for one more unit of the good. Thus all resources are compared using the same conceptual measures of value—willingness to pay.

Importance of Timing of Benefits and Costs

In most public land management actions, the benefits and costs vary over time. Many public land management actions, such as mineral leasing, will last for decades as the ore is mined. A campground may have a twenty-year lifetime. In projects such as a campground, there is a large up-front cost to build the campground, buy the tables, and so on. This is to be compared to the benefits over the twenty-year life of the campground.

But can we simply add up the dollar benefits in the first year with the dollar benefits in each of the following years? Adding a dollar's worth of benefits today to a dollar's worth of benefits in twenty years implies that people are equally willing to trade a dollar today for a dollar in twenty years. We know that is incorrect. You might suspect it is incorrect because of inflation, but that is not the problem here. Although inflation is certainly a problem, the benefits and costs of public land management actions are normally measured in real dollars that is, adjusted for inflation. Therefore the main issue is whether a dollar of camping benefits today is equal to an inflation-proofed, or real, dollar of camping benefits in the future. Put another way, would you give me a dollar today if I gave you a written guarantee to give you a dollar (fully indexed in value for inflation) twenty years from now? Chances are you would not treat a dollar today and an inflation-indexed dollar twenty years from now as equivalent. People are unwilling to trade a dollar now for an inflation-indexed dollar twenty years from now for two reasons. First, they have a time preference for consumption. Other things equal, people prefer their enjoyment or benefits now rather than later. This explains why they pay an interest premium to borrow money from banks or on their credit cards to consume now and pay later. Second, if a person is to forgo consumption benefits today, he or she can invest the money in a productive enterprise that will yield a net return. That is, people deposit money in a bank or loan money to a business that will pay them a positive rate of interest because the money is being put to a productive use.

Discounting Process

In some ways the interest rate reflects both humans' ''time preference'' for money and the rate of return that could be earned by investing this money. Most people are familiar with the concept of compound interest rates, whereby money put in the bank today will grow at some compound rate of interest to a larger sum of money in the future. For benefit-cost analysis, where we wish to take a future flow of monetary benefits and figure out the value in the present period, we use a process that is just the reverse of compounding, called *discounting*. This takes those future (real or inflation-indexed) dollar benefits and costs and computes their *present worth* or *present value*. That is, how much benefit does $10 worth of camping twenty years from now provide society today?

An example may help illustrate this process. If the interest rate is 7.125%, this implies that $1 of benefits ten years from now is worth 50 cents to us today. How did we figure that? Well, if I put the 50 cents in the bank today at 7.125%, I could take out $1 in ten years. Thus at 7.125% interest, 50 cents today and $1 in ten years are equivalent in terms of their present worth or value to me today. In some sense that is all there is to it. If we know what interest rate people use to be indifferent between present and future consumption (or the rate of return that could be earned on funds invested), then we can convert any future benefit into an equivalent present value. We can do this for the entire stream of benefits and costs over the life of some public land project and determine if the present value of the benefits is worth the present value of the costs.

The formula for calculating the present value of any dollar amount (B_t) to be paid or received t years from now is

$$\text{PV of } \$B_t = [1/(1 + r)^t] \times \$B_t$$

or present value of a dollar in year t is equal to $1 \div (1$ plus the interest rate, raised to the number of years t). Most calculators or spreadsheet programs have financial functions that allow the calculation of present values or present worths for a series of benefits or costs over time. For example, at 10% interest, the present value at year 0 of $500 a year in benefits and an initial investment cost of $2,000 with annual operation and maintenance costs of $200 a year would be calculated as shown in Table 6.5.

The net present value can be computed by either subtracting the present value of benefits from present value of costs or discounting net benefits (undiscounted benefits minus costs). As can be seen in this example, either approach to computing the net present value results in the same answer, namely, $207. Also note the error in calculating net benefits that would have resulted had the undiscounted benefits of $500 a year times ten years ($5,000) been subtracted from the costs of $3,800. This would state that the campground was worth $1,200, because it ignores the differential timing of costs compared to benefits.

Figure 6.8 provides a graph of the distribution of present value of $1 received in each of twenty years at a 4%, 7%, and 10% discount rate. Note the rapid drop-off at 10% as compared to 4%. As will be discussed more in detail later, these are the discount rates used by the Forest Service, USFWS, and the BLM, respectively. As one can see, cutting the discount rate from 10% to 7% results in a near doubling of the present value of a dollar to be received in year 20 (increasing from 15 cents

Table 6.5
Comparison of Discounted and Undiscounted Benefits and Costs @ 10%

| Year | Discount Factor | Benefits | | Costs | | Net Benefits |
		Nominal	Discounted	Nominal	Discounted	Discounted
1	0.9091	$500	454.55	$2,000	1818.18	− 1363.64
2	0.8264	$500	413.22	$200	165.29	247.93
3	0.7513	$500	375.66	$200	150.26	225.39
4	0.6830	$500	341.51	$200	136.60	204.90
5	0.6209	$500	310.46	$200	124.18	186.28
6	0.5645	$500	282.24	$200	112.89	169.34
7	0.5132	$500	256.58	$200	102.63	153.95
8	0.4665	$500	233.25	$200	93.30	139.95
9	0.4241	$500	212.05	$200	84.82	127.23
10	0.3855	$500	192.77	$200	77.11	115.66
Discounted Sum			$3,072		$2,865	$207
Net Present Value						$207

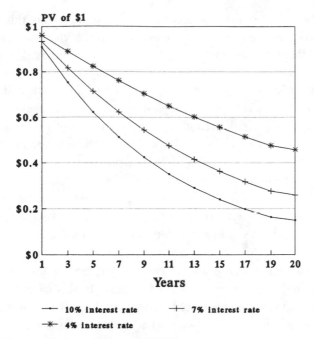

Figure 6.8. Present Value of $1 over 20 Years

to 26 cents). Reducing the discount rate from 7% to 4% results in a much slower drop in present values relative to the nominal dollar value when benefits are actually received.

Selecting the Interest or Discount Rate

The selection of the appropriate discount rate is of both philosophical and practical importance. There is an almost ethical or philosophical statement in using high discount rates: the present matters a great deal more than the future to us. If everyone was as happy to have a dollar's worth of consumption today as to have it in ten years, this would imply a zero discount rate. However, since money invested in productive enterprises today has the potential to yield a net return in the future, even an indifference of present for future consumption is not enough to yield a zero discount rate to society. However, Lind (1990:21) cautions that discounting is most defensible over time periods not exceeding a typical person's lifetime (about seventy years in most developed countries).

Components of Observed Interest Rates

Economic theory provides clear guidance on selection of discount rates in an idealized economy of zero financial transactions costs, zero risk, and no taxes. In this case the time-preference value of money (trade-off of present for future consumption) equals the rate of return that could be earned by investing this money. Once real-world distortions are added in, these two measures of the discount rate diverge. To understand this it is important to realize that the observed market rate of interest reflects four factors: (1) the real or true rate of discount; (2) a percentage return to compensate for inflation; (3) a percentage return to reflect risk that the borrower might not repay; (4) tax rate. At its simplest, if inflation is expected to be 5% over the future, there is a 2% risk premium, and the real rate of return is 6%, then the market interest rate before taxes would be 13%. If the tax rate is 25%, the after-tax interest received is only 9.75% (13% × 75%).

In general, we would want to use the full 13%, since taxes are just a transfer of the rate of return from the individual to the government (and hence society) and the taxed part of the rate of return is actually not lost. If we keep everything in real or year 0 prices, then we would not include anticipated inflation, so a real discount rate would be just 8% (6% + 2%). Some economists have argued that the federal government is so large that it is able to pool risks across the thousands of projects it does. As such, the risk premium should be zero. In this case the real discount rate for public land investments would be 6%, just the real rate of return. We now turn to the economic theory underlying what real rate of return should be used.

Theories for Determining the Discount Rate

There are several competing theories on determination of what the discount rate should be, given the many real-world distortions. One of the most prominent theories dates back to a macroeconomic view of a national economy as a closed economy (i.e., an economy where imports and exports of both goods and capital are nonexistent or minimal). A simple macroeconomic model of such an economy requires that the sum of consumers' savings and taxes equal government spending and private investment. In this economy if the government wishes to spend more money to invest in a public project, one of the other three elements must change accordingly. If the government borrows the money for the project, this simple model argues that private investment must fall by the amount of money the government must borrow (i.e., complete crowding out of private investment). If the government raises taxes to finance the project, then consumption and savings will fall by the amount of the tax increase.

Given these relationships between government financing of a public project and displacement of economic activity elsewhere in the economy, one group of economists argues that the discount rate should emphasize the opportunity costs in the private sector (forgone investment, consumption, and savings) that results from diverting the funds to the public sector (Baumol 1978). The opportunity costs in the private sector are made up of two components: (1) the rate of return on displaced private investments and the value of forgone consumption. The rate of return on investments can be obtained from financial reports of companies. The value of forgone consumption (for small changes in consumption) can be inferred from the interest rate that people would earn on their savings. That is, as consumers decide whether to consume their last dollar or save their last dollar, they compare the value of consumption to the interest rate they would earn on savings. The overall social discount rate is the weighted sum of the private investment rate of return and the interest rate on savings. The weight is determined by how much of the money to be transferred to the public sector comes from displaced private investment and how much comes from displaced savings/consumption. This ultimately depends on how the government raises the money for the public lands management project. If special financing legislation, dedicated trust fund dollars, or specific excise taxes are used, then, in principle, a public lands specific discount rate could be calculated. In practice this is unusual, and the weights would come from the proportion of the government's budget that comes from individual income taxes, corporate income taxes, other business taxes, and other consumer taxes received by the government.

Today, however, most national economies are part of an open world economy where substantial exports and imports (including foreign capital) make the closed-economy view somewhat misleading (Feldstein 1985). For no longer must private investment be reduced dollar per dollar with the increase in a deficit-financed public project. Directly or indirectly, foreign owners of capital can meet the new increased demand for funds arising from the government's project. In an open world economy there is very little crowding out of domestic private investment (Lind 1990:15). Consequently, Lind's recent reassessment of determination of the discount rate recommends that the appropriate government discount rate (in open world economies) be the interest rate the government must pay when it borrows (1990:25).

Lastly, another group of economists stresses the "social time preference" in determining the discount rate. In particular, they start by viewing the discount rate as a measure of people's preferences for present versus future consumption. They then introduce a public goods or beneficial externality argument. This argument is that individuals care more about future generations than their private savings decisions might indicate. Acting individually, their savings decisions have little effect on the welfare of future generations; however, collectively, people realize their decisions affect the future wellbeing of society. If all people would save more (and consume less) for future generations, we would all be better off. This theory is an argument for a social discount rate lower than the individual savings rate to reflect the collective concern for future generations.

What Public Land Agencies Use

It often is impractical for federal agencies to compute their own discount rates on each and every public land management project. Instead, federal policy sets discount rates that agencies must use. Much of the federal government (including BLM) uses a real rate of interest of 10% set by the federal Office of Management and Budget (OMB). OMB established this rate using the theory that the appropriate discount rate relates to the opportunity cost view particularly, the forgone rate of return on investments in the private sector. Since this discount rate is a real rate, it does not contain any accounting for inflation. Therefore the 10% rate is relatively high.

OMB permits the U.S. Fish and Wildlife Service and water-related federal agencies (U.S. Army Corps of Engineers, U.S. Bureau of Reclamation, Soil Conservation Service) to use a discount rate set by the U.S. Water Resource Council using a formula specified by Congress. This discount rate reflects the interest rate cost to the federal government of borrowing money—specifically, that on long-term Treasury bonds. This rate has averaged 7.8% during the 1980s. This rate is not only tax-free, since the individuals do not to pay federal income taxes on it, but also risk-free. This is a nominal rate of interest, since it includes a premium for inflation. However, because most federal agency benefit-cost analyses are not allowed to adjust future benefits and costs for inflation (i.e., they do them in real dollars or project period values), it is incorrect to include inflation in the discount rate. This is another example of the "Congress told us to do it that way" dilemma that federal agencies often find themselves in. Nonetheless, this U.S. Water Resources Council discount rate approach does accord (with the exception of being a nominal rate) with Lind's theoretical development that in an open world economy, the government borrowing rate is also the appropriate discount rate.

The Forest Service uses a 4% real discount rate, which was reached after analyzing the rate of returns on capital and corporate bond rates. This rate correctly includes taxes (i.e., is a before-tax rate) but excludes risk (Row, Kaiser, and Session 1981).

Although the Forest Service discount rate is probably closer to the social time preference rate and the correct long-term private rate of return, there is another source of error in comparing the economic efficiency of public land projects calculated using different discount rates. For example, the BLM might reject a wildlife habitat improvement project that would yield only an 8% rate of return because this return is below its 10% discount rate (i.e., what the funds are assumed to yield in competing investments). The Forest Service might accept a timber stand improvement project that yields a rate of return of 5% because it exceeds their 4% discount rate. Of course, what happens here is that the nation forgoes a BLM project offering an 8% return for one that offers only a 5% return. Thus it is at least as important to have a consistent discount rate across all public land management agencies as it is to have one that is theoretically correct. Although there does seem to be movement toward a common real rate of 10% in the federal government, this process needs to be accelerated so that all federal agencies (transportation, occupational safety, and so on) use the same rate.

We will now turn to a discussion of how to use discounted benefits and costs in determining an aggregate measure of the economic efficiency effects of a particular public land management action.

Alternative Benefit-Cost Decision Criteria

Once the analyst has measured the benefits and costs over the life of the project for each affected resource output, the next step is to summarize this information. For the information to be most useful for decision making it is helpful to compare the sum of the discounted benefits produced by the alternative to the discounted costs incurred.

The criteria and summary measures that are most defensible and commonly used include (1) net present value, (2) benefit-cost ratio, and (3) internal rate of return. We will discuss each of these individually and then discuss under what conditions they yield exactly the same ranking of alternative policies or projects.

Net Present Value (NPV)

As the name implies, net present value (NPV) is the present value of the benefits minus the present value of the costs. The difference is the net gain adjusted for the timing of benefits and costs. The units of measurement are present worth of dollars in the base year in which all the benefits and costs are figured.

The definition of NPV can be understood by reference to equation (6.1):

$$\text{NPV} = \frac{(B_1 - C_1)}{(1 + i)^1} + \frac{(B_2 - C_2)}{(1 + i)^2} + \frac{(B_3 - C_3)}{(1 + i)^3} \cdots + \frac{(B_t - C_t)}{(1 + i)^t} \quad (6.1)$$

where

B_t = benefits in time period t
C_t = Costs in time period
i = interest rate
t = time period

As equation (6.1) illustrates, the benefits in each year are subtracted from the costs in each year and this difference is discounted back to the present time period (t = 0). If the resource management action's present value of benefits exceeds the present value of costs, then NPV will be greater than zero (i.e., positive). In general, the larger the NPV is, the more net benefits society realizes from the resource management action.

If the analyst is using NPV as a decision rule, then only projects with NPV greater than zero are accepted as economically efficient. If a project has a negative NPV, this means society gives up more than it gets over the life of the resource management action. In many respects, NPV is a discounted measure of whether the resource management action represents a potential Pareto improvement over the life of the project. Taking benefits as measures of gains and costs as a measure of losses, if NPV is greater than zero, then the gainers over the life of the project could compensate the losers over the life of the project and still have some benefit remaining.

Benefit-Cost Ratio (BCR)

The benefit-cost ratio (BCR) takes the same information used in equation (6.1) but expresses it as a ratio. Specifically, BCR equals the present value of the benefits divided by the present value of the costs.

Equation (6.2) presents this relationship:

$$\text{BCR} = \frac{(B_1/(1 + i)^1) + (B_2/(1 + i)^2) + (B_3/(1 + i)^3) \ldots + (B_t/(1 + i)^t)}{(C_1/(1 + i)^1) + (C_2/(1 + i)^2) + (C_3/(1 + i)^3) \ldots + (C_t/(1 + i)^t)}$$

(6.2)

$$\text{Benefit Cost Ratio} = \frac{\text{Present value of benefits}}{\text{Present value of costs}}$$

Internal Rate of Return (IRR)

The internal rate of return (IRR) is the percentage return on money invested in the project that equates the present value of the benefits with the present value of the costs. In essence, IRR is the earning power or interest earned when the money is invested in the project under study. Mathematically, IRR is that interest rate that results in the NPV of the project equaling zero. Thus the IRR formula solves for i in equation (6.3) such that

$$\text{NPV} = 0{:}(B_1/(1 + i)^1) + (B_2/(1 + i)^2) \ldots + (B_t/(1 + i)^t) -$$
$$[(C_1/(1 + i)^1) + (C_2/(1 + i)^2) \ldots + (C_t(1 + i)^t)] = 0.$$
(6.3)

Unlike the BCR or NPV, which can easily be computed on even a simple calculator, determining the IRR involves a series of relatively complex calculations. As such, it is best performed using financial calculators that are programmed to calculate IRR or using a computer software program designed to calculate IRR. By itself this added computational burden is not a valid reason for reluctance to use IRR (there may be other technical reasons, which will be explained later).

To use the calculated IRR to determine the economic efficiency of a project involves a simple comparison. The calculated IRR is compared to the interest rate at which the agency or society can borrow money.

As will be recalled from our earlier discussion, one view of the interest rate is the opportunity cost of forgone rates of return in society's next-best investment. Thus the IRR decision rule is to recommend any project that generates an internal rate of return greater than the interest rate.

Does It Matter Which Discounted Measure Is Used?

A comparison of the three formulas for NPV, BCR, and IRR indicates that all three use the same information. Therefore one would expect that these measures would yield the same conclusions on which projects should be adopted and which should be rejected. However, this is only true under certain conditions.

The conditions under which all three measures will yield the same conclusions about the desirability of projects are

1. Either there are no budget limits that preclude adopting all economically efficient projects that pass their respective decision rule or the size of projects in terms of investment costs is similar.
2. The projects or policies are not mutually exclusive. That is, each project could be implemented without precluding implementation of another.
3. Either the projects being compared have a similar length of time (i.e., economic life) or the reinvestment opportunities yield the same returns as the original project.

If these conditions are not met, what appears to be the best project alternative will vary, depending on which discounted measure (NPV, BCR, IRR) is used to evaluate the alternatives. For example, table 6.6 involves violation of both aspects of condition 1 and part of condition 3. First, the projects have substantially different investment requirements. If we add to this the budget constraint that society has only $10 million to invest, then we must choose between the projects rather than adopt both (which is warranted, since both are economically efficient, regardless of whether one uses NPV or BCR). In addition, assume we cannot find reinvestment opportunities that would yield the same return as we get from doing several project A's. (Thus the default assumption is that the remaining $8.3 million of the original $10 million not invested under Alternative A would earn just the discount rate.) Under these circumstances, which project is best to select depends on whether we rank by NPV (choose Project B) versus rank by BCR (choose Project A). Since we are likely to find situations where governments have very limited budgets to devote to public projects or where nature makes the projects or policies mutually

Table 6.6
Comparison of Benefit-Cost Ratio to Net Present Value
(Millions of Dollars)

Project	PV of Benefits	PV of Costs	NPV	BCR
A	4	1.7	2.3	2.35
B	16	10	6	1.6

exclusive, it is important to know the advantages of each of the three discounted measures of project worth.

Advantages of NPV as a Theoretically Correct Measure of Net Benefits

We can illustrate the merits of NPV relative to BCR by reference to table 6.6. When the preceding conditions 1–3 do not hold, the choice between Plan A and Plan B becomes clear: does society prefer $6 million in net benefits or $2.3 million in net benefits? If all the opportunity costs are correctly measured in the present value of costs and the discount rate accurately reflects the forgone investment opportunities, then Plan B is preferred. People prefer more ($6 million) to less ($2.3 million) if these are mutually exclusive projects and we cannot invest in several identical Project A's. Our theory of what constitutes an improvement in well-being is written in terms of maximizing the total amount of the net gain, not in terms of maximizing the rate of return. A very high rate of return for which we are restricted to collecting on only a $1 investment is not attractive if it means giving up a larger total return on some other investment. (Note that if the remaining $8.3 million not invested in Project A just earns the discount rate, then this would pull down the BCR of Project A to 1.23, again demonstrating the superiority of Project B.)

As Boadway (1979:176) discusses, maximizing NPV is a necessary condition for a project or policy to result in a potential Pareto improvement. Maximizing the BCR is not strictly related to attainment of potential Pareto improvement, since the BCR is the average rate of return. A high average rate of return tells us little about whether the total benefits are maximized, but it is consistent with a project that is not large enough (marginal benefit from expansion greater than marginal cost of expansion (i.e., MB > MC). To attain the maximum gain, marginal cost must equal the marginal benefit (MB = MC). Thus NPV is a useful criterion for both determining the optimal scale of a resource management action and timing implementation. The BCR is not an unambiguous indicator of either the timing or the scale.

Advantages of BCR

The BCR does have some advantages. Some authors claim (Sassone and Schaffer 1978:21) that if one must rank many projects within a budget constraint, the BCR (and calculating incremental BCRs) is the appropriate procedure. In addition, it is easily understood by the public because it is expressed as dollars of benefit per dollar of cost. Thus it has a nice, intuitive appeal. Lastly, in comparing projects where one does not want the scale of the project to influence opinions about desirability, the scale of the projects does not affect the BCR.

Merits of IRR as a Measure of Discounted Net Benefits

So far we have focused on the relative merits of NPV and BCR. However, the IRR has a similar set of advantages and disadvantages. Because the it can be expressed as a return per dollar of investment, like the BCR, the public can easily understand it and is unaffected by the scale of the project. However, since it also shares BCR's drawback of using a rate of return rather than total return as the decision rule, it too is a poor indicator of which project provides the greatest total gain to society. What is worse, there may exist several interest rates that set NPV at zero. Depending on the pattern of costs and benefits over time, there may be two, three, or more interest rates (i) in equation (6.3) that are possible answers. Unfortunately, the analyst is likely to be completely unaware of which interest rate has been solved for.

However, the IRR has one advantage not shared by BCR and NPV: the ability to make comparisons between alternative projects or policies without first having to specify an exact discount rate. In essence the IRR formula solves for the discount rate (i.e., IRR) that would just make the present value of benefits equal to the present value of costs. If the actual discount or interest rate is lower than the internal rate of return, then this project will provide net benefits in excess of its opportunity cost. Hence this project or land management option would be acceptable on economic efficiency grounds. However, in many instances an agency or decision maker need not state the exact discount rate it is using. As long as the project or policy generates a "reasonable" return (given the implicit discount rate the agency is using), the project can be accepted without ever having to declare to the analysts and others the exact discount rate being used. The practical advantage of such an approach should be obvious, given the earlier discussion about the difficulty in getting universal agreement on what should be the appropriate discount rate.

However, as discussed earlier, most agencies have their discount rate specified, so this feature of IRR is not always compelling. In addition, the decision criteria (i.e., NPV, BCR, and IRR) of many agencies are specified as well. Agencies such as the Forest Service, Army Corps of Engineers, and USDA Soil Conservation Service generally rely on the NPV, although BCRs are also computed as supplemental information.

In general, using the NPV as the benefit-cost analysis decision criteria has a great deal to offer. It is the most theoretically correct measure in the sense of being most consistent with the conceptual foundation of BCA. Moreover, it does not suffer the multiple solutions problem that can plague IRR calculations.

Steps in Performing a BCA

Five major steps must be followed in performing a complete BCA, each of which has many components:

1. *Define the problem and the analysis environment.* This step involves five components:
   a. Develop the range and composition of alternative solutions to the problem (i.e., what management practices and inputs will be used).
   b. Determine the likely future without the project or the consequences of the no-action alternative.

  c. Select which accounting stance will be used to measure benefits.
  d. Determine the overall study constraints in terms of time, personnel, and budget available.
  e. Choose the discount rate to be used.

2. *Design the analysis.* This step involves five components:

  a. Describe problem structure. Are alternatives mutually exclusive? Is there an overall project investment budget constraint?
  b. Select the BCA decision criterion. Is NPV or BCR or IRR to be used as the summary criterion?
  c. Identify all primary benefits and costs that must be included in the analysis.
  d. Choose the variables for sensitivity analysis. That is, determine which variables are both fairly uncertain and likely to have a significant impact on the economic feasibility of the project (i.e., on the size of NPV, BCR, or IRR).
  e. Develop a data collection plan.

3. *Collect the data.*

4. *Perform the benefit cost analysis outlined in step 2.*

  a. Quantify benefits and costs.
  b. Express as many benefits and costs in dollar terms as is consistent with accounting stance chosen.
  c. Compute the BCA decision criterion (i.e., NPV, BCR, or IRR) chosen in step 2.
  d. Perform the sensitivity analysis.

5. *Present results.*

  a. Use a form understandable to decision makers.
  b. Use a form that highlights the merits of each alternative available to solve the original problem.

Clearly, performing a comprehensive BCA is a great deal of work. However, this BCA framework can be tailored to the magnitude of the resource problem to be analyzed. As will be illustrated in chapters 9, 10, 12, and 13, this framework has been implemented frequently by a small group of people to evaluate a small number of public land management alternatives. However, if the resource decision involves an irreversible commitment of a relatively unique natural resource, then it is often socially optimal to devote the thousands of dollars of analysis effort to ensure that the resulting solution is the best one available.

Relationship of BCA and Planning.   It is worthwhile to note the similarities between the steps in the rational-comprehensive planning process and those in performing a benefit-cost analysis. Both share a common core of evaluating alternative solutions to a given problem. As such it is easy to see how benefit-cost analysis and public land management planning can be easily integrated together into one process.

Valuation of Natural Resources from Public Lands

Valuation of Marketed Commodities

As was discussed earlier in this chapter, one can often use the market price of a good to value small changes in the supply of that good. Nonetheless it is important to understand how that market price is determined and what it represents in the case of resources produced from public lands. Although many of the commodities sold from public lands have prices, they are not always expressed in the same way as traditional market prices. Most goods, such as items of clothing, have a price per unit posted in the store, and any buyer can purchase as many or as few items as he or she wants at that price. On the other hand, timber is not often sold from public lands with an explicit posted price per board foot in which many different buyers can purchase as little or as much as they want at that price. Instead the cutting rights to an entire area of forested land are offered for bid, and the highest bidder is awarded the rights to cut the entire area. The same is true for mineral leases. Firms cannot buy 5 tons of coal at a price per ton from a tract of public land; rather, they must bid on the mineral rights to areas of land as large as several hundred acres.

Other public land commodities are not themselves sold in formal markets; rather, they are often exchanged as part of a package of natural resources to a select group of buyers. For example, unlike private and some state lands, where grazing privileges are auctioned off on a pasture-by-pasture basis, on most federal land grazing privileges are not put out to bid. Rather, only "qualified" ranchers (i.e., those owning nearby ranches that are considered "base property") can request these grazing privileges. In most cases these privileges are retained by those ranchers who acquired the historic grazing privileges several decades earlier or by whoever bought the entire base property. That is, one cannot buy one hundred animal unit months of public land grazing in a formal market. Often the only way to acquire the grazing privileges is to buy a ranch that has them. To make valuation even more difficult, the price the federal government charges for these privileges is set by an administrative formula that results in a grazing fee of about one-third the market clearing price where comparable bidding exists (Forest Service 1990a).

Much of this discussion about market structure for public land grazing also applies to the valuation of water from western public lands. Water withdrawn for irrigated agriculture and domestic and industrial use is often based on historic appropriation, sometimes characterized as "first in time, first in right." Once again, historic users have priority over more recent uses. Formal market exchanges of water rights are beginning to occur more frequently, but they are not frequent enough to constitute what we might call a competitive market. Certain water "uses," such as in-stream flow, pollution dilution, fisheries habitat, and so on, also have many public good characteristics. As discussed in chapter 3, even the most competitive markets will generally fail to recognize the full value of these public good attributes of water. With water there is often an added challenge, because, unlike grazing, which takes place directly on the public lands, it is often more difficult to identify which public lands produced a given amount of water that is consumed several hundred miles downstream from several ownerships of public land.

Nonetheless, as the following sections will illustrate, there is a well-developed and conceptually sound foundation for valuing natural resources from public lands. This is true whether the natural resource's values are revealed directly through markets or through quasi-markets. It is also true whether the values are inferred from consumers' behavior in visiting recreation sites or in response to surveys about the value of unique natural environments. It is hoped that by the time the reader has finished these sections, he or she will avoid ever thinking that somehow economists assign values to natural resources. Economists, with the help of other social scientists and statisticians, merely record and organize information on people's behavior to quantify in dollars the values people already possess. Economists no more create value for these commodities than did scientists create subatomic particles when they first peered through an electron microscope or astronomers created distant galaxies when they first peered through more powerful telescopes. Like these subatomic particles or distant galaxies, the values have always been present; only recently have we been able to measure them accurately.

Common Methodological Approaches to Valuation of Public Land Resource Commodities

In valuing timber, animal unit months (AUMs) of livestock forage, and water and minerals produced from public lands there are at least four common approaches: (1) market or transactions evidence; (2) demand estimation; (3) residual valuation; (4) change in net income (Davis and Johnson 1987; Gray and Young 1984).

Market or transactions evidence requires observation of similar sales of resources in competitive markets to infer a value for the resource of interest in this particular public land setting. For example, to value AUMs of forage from public lands, the analyst might look at bids for AUMs on private or state land leases. The key requirement for accurate valuation with this approach is finding comparable sales in terms of characteristics of the resource in the private sale to the public land case.

A demand curve approach can be employed if there is a large enough number of past sales or a data series of prices and quantities sold over time or across the country to estimate a demand function statistically. This has been performed for regional and national timber demand (Adams and Haynes 1980) and for municipal demand for water (Young 1973, Foster and Beattie 1979).

Residual valuation involves subtracting from the total value of the output produced all the costs of inputs except the ''unpriced'' public land input, such as water or forage. The value remaining after all other costs (including a normal profit, return to management, capital, and so on) is then attributed to the unpriced or underpriced public land input. Thus the irrigation value of unpriced water flowing from a National Forest is the value of the agricultural crop produced minus all the priced inputs, such as land, seed, fertilizer, machinery, labor, and management. The key assumptions here are that all other inputs are priced and that total value of output can be apportioned according to the marginal product of the inputs (Gray and Young 1984:171). This approach is commonly used to value water and timber and will be discussed in more detail later.

Change in net income involves simulating a firm's or producer's net income with and without additional quantities of the public land resource. For example, the value of additional AUMs of forage to a rancher is frequently calculated by simulating how the ranch enterprise would modify its herd size or ranch operation if it had an additional one hundred AUMs of public land forage at a given time of the year in its current grazing allotment. Often this simulation is performed using ranch budget data and linear programming models.

Given these general techniques, it is worthwhile to examine briefly how they are applied to calculate economic values for each of these marketed or quasi-marketed resources. Once we have done this, we will turn to valuation of nonmarketed uses of such public land resources as recreation, fisheries, wildlife, wilderness, and so on.

Valuation of Timber

Timber on public lands is often referred to as stumpage. It is the trees on the stump that are being valued. The Forest Service generally uses the residual-value method for appraising the standing timber in the National Forest prior to sale. As mentioned earlier, this method starts with the value of the an intermediate product, such as lumber, and then subtracts the milling and logging costs. We shall sketch this process, starting at the mill and moving back to the forest itself.

Specifically, prices that lumber mills receive for finished lumber of different types of wood (Douglas fir, cedar, and so on), different grades (e.g., qualities), and different dimensions (width, lengths) are studied. This is the value of the output to the lumber mill.

Now we must begin to subtract the costs of milling (i.e., machine time and labor) as well as sorting and storing logs at the mill. Next, the transportation costs to move the logs to the mill must be subtracted. The actual costs of the timber harvest must also be subtracted. This includes cutting the trees down (e.g., felling the tree) and skidding the trees to the landing or logging road spur. The costs of loading the logs onto the truck must be included. Finally, there is the considerable expense of building the roads to the timber sale area, laying out the timber sale itself, and finally any preventative measures before the sale or mitigation after the sale to protect other resources from damage during or after logging.

There has been considerable debate about whether the complete cost of building the roads should be charged to the timber sale, because sometimes the access road benefits other resources, such as roaded recreation, livestock grazing, and so on. We will discuss this issue of joint costs and joint benefits in a case study in chapter 9. To muddy the cost accounting further, federal agencies, such as the Forest Service, frequently offer a ''purchaser road credit'' to the firm to build the road. This does not change the cost, and as long as it is netted out, no error results. Unfortunately, sometimes the Forest Service shows stumpage prices that include the purchaser road credit, substantially inflating the net economic value of the timber (Forest Service 1990a:B-9).

In sum, the value of the lumber would be subtracted from the sum of these milling, harvesting, road-building and transportation costs to arrive at a residual or stumpage value. This value is sometimes called the *appraised value* of the stumpage, as it often reflects the minimum price the federal agency must receive. The costs are based on firms with average milling and harvesting efficiencies and with a provision for a reasonable margin for profit and risk.

This residual valuation procedure is used not only by federal agencies to arrive at appraised stumpage value but by the private firms to determine how much they should bid on the timber offered for sale. Differences between what a specific firm bids and the appraised value are often a result of the firm having a greater efficiency than the average firm or a willingness to accept slightly less than the average profit on the sale.

Note that these timber sales yield a value per sale. It is usually not a value per board foot for specific tree species, which is what is frequently needed for benefit-cost analysis or forest planning optimization models. Sale bids are normally converted into per-board-foot valuations of specific tree species by computing a tree species specific weighted average bid from the overall sale. For example, a bid of $100,000 for a sale with 4,000 mbf of Douglas fir and 1,000 mbf of ponderosa pine might be apportioned based on volume as well as relative end lumber product prices. Thus it is worthwhile to remember that these ''prices'' per board foot are derived and are not a typical posted price, whereby buyers can purchase as few or as many board feet of a given tree species on the stump as they would like. Although these derived prices are often used as normal market prices, they may reflect more of an average producer surplus per unit output than a typical price at the margin. These stumpage prices are still directly useful in timber valuation for benefit-cost analysis and in forest planning optimization models. However, they are not the pure measure of market price per unit that they are often made out to be. For a detailed discussion of valuing stumpage see Davis and Johnson (1987).

Valuation of Livestock Forage on Public Lands

Three techniques are common for measuring the economic value of public land forage to ranchers. The *market/transactions evidence approach* is frequently used when comparable sales of forage from nearby private or state lands can be found. In 1984 the Forest Service and BLM joined in an attempt to arrive at such comparable rental values for private and state pasture (Tittman and Brownell 1984).

This report also adjusted these prices for differences between public land grazing conditions and these private and state lands.

The *change in net income* is another common approach that is used with ranch budget data. The change in profit or net income is calculated from a change in public land forage supplies. This has been the basis of many early estimates of public land forage. Since the mid-1970s the use of linear programming models with ranch budget data has been common in estimating the contribution of public land forage to rancher net income. As will be discussed more in chapter 9, the ability of these linear programming models to calculate a "shadow price" that reflects the change in value associated with one more unit of a constrained input such as public land forage is a very desirable feature of these models. Bartlett (1982, 1984), Wilson and colleagues (1985), and Hahn and colleagues (1989) have all used the linear programming approach to develop estimates of the marginal value of another AUM for the Forest Service and BLM's land management plans and EISs.

Valuation of Minerals from Public Lands

The residual valuation approach is frequently used in determining the net economic value of minerals on public lands. Much like the timber appraisal method, this technique starts with the product price as given. This is especially appropriate for minerals where the prices are set in national and international markets (Vogely 1984:100; Forest Service 1990a:B-15). The costs of ore or mineral processing, mining, transportation, and a normal rate of return are subtracted from these prices to yield a net economic value of the minerals in the ground. The U.S. Bureau of Mines provides estimates of the cost of milling and mining different ores. In some cases where competitive bidding is used to sell the mining rights for certain leasable minerals such as coal or oil, the bids can be used as a source of information to derive the net economic value of the mineral resources in the ground. Note that, like timber, these mineral "prices" per unit derived by either approach reflect an average price over some lease area or deposit rather than a posted price per unit. For a more detailed discussion of valuation of minerals see Vogely (1984).

Valuation of Water

Valuation of water begins our transition from marketed natural resources to non-marketed natural resources. The first challenge in valuing water flows from public lands relates to determining which of the many uses this incremental flow will be allocated to. Will the additional water be used in irrigated agriculture, used municipally, or left in the river to add to in-stream flow? Additions to existing water supplies are usually assumed to go to the lowest valued use of water, such as irrigated pasture or irrigated agriculture. The reason is that higher-valued municipal and industrial users can normally purchase water from these lower-value uses. This is not always the case, however, as rigidities in the prior appropriation doctrine of water law give senior rights to the first users rather than to the highest-valued users.

After determining which water uses are most likely, the next step is to determine which of the four techniques to use. This is frequently done based on the type of water use to be valued. If the water use is irrigated agriculture, a change in net income or residual valuation approach is often used. Here farm budgets are used to simulate the increase in crop values stemming from additional water. From the change in crop values the cost of putting this additional water to use and of such associated farm inputs as additional seed, fertilizer, tractor time, and so on, are subtracted to yield the change in net income.

If the use of water to be valued is municipal, then demand estimation or market transactions will frequently be employed to calculate a value of water in this use. For a discussion of valuing water see Gray and Young (1984) or Colby (1987).

Of course, if the additional water is to be left in the stream to enhance fisheries or recreational boating, then in most areas of the United States we must rely on slightly different techniques for valuation of water in these cases. We say *slightly different* because the first technique discussed falls basically under the category of demand estimation. This technique is called the *travel cost method* and will be discussed in the next section.

Valuation of Nonmarketed Natural Resources Such as Recreation, Wildlife, Fisheries, Wilderness, and Rivers

Why Nonmarket Valuation?

As we have seen throughout several earlier chapters of this book, there is a big difference between a financial cash flow and economic value to society. This distinction arises for public goods and for many uses of natural resources that are not traded in markets. Although the economic forces underlying nonmarketed natural resources such as sport fish, wildlife, and wilderness areas are similar to those economic forces for any natural resource or good, there are some important differences. The nonrival and nonexcludable nature of consumption of pure public goods discussed in chapter 3 is a significant one. The fact that the producer cannot exclude nonpayers from consuming public goods almost always makes optimum supply financially unprofitable, even if it is economically viable to society. Unfortunately, without an explicit market, direct observation of willingness to pay (i.e., benefits) is made difficult. Thus there is the need for techniques that reveal the demand and value for publicly provided natural resources.

The other need for valuation of natural resources is that, institutionally, most western states have taken ownership of sport fish and wildlife out of the free market. Most game animals are owned by the state and cannot legally be bought and sold like livestock. In addition, the state does not ration access to wildlife based on market-determined prices. In the case of Wilderness, Congress has specifically instructed all federal agencies not to charge for access to Wilderness areas, let alone establish a market for Wilderness permits.

Thus although society has chosen not to allocate access to these resources through the market, economic forces are still present. That is, there is still scarcity of these natural resources, excess demand at these zero or below-cost prices, and competing marketed uses for these resources (i.e., forage, demanded by ranchers for livestock, instead of wildlife; timber rather than wilderness recreation; and so on). Given the scarcity, excess demand, and competing demand, allocation decisions are still required. If society is interested in receiving the most benefit possible from these nonmarketed resources, some attention to benefits and costs is necessary. As will be recalled from chapter 4, net economic benefits (benefits minus costs) is not the only criterion for allocating natural resources on public lands, but it is one of the criteria emphasized in many public land laws.

Travel Cost Method for Estimating Recreation Demand and Benefits

The first method we wish to discuss is one limited to valuing recreational uses of natural and environmental resources. This technique falls into the category of demand estimation briefly mentioned in the previous category on valuation of marketed commodities. After discussing this travel cost method we shall discuss a method that can value not only recreation but also the existence and bequest of natural and environmental resources to current and future generations.

What Makes Recreation a Little Different

Unlike marketed goods that we ship to the consumer, with recreation the consumers must take themselves to the recreational resource. Therefore people face substantially different market prices for a recreation site, depending on where they live. As a result, we can trace out a demand curve for the recreation site and then calculate the net willingness to pay or consumer surplus from it.

Calculating a Demand Curve Using the Travel Cost Method

To calculate a demand curve for recreation at a specific site we first recognize that trip costs are prices. Figure 6.9 provides a schematic map of four cities where rafters live who visit a river shown in the center of the map. The map shows the trip costs (transportation costs and travel time costs) to visit the river from four cities of equal population size. Rafters in each city face a different "price" to float this river. As a result, we find that rafters in the close city (D) take more trips than those from the more distant city (A). This is as one would expect, holding other factors, such as income and substitutes, constant across cities.

Although there is a substitute river within the market area boundary, trip frequency behavior by rafters demonstrates that for rafting they prefer this river to the substitute site. This preferred river may have bigger rapids, or more dependable flows, or be closer to some rafters (i.e., have a lower price).

Recreation Market Area

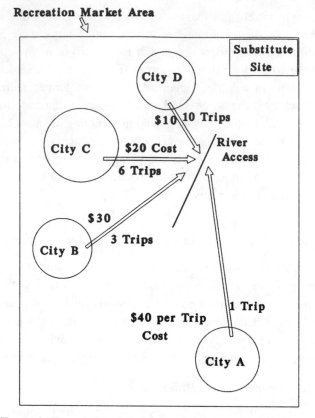

Figure 6.9.   Example Travel Cost Model Spatial Recreation Market

Figure 6.10 shows how the dollars per trip (shown on the vertical axis) and the number of trips per city (on the horizontal axis) from each of the four cities allow one to trace out a demand curve for this river. That is, each trip cost and each number of trips per city is a price-quantity combination. If we have adjusted for other socioeconomic differences such as income (by including them as demand shifters), these price-quantity points trace out a single demand curve.

Figure 6.11 illustrates how we use this one demand curve to calculate the consumer surplus rafters in each city receive when they visit this site. In figure 6.11 we have repeated the same demand curve four times, once for each city. This makes sense because they all visit the same river and consume essentially the same recreational experience. Since rafters located in each of the four cities face a different trip cost or price, the consumer surplus area is different for each city. We calculate the consumer surplus triangle as the area under their demand curve but above their current trip cost or price. Those living close to the site buy more rafting trips and pay less per trip. Hence they receive a much larger consumer surplus. In other words, they would be willing to pay more than those living further away to have access to the river or to prevent it from being dewatered or developed, other things remaining equal.

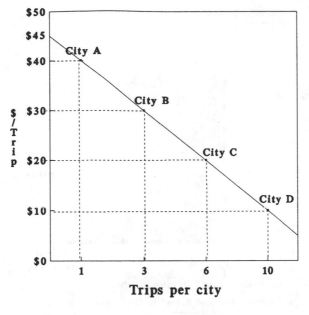

Figure 6.10.   Example Travel Cost Model (TCM) Demand Curve

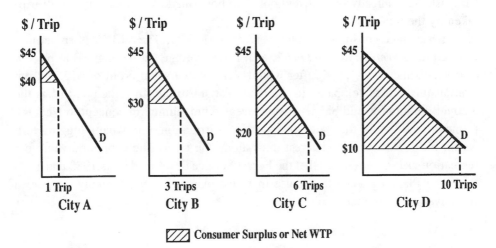

Figure 6.11.   Calculating Consumer Surplus from TCM Demand Curve

Calculation of Total Site Consumer Surplus

The total rafter benefits is found by adding up all the trips from the four cities (e.g., twenty trips) and adding up the shaded consumer surplus areas under each city's curve. This total consumer surplus or net willingness to pay for this site is shown in figure 6.12. The vertical axis on this figure is the added dollars of willingness to pay or consumer surplus over and above the rafters' actual expenditure. Thus

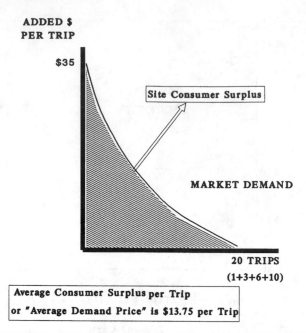

Figure 6.12.   Total Recreation Site Consumer Surplus from All 4 Cities

trip costs have already been netted out. The horizontal axis is the sum of all trips taken by the four cities.

The average consumer surplus per trip is $13.75. The $13.75 is sometimes referred to as *average demand price,* as it is the average net willingness to pay for a trip. This means a typical rafter would receive a personal profit of $13.75 from visiting this site as compared to the next-best substitute. In principle, this is the amount the rafter would bid per trip in a sealed bid, similar in concept to what the farmer would bid for the water in the river for irrigation or to what a logging firm would bid for the right to harvest trees along the river. The economic value per recreation visitor day (RVD) that the Forest Service developed in its 1985 and 1990 Resource Planning Act EIS for use in forest planning was originally developed from average consumer surplus estimates.

Basic Assumptions of the Simple Travel Cost Method

Single-Destination Trips.   Like any model, the TCM requires a few assumptions to permit valid estimation of economic values. The most important assumption is that recreationists take single-destination trips. In terms of our simple map in figure 6.9, this means that recreationists go from their city of residence to the river and return directly home. If this occurs, then it is accurate to measure the round-trip distance from home to the site as a measure of the price of visiting the site. If they visit other recreation sites on that same trip, then it makes measurement of the price of visiting any one of the sites difficult; these visitors are often excluded from the estimation of the demand curve (but the average consumer surplus is applied to them at the end). The alternative contingent valuation method described later can be used to value such multidestination trips. Thus the number of visitors on multiple-destination trips must be researched before the TCM can be applied. In general, for many public-land campgrounds, Wilderness Areas, rivers, Wildlife Refuges, hunting trips, fishing trips, hiking areas, National Recreation Areas, and even some National Parks, 80% or more of the visitors do take single-destination trips.

The second major assumption is that there is sufficient variation in travel costs to trace out a demand curve. In terms of figure 6.9, we needed rafters visiting from cities located at different distances to trace out a demand curve. If all the rafters came from cities located exactly the same distance, all we would obtain is an estimate of one point in figure 6.10. With one point there is no way to statistically (or graphically) trace out a demand equation. Once again, this is usually not a problem for public-land recreation areas. They are located close to a few small towns and at varying distances from many cities and counties.

The third assumption is that travel itself has no benefits. That is, we assume the trip cost is incurred to gain access to the benefits of the recreation site. If people enjoy the travel en route, then part of the travel cost is paid for the en route sightseeing and is not solely a sacrifice made to gain the benefits of the recreation site. If people do enjoy the driving en route, then our estimate of recreation site benefits will be overstated.

The empirical evidence on this assumption is mixed. For short drives to infrequently visited sites in such scenic areas as Colorado or the Pacific Northwest there does seem to be some sightseeing value during the first hour of the drive. After the first hour or so, the disutility of travel time begins to dominate. In hot and arid climates, such as that of New Mexico, even the first hour of travel seems to have a disutility. Of course when recreationists make frequent trips up the same roads to visit the same sites it is reasonable to assume there is probably little sightseeing value. Nonetheless, this is an assumption that with properly worded questionnaires can be empirically tested for the recreation site in question. (See Walsh and colleagues [1990] for a discussion of such questions and techniques.)

When the assumptions of the TCM are met, the method is expected to yield valid estimates of visitors' net willingness to pay (WTP). When they are not met, the method should generally not be applied.

Two Ways the Travel Cost Method Demand Curves Reflect Presence of Substitute Recreation Sites/Opportunities

The actual number of trips taken by rafters to the river in our example reflects the presence of other substitute rivers. If those other rivers were not present and this were the only river, more rafters would visit this river. In figure 6.13 the influence of other raftable rivers is reflected in the lower number of trips being recorded to the site on the horizontal axis.

The vertical intercept of the demand curve is shifted downward because of the availability of substitute rafting rivers. This means the maximum amount a rafter

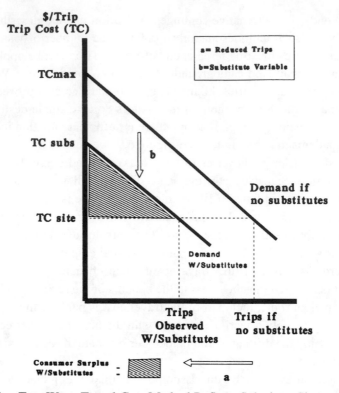

Figure 6.13.   Two Ways Travel Cost Method Reflects Substitute Sites

would pay to visit this river is capped at the trip costs to similar-quality (as perceived by the rafter) rivers. In the travel cost method this downward shift in the demand curve is reflected by a substitute index variable. If this index rises when the number of other rivers accessible to a given city increases, then the variable will have a negative sign in the demand equation.

The net result is that the consumer surplus calculated reflects the presence of substitutes. In essence the consumer surplus obtained from a TCM demand curve that includes a substitute variable is the net willingness to pay for this preferred site over and above the next-best substitute. That is, the benefits are the additional personal profit realized by having access to this site as compared to a less preferred or more expensive one.

These examples reflect one of the key advantages of the TCM: its ability to generate values that are based on actual recreation behavior observed. In addition, another advantage of TCM is that it can often be performed using existing data, such as campground fee receipts, recreation permits for hunting, fishing, wilderness and rafting, and simple license plate surveys. Thus the TCM is a relatively inexpensive method to calculate site-specific recreation values.

Two detailed examples of the application of the travel cost method to fishing and hunting are provided in chapters 9 and 10, respectively. These two examples also illustrate how the TCM demand equation can be statistically estimated to incorporate fishing and hunting quality. Since this quality is often related to that of the habitat, this helps to make the TCM demand equation into a quasi-bioeconomic model. (For more details on the travel cost method see Ward and Loomis [1986] or Walsh [1986].)

Contingent Valuation Method

The contingent valuation method (CVM) is a survey technique that constructs a hypothetical market to measure willingness to pay (WTP) or accept compensation for different levels of nonmarketed natural and environmental resources. The method involves in-person or telephone interviews or a mail questionnaire. CVM not only is capable of measuring the value of outdoor recreation under alternative levels of wildlife/fish abundance, crowding, in-stream flow, and so on, but is the only method currently available to measure other resource values, such as the benefits the general public receives from the continued existence of unique natural environments or species.

The basic notion of CVM is that a realistic but hypothetical market for "buying" use and/or preservation of a nonmarketed natural resource can be credibly communicated to an individual. Then the individual is told to use the market to express his or her valuation of the resource. Key features of the market include (1) description of the resource being preserved, (2) means of payment (often called payment vehicle), and (3) the value elicitation procedure.

Value Elicitation Format

The value elicitation procedure can be one of three types. First, there is an open-ended willingness-to-pay question format that simply asks the respondent to state his or her maximum WTP. This was one of the early forms of CVM and is still the simplest. The answer to such open-ended questions is the respondent's WTP in dollars.

Another procedure is the closed-ended "iterative bidding" type of question, where the interviewer states an initial dollar amount and the respondent answers whether or not he would pay. If a yes is elicited, the amount is raised and this process is repeated until a no is recorded. The highest dollar amount receiving a yes is recorded as the person's maximum WTP. If a no is recorded on the first dollar amount, the amount is lowered until the first yes is achieved. In this iterative approach, poor selection of the starting value may influence the respondent's final reported value. This starting-point problem can often be minimized by selecting the initial bid amounts around the mean of WTP from an open-ended pretest, or at least avoiding extremely high or low starting points.

Recently, a third approach, called dichotomous choice, or "referendum," has been developed in which the respondent answers yes or no to one randomly assigned dollar amount chosen by the interviewer. The dollar amounts vary across the respondents. In this way the entire sample is asked to pay a variety of dollar amounts. Then the analyst can calculate the probability of a respondent answering yes or no to a given dollar amount. From this probability the statistical expected value of WTP can be calculated. Chapter 9 provides an example of the dichotomous-choice CVM approach.

This dichotomous-choice approach has several advantages. Its "take it or leave it" format matches the market where individuals either buy or do not buy at a given price. In addition, when cast as a referendum, it matches how people vote on many public programs. That is, they vote yes or no at a specified tax price. In general, the dichotomous-choice approach avoids the starting-point problem associated with iterative bidding. Unlike the open-ended WTP questions, where some respondents have difficulty providing an exact dollar amount of their WTP, in the dichotomous-choice format respondents must simply determine if their value is greater or less than the dollar amount they are asked to pay. Then, much like the TCM, the analyst uses statistical analyses to infer how much more their WTP is than the amount they are asked to pay. For more details on these three approaches see Schulze and colleagues (1981); Cummings and colleagues (1986) and Mitchell and Carson (1989).

Means of Payment

A respondent can be asked to pay his or her WTP in several ways. The particular means should be tailored to match the characteristics of the hypothetical market the analyst has set up. For preservation of wilderness areas, rivers, or nongame wildlife habitat a payment into a trust fund similar to "nongame check-offs" on many state income tax forms is often used. Other payment vehicles include utility bills, higher taxes, or higher product prices. In selecting the payment vehicle the analyst should pretest different payment vehicles to see which ones yield the fewest emotional responses that are clear protests to the means of payment. For whatever payment vehicle is ultimately chosen, it is still important to use a check question to determine if people responding with either zero WTP or consistent no's in a dichotomous-choice format are people with true zero WTP due to lack of interest or income or if they are protesting some feature of the hypothetical market. For example, people will sometimes respond that some other group should pay or that government should reallocate more of its existing taxes to supplying this good. Either of those answers indicates a positive value for the resource but a concern over the market portrayed in the survey. These protesting responses are normally omitted from the final analysis of WTP.

Description of Resource to Be Valued

The natural or environmental resource to be valued can be described to the respondent in words, drawings, photographs, charts, maps, and so on. Often some combination of these approaches is used, depending on the nature of the environmental change being communicated. For changes in air quality or water quality many times photographs are used. For changes in lake levels, river levels, or acreage of wilderness, simple maps or drawings are frequently used. For changes in hunting or fishing success, a simple narrative will work because hunters and anglers are often quite familiar with the resource to be valued. The key in any of these cases is to provide a short and neutral description of the resource. The regional or national significance of the good can also be described as part of the survey.

These three components (the elicitation procedure, means of payment, and resource description) are combined into a survey instrument. The wording of the question needs to be very matter-of-fact: if respondents pay the money, they get preservation of resource; if they do not pay, they do not receive the preservation of the resource. An example of CVM for valuation of elk hunting is given in chapter 9.

Advantages of CVM

CVM has at least three advantages. First, the method will work to value recreation, even in the face of most visitors taking multiple-destination trips. Therefore CVM can be used to value recreation areas that are visited as part of a longer trip to numerous sites. The second advantage of CVM relates to use of its hypothetical nature as an asset. That is, sometimes decision makers need information about how people value a variety of irreversible public land management scenarios (e.g., the value of maintaining a free-flowing river instead of damming it or of keeping an area free of roads and not logging it). These situations call for information on intended behavior about possible alternatives rather than past behavior if the information is to be timely in assisting the decision maker.

One of the strongest advantages of CVM may lie in valuing those resources for which recreation is but a small part of the social benefits. For example, many unique natural environments and species are rarely visited or seen by people. Yet many people derive a substantial enjoyment or satisfaction from knowing that unique areas exist or that such species as spotted owls, whooping cranes, wolves, condors, and grizzly bears are still in the wild, even though they may never see them. This existence value is often reflected in a person's contribution to conservation organizations to save these species or natural environments. In addition, some people are willing to pay to preserve areas for future generations. This bequest value is an additional motivation. Lastly, some are willing to pay a premium over their future-use value as an insurance premium to ensure that natural environments or wildlife species exist so they can visit them in the near future. This is sometimes referred to as an *option value*.

These option, existence, and bequest values capture much of what is called the public trust values of natural resources. Others refer to these three values as preservation values or off-site or nonuse values. In any case, CVM is presently the only method for measuring these values in dollar terms. Thus in cases where the public land decision involves relatively unique environments or irreversible decisions, these values are often empirically important. A CVM study of these preservation values for protecting additional roadless areas in Colorado found that these values made up about 50% of the total value to society (Walsh et al. 1984). That is, these values were at least as large as the recreation values provided by protection of areas as Wilderness. (For more discussion of the use of CVM for valuing existence values see Mitchell and Carson [1989].)

Assumptions of CVM

There are two primary assumptions of CVM (Dwyer and colleagues 1977). The first is that a respondent does have a well-thought-out value for the natural or environmental resource of interest. That is, he or she has well-defined preferences for this natural resource relative to other resources and other goods he or she buys. The second is that this value can be elicited accurately from the respondent in a survey or interview. Specifically, the survey can provide the context for the re-spondent to reveal their true values.

Both of these assumptions are often quite reasonable. With regard to the first assumption, the more familiar the respondent is with the resource in question, the more likely the assumption is to be met. For example, hunters and anglers asked questions about fishing quality or deer populations will likely have a well-devel-oped value. After all, they have repeatedly traded money in the form of travel costs (and hence given up other goods) to visit the site to hunt or fish. The same is true for many other recreational users. For CVM questions dealing with preservation of resources the respondent may have only read about or visited in the distant past, the degree of conformity to the first assumption is somewhat less. Nonetheless, much like new product market testing, economists believe that a respondent can make a judgment about how worthwhile this resource is compared to ones currently being consumed. Especially if the choice is framed in a referendum context, this decision is similar to voting on ballot items to protect some area as a public park.

With regard to the second assumption, much effort in survey design has gone into construction of the hypothetical market that will elicit honest answers. Drawing on past research and pretesting, a credible hypothetical market or referendum can often be constructed that does not provide incentives to respondents to bias their answers. Without any obvious incentives to bias their answers, truth telling seems to be the dominant behavior in CVM survey responses. We now turn to the issue of whether these respondents' truthful answers in the survey (known as intended behavior) would translate into actual behavior if a real market with real money was present.

Validity and Reliability of CVM

It is reasonable, of course, to question the accuracy of answers where hypothetical, not real, money is involved. Would people really pay the dollar amounts they state in these surveys? The empirical evidence to date indicates that when we ask about WTP (rather than willingness to accept), people would pay approximately what they state in the surveys. This conclusion is based on several comparisons of real cash markets with hypothetical markets used in CVM. One comparison was between housing price differentials in areas of Los Angeles with less air pollution and how much people said they would pay for the same reduction in air pollution. The study by Brookshire and colleagues, published in the 1982 *American Economic Review,* showed that stated WTP for a reduction in pollution was statistically less than what people had actually paid.

The second series of CVM validity experiments was carried out by Bishop and Heberlein (1979) and Welsh (1986) in Wisconsin. Bishop and Heberlein compared how much goose hunters said they would pay for a goose hunting permit with the amount of cash a similar group of goose hunters actually accepted for their goose hunting permits. The actual cash value of the goose hunting permit was $63, whereas the stated WTP was $21.

In more recent work on deer hunting permits, the Wisconsin Department of Natural Resources allowed Welsh (1986) to buy and sell up to seventy-five deer permits. Two markets were set up to allow hunters to buy them. One was a real market, where hunters without permits were offered actual hunting permits at a specific price and asked if they would buy at that price. The other was a contingent-value market, where the same price was quoted to a different sample of hunters and the transaction did not involve real money and no permits traded hands. The results indicate the contingent-value market obtained values that were generally 25% higher than the actual cash ones (Welsh 1986).

In a test of the reliability of households' WTP for recreation, option, existence, and bequest values for preservation of a unique wild bird habitat (called Mono Lake), Loomis used the test-retest method. Here the same households were resurveyed eight to nine months after their first survey. The surveys were identical and people were asked to give their value in response to both open-ended WTP and dichotomous-choice questions. For both the visitor and general public samples, using both the open-ended and dichotomous-choice question format, CVM WTP values proved to be reliable. Specifically, there were no statistical differences between the WTP given in their first survey and that in the second. In general, this indicates the answers are thoughtful responses that are relatively stable over time. (See Loomis [1989] and [1990] for more on this reliability study.)

The essence of all these comparisons is that respondents do attempt to give their true value of the resource in a CVM survey. The behavior exhibited and statements of value appear at some times to understate the value and at other times to overstate the value slightly. The degree of overstatement, when it occurs, seems to be reasonably small. Based on these studies, it appears that one can have some confidence that statement of WTP elicited in contingent-value surveys bears a close resemblance to the behavior that would be observed if the situation described in the survey arose in a real market.

One of the biggest unresolved issues in CVM may relate to a common problem in much of benefit-cost analysis: focusing on one resource at a time. This is common when estimating demand for most market goods, such as timber, beef, coal, cars, and so on. That is, the demand curve for one good is estimated with a limited set of substitute prices included. In doing so, economists appeal to the view that consumers engage in two-stage budgeting: first, they optimize their overall budget or income across broad categories of goods, such as food, clothing, recreation, shelter, transportation, and so on. Then they optimize their consumption choices within each category. In essence, consumers substitute among goods within these groups much more than across the groups. Thus demand is estimated, including the substitute goods within the bundle, but not across the bundles.

Although we accept this simplification for market goods, there has been a concern that when people use separate CVM surveys to value each natural resource, they are not explicitly accounting for many of these substitution effects when giving WTP responses to CVM surveys. More to the point, we need to ask whether we get the same estimate of WTP for a species if we ask about it by itself or if we ask about it as one of one hundred other species. In some sense this is an unrealistic scenario from both a policy analysis and CVM survey standpoint. Most public land management or policy issues deal with a few species or resources at a time. Nonetheless, recent advances in CVM make possible valuations of one resource within the context of multiple species/resources.

Use of TCM and CVM by Federal and State Agencies

Both TCM and CVM are widely accepted by government agencies for valuing both recreation and other nonmarketed benefits of environmental resources. TCM and CVM have been recommended twice by the federal U.S. Water Resources Council (1979, 1983) under two different administrations as the two preferred methods for valuing outdoor recreation in federal benefit-cost analyses. Recently, the U.S. Department of Interior (1986) endorsed both as preferred methods for nonmarketed natural resource damaged by oil spills and other toxic events.

In January 1993, a blue-ribbon panel that included two Nobel-prize winning economists concluded that CVM was reliable enough for use in both administrative and judicial proceedings involving establishing the economic losses to society from oil spills.

The U.S. Bureau of Reclamation and National Park Service relied on CVM to value in dollar terms the recreational fishing and rafting effects of alternative hydropower water releases from Glen Canyon Dam into the Grand Canyon. The Montana Department of Fish, Wildlife, and Parks relied on a CVM survey of the benefits of viewing and hunting elk when justifying its purchasing additional elk winter range outside of Yellowstone National Park. Several other state fish and game agencies, such as those of Arizona, California, Idaho, Maine, Missouri, Nevada, and Oregon, use the travel cost method and contingent valuation methods for valuing fish- and wildlife-related recreation.

As can be seen from the discussion of valuing both marketed and nonmarketed natural resources, there is a well-developed and conceptually sound basis for derivation of such values. In most cases, these values stem from people's actual behavior in buying products made from natural resources or from visiting recreation sites. The techniques discussed in this chapter simply allow the analyst to measure those values in dollar terms so as to compare benefits to costs of alternative uses of those natural resources. In this way, the public and the manager can determine if resource allocations or management actions are economically efficient or not. As we have stressed repeatedly, economic efficiency is not the only criterion, but neither should it be ignored.

Another View: Criticisms of the Conceptual Foundations of Benefit-Cost Analysis

Compensation Paradoxes

Since with the potential Pareto improvement the compensation is not actually paid, three major concerns have been voiced about reliance on BCA for public decision making. The equity issue has already been dealt with in the first part of the chapter and will not be repeated here. The second major concern results from the possibility that implementing the new policy might change the relative prices of goods and income levels that influence individuals' WTP. For example, a move from a situation where salmon are scarce and hydropower plentiful to an alternative situation where salmon are plentiful and hydropower is scarce could change relative prices of salmon and hydropower. In addition, the income of commercial fishermen relative to owners and workers in businesses that use a large amount of electricity would probably change. If the effect of the policy was so widespread that relative prices or incomes changed substantially, then the desirability of the policy might change, depending on whether we evaluate it using the original set of prices and incomes or the set occurring with the change. This is what is referred to as a *compensation paradox.*

There are a couple of responses to this concern. First, most public land policies or natural resource management actions that are typically evaluated make such small changes in the market supplies of goods or factors of production that relative prices or overall factor income shares rarely change. Second, in comparing an inefficient bundle of resource uses to an efficient bundle, the compensation paradox may not occur. (The interested reader should see Just, Hueth, and Schmitz [1982:38] for a more detailed discussion regarding the resolution of the compensation paradox.)

Theory of Second Best

The purpose of BCA is to recommend projects, policies, or programs that make people better off (i.e., represent an improvement in economic efficiency). In BCA we evaluate the economic efficiency of just the resources affected by the proposed resource management action or project. If the BCA indicates that we should allocate more forage to deer from livestock, then implementing such a recommendation would be thought to improve economic efficiency of the entire economy. That is, we have improved economic efficiency in use of three resources (forage, deer, and livestock) and not negatively affected any other resources elsewhere in the economy (e.g., steel production, housing construction, automobiles, and so on).

But the theory of second best postulates that if there are inefficiencies elsewhere in the economy, then improving economic efficiency in one part of the economy will not necessarily increase social welfare. In other words, reducing economic inefficiency, as long as there are still some inefficiencies remaining, may not represent an improvement.

If this theory of second best is universally true, then we could not reach any conclusion on the desirability of any single project based on the information contained in a BCA as long as other industries or resource uses remained economically inefficient. Fortunately, reevaluation of theory of second best has shown that we need not be so pessimistic about our chances of recommending economically efficient uses of natural resources in an otherwise inefficient world. First, Mishan (1981:293) indicates that any time we can produce the same bundle and level of output for less cost, this represents an unambiguous gain to society. Second, as Davis and Whinston (1965) have shown, if there is separability or independence between economically inefficient sectors of the economy and the ones under study in the BCA, we may be able to make unambiguous statements about improvements in economic efficiency. Since there often is little connection between public natural resources, we will be evaluating numerous other sectors of the economy (e.g., clothing, electronics, automobiles, and so on), concerns over second best need not preclude us from concluding that increased efficiency in natural resource use will improve social well-being.

7

Regional Economic Analysis and Input-Output Models

Need for Regional Economic Analysis

The need to quantify the change in local income and employment related to public land management actions has been referred to frequently in chapters 4 and 6. In terms of chapter 4, addressing the distributional-equity criteria often requires some information on changes in local jobs and income (i.e., wages, rents, business income). That is, does a particular public land management action affect local economic activity? Is the change in local economic activity relatively small in magnitude compared with the scale of the local economy (i.e., a loss of one hundred jobs in a county with ten thousand workers is minor compared with a loss of one hundred jobs in a county with five hundred workers)? More importantly, is much of this change in local employment concentrated in one or two industrial sectors (e.g., mining). A loss of one hundred jobs spread evenly throughout the retail, transportation, and food service industries may be less problematic to a local economy than a loss of the same number of jobs concentrated in one particular industry. Although the relative weight to give gains and losses in jobs in different industries may often be a value judgment (i.e., does the gain of one fishing guide offset the loss of one truck driver?), the agency managers and the public have a right to know what the change in jobs is by industry from some public land management decision.

This right to know is exemplified in the National Environmental Policy Act's requirements for preparation of an Environmental Impact Statement (EIS). Specifically, what are the effects on the human environment in terms of employment? Thus a regional economic analysis is a common feature of most EIS's performed by nearly all federal land agencies.

It also makes good sense for a manager to pay attention to local income and employment effects. Although we demonstrated in chapter 6 that local gains and losses in employment are nearly always transfers of economic activity at the national level (and hence should be omitted from a calculation of benefits and costs), these changes in local income and employment sometimes take on symbolic proportions in public land management debates. Opponents of proposed public land management actions, such as Wilderness designations or reductions in livestock grazing, will often raise the specter of massive reductions in local employment. Conversely, proponents will often counter with rosy predictions of gains in employment in other industries that will benefit from the proposed action.

Therefore there is much to be gained by the agency performing an objective regional economic impact analysis of what the employment effects are likely to be. This helps add an element of realism and sometimes demonstrates that local employment concerns are a smoke screen for real reasons for local opposition or support.

The objective of this chapter is to provide an understanding of the basics of regional economic analysis and the types of models used. With this background the reader will be much more able to judge the soundness of the many income and employment estimates commonly tossed around in public land debates.

Concepts of Regional Economic Analysis

Economic Linkages and Leakages

The easiest way to grasp the notion of regional economic analysis is to think about how a local economy might react to an agency offering a mining lease on public land adjacent to this local economy. By *local* we often mean a county or group of counties directly linked to or influenced by management decisions on adjacent federal land. The development of a mineral lease offered by the federal agency will have both direct and indirect effects that we sometimes group under the headings *backward* and *forward linkages* in this local economy. For example, the leasing of coal will directly affect the local economy, as miners, managers, and truck drivers are hired (locally and/or from outside the region). These people work directly at the site; hence their wages directly affect the local economy. (The importance of the source of these workers is discussed later.)

But building and operating a new mine takes many *inputs,* such as mine timbers, explosives, heavy equipment, and other tools and supplies. To the extent the local economy can meet some of this increased demand for inputs, there is an additional stimulation of local industry. This stimulation to supply inputs is known as the *backward linkages* and gives rise to *multiplier effects,* which are discussed later in this chapter.

In addition, the raw coal taken from the ground must often be crushed, processed, and washed before it can be delivered to end users, such as power plants or steel mills. If such processing is done in the local economy, this too stimulates the demand for local resources. Processing equipment must be operated and maintained, and private land must be rented for the processing center. Hence additional workers must be employed in the coal-processing sector, and of course supervisors and managers must be hired as well. This stimulation of final processing of outputs creates the *forward linkages* to a local economy.

These potential linkages are displayed in figure 7.1 for a mine or a timber sale in a local economy that would contain both the forward and backward linkages. The direct effect of the mine is located in the middle of figure 7.1, with the backward linkages to supporting input sectors at the bottom of the figure. The top of the figure represents the forward linkages of the processing sectors.

If a local economy either is so well diversified industrially or has the appropriate land, water, and other factors of production to supply these backward and forward linkages, mineral leasing or continuous timber harvesting will maintain a large amount of local economic activity.

On the other hand, most isolated rural economies adjacent to public lands do

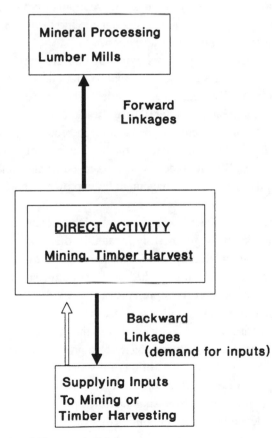

Figure 7.1.   Regional Economic Linkages

not have the diversified economic base to supply most of the inputs or the processing of final outputs. The specialized machinery is often manufactured in large industrial cities in far-away regions of the country or abroad. Thus the local economy gains little when a mining company buys one of these machines. The same is sometimes true of other inputs. Consequently, the analyst must be very careful to scrutinize which required inputs are likely to come from local production (and hence represent a backward linkage) and which are imported from outside the local economy. Although the potential exists for a new local demand for these inputs to make presently nonexistent local production economical, it is important not to let local optimism overwhelm common sense. Often production of these inputs is subjected to economies of scale so large that a few plants supply the entire national demand for these inputs. Consequently, one local mineral lease may not make local production profitable.

It is worth reinforcing the concept of *leakages* and *value added,* which will be discussed several times in this chapter. If some of the supplies needed by the mine are not produced locally (i.e., they are produced by industries outside the local economy) but the mining company purchases these inputs from local retailers rather than directly from the factory or wholesalers outside the local economy, then there is some local economic gain on this transaction.

Specifically, if imported explosives worth $1,000 are purchased from a local retailer, the gain to the local economy is equal to the retail markup if the store is owned by a local owner. The retail markup reflects the contribution of local labor, capital, management, and land to order, unload, stock, and sell the imported explosives. The payments to local factors of production in processing this sale are often called the *local value added.* Of course, if the store is part of a nationwide chain, the only part of the local value added would be to labor and land (e.g., rent), as returns to capital and upper-level management flow to owners and investors outside the local economy.

This logic carries across nearly all sectors of a local economy. Hence spending by recreationists in a sporting goods store located in the local economy often just translates to local value added associated with the retail margin. That is, purchase of film or gasoline produced outside the local economy simply adds the retail margin to the local economy, if the store is locally owned.

The chain of reasoning used with backward linkages applies when evaluating the likelihood of forward linkages associated with the processing sectors. The analyst must ask whether the local economy has such a processing sector. If not, would such a sector likely emerge from the addition of this one lease? If the lease promises a large enough supply for a length of time equivalent to the economic life of new processing facilities, then new processing facilities might develop in response to the lease. Again, it depends on the economies of scale in processing relative to the locational and other advantages enjoyed locally.

Indirect Rounds of Related Economic Activity

Once the direct spending and payment of wages occurs in the local economy, more ripplelike effects are felt in other sectors. For example, for every ton of coal mined, the backward linkages suggest the mining company requires several dollars worth of inputs. This stimulates the industrial sectors supplying those inputs. As those support industries are stimulated, they also need more inputs. That is, if demand for explosives goes up, the firms that manufacture them must hire more labor and buy more raw materials. More underground mining requires more timbers to support the shafts. This stimulates the forest products sector to harvest more trees and lumber mills to process the timbers. Finally, trucks and fuel are needed to transport the timbers to the mining firm.

Role of Imports in Determining the Size of the Total Effect

The amount of stimulation of economic activity depends on how self-sufficient the local economy is in supplying the labor and capital necessary to support the expansion of basic industries. One critical factor worth investigating is the supply of labor inputs to a new or expanding industry.

The demand for more labor arising from a new project is normally met from one of four sources: (1) hiring previously unemployed workers in the local economy; (2) hiring new workers from outside the local area who now relocate their households to this new area; (3) taking away workers from existing jobs in the local economy by offering higher wages in the expanding industry; (4) hiring temporary workers who maintain their households outside the local economy. Determination of the direct change in local income and employment in each of these cases may be somewhat different, although as an approximation in cases 1–3, analysts often simply count the number of new jobs created, ignoring the possibility that the three different sources of supply of labor may affect true gain to the local economy. This is often a reasonable simplification to make unless the circumstances surrounding a particular development project clearly contradict conditions 1–3 (e.g., it is a short-term but large-scale construction project, such as a power plant, that requires specialized labor services not found in the area). When case 4 arises, much of the wage income is lost to the local economy, because leakages are substantial.

An Example of Stimulation of Economy from Direct Effects and the Resulting Total Effects

For starters we will estimate the annual change in economic activity associated with operation of a new campground. Assume that all the visitors come from outside the local economy (i.e., people who live adjacent to the National Forest don't go there to camp). To keep this example simple we will concentrate on just one sector, the retail sector.

Suppose each camper spends $10 at the local sporting goods store for film, additional ice for the ice chest, and other miscellaneous supplies. If there are one thousand campers at this campground each year, this means $10,000 of new demand on the retail sector of the local economy.

The immediate effect of the first round of spending results in hiring more part-time workers to order more supplies, stock shelves, ring up sales, and wait on customers. In addition, the local owner makes his or her profit on the additional $10,000 of sales. The local store owner must also buy more price tags, cash register tapes, and so on, from the neighborhood stationery store. But the store owner, of course, also pays the wholesale distributor (which we will assume is outside the local economy) more money to bring in additional film, ice, and other supplies bought by campers that are not manufactured locally.

The first-round effect is illustrated in table 7.1, where it is assumed that local value added is 40%. Thus $4,000 of the $10,000 in visitor spending remains in the local economy to pay for local retail inputs. The composition of the $4,000 in the local economy would include about $600, or 15%, as a trade service sector payment (e.g., marketing sector) and about $200 to $400 (5% to 10%) as local transportation charges; the balance would be direct payments to local producers (e.g., farmers, manufacturers, services, and local government agencies).

But this $4,000 received as additional income by workers and business owners stimulates another round of spending. That is, part of the $4,000 of additional income will be spent by the recipients of this income. The extent of this second round of spending depends on the propensity to consume locally (McKean 1983).

Table 7.1
*Illustration of Direct and Indirect Effects
and Multiplier*

Round	Total Spending	=	Leakage	+	Local Value Added
1	$10,000		$6,000		$4,000
2	4,000		2,800		1,120
3	1,120		806		314
4	314		226		88
5	88		63		25
6	25		18		7
7	7		5		2
8	1.93		1.39		.54
Total					5,555
Multiplier = total/1st round					1.3888

If the local workers and business owners buy most of their consumer goods and make most of their investments locally, then a large percentage of this $4,000 will be spent in the local economy. For example, if 70% of the gain in local income is spent or invested locally, the second-round change in income is $2,800 ($4,000 × .7).

To keep the example simple, let us assume that the same retail markup applies to this spending. Therefore of the $2,800 spent locally, 40% is retained in the local community as local value added. That is, 40% is payment to local factors of production, with the remaining 60% representing imported inputs or purchases from wholesalers outside the local economy. The 40% of $2,800 is $1,120. So this is the gain in local income (wages and business income) realized from this second round of spending.

But of course this $1,120 gain in local income to workers and business owners in this local economy stimulates further spending by these recipients. If the same pattern of 70% local consumption with a 40% value added continues, then $784 of the $1,120 would be spent locally and $313.60 would become local value added or income. This $313.60 of additional income continues to stimulate spending by the local recipients. If the same pattern continues, this $313.60 results in approximately $88 of new local income. Following the same pattern as before, the $88 results in about $25 of additional income to workers and business owners. If the same spending patterns continue, this round results in $7 of additional income to the local economy. Finally, this $7 results in about another $2 of local income in the final round. Table 7.1 adds all these rounds up and calculates the total effect of $5,555. The total effect could be calculated from the first-round effect by applying the income multiplier derived in table 7.1. *The income multiplier is the ratio of the total effect to the direct or first-round effect.* In this case the income multiplier is 1.388. This means that each initial $1 of local income generates an additional 38 cents of income, for a total effect of $1.38.

This multiplier can be analytically derived by the following formula:

$$\text{Income multiplier} = 1/1 - (\%C \times \%VA) \qquad (7.1)$$

where

$\%C$ = percentage of income spent locally

$\%VA$ = value added by local businesses (i.e., 1 − the percentage of imported inputs or goods used by local businesses)

In our example, $\%C$ is 70% and $\%VA$ is 40%. Therefore the income multiplier is $1/(1 - (.7 \times .4)) = 1.3888$, the same number we obtained from adding up the repeated rounds of spending.

These repeated rounds of additional spending generated from the initial increase in final demand constitute another way to define the multiplier. If we define the *initial* increase in final demand as the *direct effect* and the subsequent rounds of spending as the *indirect effects,* then the multiplier can also be defined as follows:

$$\text{Multiplier} = (\text{direct} + \text{indirect})/\text{direct} \qquad (7.2)$$

The analytic formula for an income multiplier highlights the importance of local consumption and leakages in the multiplier. The greater the local percentage consumption or the more self-sufficient local industry is in meeting local demands, the greater the income multiplier.

Generally, businesses in remote rural towns import a high percentage of their inputs, and many consumers make large purchases out of town. Thus value added or income multipliers of 1.25–1.75 are common in these rural areas surrounding public lands. As Coppedge (1977:4) states with regard to income multipliers, "Most estimates would fall between 1 and 2. . . . A multiplier that exceeds 2 should be subjected to critical review before acceptance or use in further analysis." Of course the larger the geographic area and size of the economy covered, the larger the multiplier would be. Thus a multiplier for "Tiny Town," California, might be 1.3, that for Los Angeles might be 1.75, and that for California as a whole might be 2.5.

A Simple Model of a Local Economy

Now that the reader has an intuitive understanding of the income multiplier and regional economic linkages, it is time to develop more formally the analytical methods used to calculate the multipliers for all sectors of the economy. To start with we will consider two views of a given local industry. The first is based on output; the second is based on use of inputs.

The output of a given industry can be used in one of five ways. A given industry's output can be directly consumed by households (C), used for an investment in capital goods (I), consumed by government agencies (G), used as an input by other local industries (x_j), or exported to any one of these uses outside the local economy (E). Equation (7.3) presents this relationship for the output (X) of industry i:

$$X_i = \sum_{i=1}^{n} x_{ij} + (C_i + I_i + G_i + E_i) \qquad (7.3)$$

where x_{ij} is the amount of industry i's output used as an input (or intermediate good) by industry j.

The amount $C_i + I_i + G_i + E_i$ is often referred to as the *final demands* to distinguish it from $x_{ij,}$ which are intermediate demands of the processing or producing sectors. Of course the other way to view industry i is from its use of inputs. For example, industry i might use labor, land, capital, raw and semifinished materials, and management inputs. The inputs may be obtained locally or by imports from other industries outside the region. What many regional economic models do is to trace the *factor payments* to both local owners of inputs and importers of inputs. The payment view of equation (7.3) is given in equation (7.4) as the total outlay (*X*) of industry *i:*

$$Xi = \sum_{i=1}^{n} x_{ij} + (W_i + L_i + K_i + P_i) + M_i \qquad (7.4)$$

where

W_i = wages paid to workers in industry *i*
L_i = rental payment on land used in industry *i*
K_i = rental or interest paid for use of capital in industry *i*
P_i = the profit or return to management in industry *i*
M_i = payments to imported inputs (labor, materials, capital, management) from outside the region.

As the reader has probably noticed, the left-hand-side variables in equations (7.3) and (7.4) are the same. This is because the value of the output produced by an industry (7.3) must equal the payments to all the factors of production, including profits (7.4). That is, the value of the output must equal the total cost of production plus profit.

Given the equality of factor payments and value of output, we can show that Gross Regional Income (what the local economy receives) equals Gross Regional Product (the value the local economy produces; Richardson 1972:20)). Specifically, since the right-hand side of equations (7.3) and (7.4) both equal X_i,

$$\begin{aligned} \text{Sum } x_{ij} + (W_i + L_i + K_i + P_i) = \\ \text{Sum } x_{ij} + (C_i + I_i + G_i) + (E_i - M_i) \end{aligned} \qquad (7.5)$$

The relationship shown in equation (7.5) is the building block of the *transactions table* that underlies many input-output models used to describe entire economies and calculate multipliers for all their sectors. Figure 7.2 shows a conceptual view of the three main components of a transaction table.

This conceptual transactions table is made up of three components: a processing or producing sector, a payments sector, and the final demand sector. Thus any economy can be viewed as having a core-producing sector that takes inputs and satisfies final demands. At the same time, payments to the factors of production (including taxes) used by the processing sector allow the households and the government to finance their final demands for goods and services. In this sense, the elementary circular flow in any economic system is represented in the transactions table. Table 7.2 provides a somewhat more detailed picture of a transaction table that will serve as the basis for our discussion.

This transaction table provides a "double-entry" bookkeeping view of this

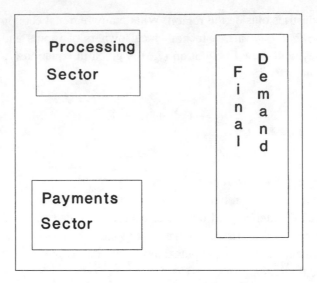

Adapted from Peterson, 1982

Figure 7.2   Conceptual Transactions Table

simplified local economy. Reading *across the first row,* industry 1 (X_1) sells itself $2 worth of its own output (i.e., produces or uses some of its own output as an input to produce more output). Industry 1 also sells $4.50 worth of its output to industry 2 and just 40 cents worth of output to industry 3. Industry 1 also sells $3.10 worth of its output directly to final consumers (e.g., consuming households or government or export out of the region). Thus industry 1's sales of output is $10.

Now, reading *down the first column,* we see a factor payments view of industry specifically, what industry 1 must buy or pay out to get the inputs it needs for its own production. Here industry 1 buys $2 worth of its own output, $4 worth of industry 2's output, and just 40 cents' worth from industry 3. In addition to these materials, industry 1 buys $1.60 worth of labor (i.e., pays wages and salaries to workers of $1.60), uses 80 cents' worth of other local inputs (e.g., land and building rent, local taxes, profits to local owners) and imports $1.20 worth of other inputs unavailable from the local economy. Thus to produce $10 in sales of total output (reading across the first row), it requires $10 in total outlays (the first column). Thus there is a balance for X_1 and every other processing sector as well.

The transactions table illustrates the economic interdependence between processing sectors themselves and the processing sectors and the rest of the economy. The magnitude of these economic interdependences is what determines the size of the income multiplier in each sector.

To estimate the multipliers from the transactions table, two additional steps are

Table 7.2
A Three-Sector Transactions Table

Sales to →		Processing Sector X_1	X_2	X_3	Final Demand	Total Output
Purchases	X_1	2.00	4.50	.40	3.10	10.00
From	X_2	4.00	12.00	2.00	32.00	50.00
	X_3	.40	6.00	3.60	30.00	40.00
Payments to: labor		1.60	20.00	15.00	4.00	40.60
Other local inputs		.80	3.00	10.00	1.00	14.80
Imports		1.20	4.00	9.00	1.00	15.20
Total outlay		10.00	50.00	40.00	70.10	170.00

necessary. First, the analyst must construct the *direct coefficients table* using the information in the transactions table. The number in this table represents the dollar amount of other inputs required from a given processing sector to produce $1 worth of output. This is the first-round effect discussed earlier. Given this definition of direct coefficients, it is easy to see how they are calculated from the transactions table: the direct coefficients are the individual matrix elements from the transactions table divided by the column total (i.e., total outlay for that industry).

For example, the technical coefficient for industry 1 buying from itself (denoted X_{11}) is $2/$10, or .20. For industry 1 buying from industry 2 (X_{21}) is $4/$10, or .40, and so forth. These technical coefficients show the first-round economic interdependence between sectors on a per dollar basis. That is, for industry 1 to produce an additional $1 of output, it must purchase 40 cents' worth of industry 2's output. This technical coefficient relating what industry 1 needs from industry 2 is often symbolically labeled a_{21}.

But, of course, for industry 2 to produce more output it must purchase inputs from other industries as well. Using the technical coefficient matrix shown in table 7.3 (next page), industry 2 must buy 12 cents of output from industry 3. This is additional stimulation to industry 3 over and above the direct stimulation received by purchases from industry 1 itself (4 cents) and is the beginning of the second-round effect.

Similar effects occur in the third round as input suppliers to industry 3 are stimulated in response to that industry's attempt to supply more input to industry 2. To calculate the multipliers, the information in the technical coefficient matrix is used along with the relationship in the transactions table. Specifically, the output of X_1 is as follows:

$$X_1 = a_{11} \times X_{11} + a_{12} \times X_{12} + a_{13} \times X_{13} + Y_1 \tag{7.6}$$

where Y equals the final demands of consumers, government, investment, and ex-

Table 7.3
Technical Coefficient Matrix

Sales to →		Processing Sector		
		X_1	X_2	X_3
Purchased	X_1	.20	.09	.01
from	X_2	.40	.24	.05
	X_3	.04	.12	.09

ports. Thus equation (7.6) is similar to equation (7.3). The key addition is that with the technical coefficients (the a_{11}, a_{12}, a_{13}) one can calculate the multipliers from a series of equations like (7.6).

We said "series of equations" because there is, of course, one equation like (7.6) for each processing sector. In our example there are three such equations. In real I-O models there are often a few hundred processing or manufacturing sectors, and hence a few hundred equations like (7.6). To calculate the multipliers more directly, the equations are algebraically rearranged and put in matrix form as

$$\mathbf{Y} = \mathbf{X} - A\mathbf{X} \tag{7.7}$$

where **Y** is a matrix of final demands and **X** is a matrix of total output of each processing sector. *A* is the technical coefficient matrix (i.e., table 7.3). To begin to solve equation (7.7) we arrange it as follows:

$$\mathbf{Y} = (I - A)\mathbf{X} \tag{7.8}$$

where *I* is an identity matrix (i.e., a square matrix with the main diagonal elements being 1). To solve for the multiplier that measures the direct plus indirect output associated with an expansion of final demand for **X,** (7.8) is rearranged one last time:

$$\mathbf{X} = (1 - A)^{-1} \times \mathbf{Y} \tag{7.9}$$

where -1 as a superscript denotes a complex matrix mathematical operation called the inverse. In this case, the $(1 - A)^{-1}$ is called the *Leontiff inverse* after the developer of the modern-day input-output model. As Walsh (1986:385) points out, solving this inverse is equivalent to computing all the iterations in rounds of spending that would occur (as we did in table 7.1) in response to a $1 increase in final demand. Table 7.4 illustrates these coefficients for our example.

The interpretation of these multipliers is that for every $1 increase in direct final demand for sector X_1's output, there will be a $2.25 total increase in X_1's output. This occurs because of the linkages between X_1 and the other sectors of the economy (here X_2 and X_3). Because of the smaller linkages of X_3 to the other sectors of the economy, the multiplier for X_3 is smaller.

Table 7.4
Direct and Indirect Coefficients per Dollar of
Final Demand

		Processing Sector		
		X_1	X_2	X_3
Purchases	X_1	1.33	.16	.02
from	X_2	.77	1.41	.08
	X_3	.15	.19	1.11
Output multipliers		2.25	1.76	1.21

Distinguishing Between Type I and Type II Multipliers

The multipliers in table 7.4 meet the definition of a type I multiplier, which is defined as in equation (7.2) as the ratio of direct and indirect effects to direct effects. But these indirect effects in tables 7.3 and 7.4 reflect only the indirect effects on interrelated processing or manufacturing sectors of the economy.

However, as discussed earlier, the additional factor payments (wages, profits, and so on) become income to households. Naturally, households will spend a large portion of this additional income. As was seen earlier, this sets off additional rounds of spending in the economy. This effect of increased consumer demand by households associated with the direct and indirect effects is called an *induced effect.* If we include this induced consumption effect in our estimate of the multiplier, we obtain the type II multiplier. The type II multiplier is defined as

$$\text{(Direct + indirect + induced)/direct} \tag{7.10}$$

Not surprisingly, the type II multiplier is larger than the type I. Often the type II multiplier is much (25% to 50%) larger than the type I multiplier. The type II multiplier is often the appropriate one to use to capture the full *long-run* effect of a *permanent* change in final demand in a local economy.

The type II multiplier can be calculated within the input-output model framework. The key change is to expand the view of the processing sector portion of the transactions table to include households as a "processing sector." This is sometimes referred to by analysts as making households "endogenous" in the model. That is, instead of being treated solely as a predetermined constant outside (i.e., exogenous) of the processing sectors, household consumption demand is determined partly by the interactions among the sectors.

Income Multipliers

Much as the business or output multipliers were determined, the income multiplier can be analogously determined in the input-output model. The type I and type II income multipliers can be defined similarly to equations (7.2) and (7.10); that is, the type I income multiplier is the ratio of the direct plus indirect change in income to the direct change in income.

The key difference between output multipliers and income multipliers is that the income multiplier applies just to the direct change in income, not to the direct change in dollar output. A frequent mistake is to multiply an increase in $1 million in final demand by an income multiplier, say, 1.5, and conclude that $1.5 million in local income will be derived. The fallacy here is that much of the $1 million increase in final demand will be met by importing more inputs into the local economy. The direct gain to the local economy is only the local value added. This local value added represents the direct or first-round income effects. Therefore it is this first-round *income* effect that should be multiplied by the income multiplier.

Employment Multipliers

Besides measuring the change in local income, the analyst and decision maker are often interested in the change in local employment associated with the direct change in final demand in the local economy. This requires that one additional piece of information be developed to be used along with the input-output model. That piece of information is the number of workers per dollar of output delivered to final demand. For example, each $1,000 of livestock production might require .03 workers.

This is the direct employment coefficient, which is in some respects analogous to the direct output or technical coefficients presented in table 7.3. Of course the analyst and decision maker are also interested in the indirect and hence total employment effect. To calculate this requires multiplying the matrix of direct employment coefficients by the direct plus indirect coefficient matrix (e.g., table 7.4). If the form of table 7.4 is used with exogenous households, then the analyst obtains type I employment multipliers. If households are endogenous, then type II employment multipliers result. The type II employment multipliers can be about one and one-half to three times larger than the direct employment effects alone (McKean 1981:39–40).

A simpler, more direct process of calculating the total employment effect would be to multiply the direct employment coefficients by the total (direct plus indirect) gross output levels associated with a project.

Application of Input-Output Models

One result of building an input-output model is a set of multipliers. Often these multipliers are used directly in quantifying the total effect (direct plus indirect) associated with some public land management action. When the full input-output model is available, it is often better to run the direct changes through the model to calculate the total effect. In this next example, we shall use the multipliers derived earlier to demonstrate how to perform simple impact analysis.

Regional Economic Impact Analysis Example

Suppose a proposed public land management action involved a timber sale offering, an increase in livestock forage, a mineral lease offering, and a recommendation of an area as Wilderness. To calculate the regional economic effects of such an action, the first step is often to multiply the physical outputs by their respective dollar values. For timber this would be the amount of wood volume times the expected wood product price after the first round of processing per board foot (Sullivan 1983:17). For mining it would be the price per unit of the mineral (e.g., price per ton for coal times the number of tons). In the case of Wilderness recreation, the "price paid" by the recreationists is their trip expenditures (travel, lodging, supplies, and so on).

In this example, suppose that the annual effect of the action was $100,000 of additional sales. The next step is to allocate that $100,000 between industries in the local economy and those outside. One approach would be as follows: First, determine which sectors of the economy will receive the $100,000 of final demand. For example, $10,000 of additional wilderness spending by recreationists would be distributed according to a typical "bill of sale" associated with that recreational activity among transportation, food, lodging, and supplies. Only the percentage of the recreationists spending that took place in the local economy would be used with the input-output model or the multipliers.

In the case of a $50,000 timber sale, the forestry industry (logging, sawmills, and so on) would receive much of this, but some might go to transportation (trucking logs to the mills) if this is not already built into the forestry industry. In our example we will assume the wood products industry includes all groups from logging road construction, logging, transportation, and milling.

If no end user of lumber (e.g., furniture manufacturing) exists in the local economy, then no forward linkages exist and the value of that lumber destined for uses outside the region would not be considered further by the input-output model. A sophisticated input-output model database will have information as to which industries exist in a predefined local impact area and which do not. If a stimulated industry is not present in that local economy, the input-output model will not calculate subsequent rounds of local impacts associated with that change in final demand. In the simple case where the analyst is applying the multipliers directly, the multiplier would not be applied to that portion of the change in final demand for an industry not located in the local economy.

In the case of increasing livestock forage, the additional beef to be produced from the additional forage is multiplied by the price per pound at the first exchange level (i.e., the rancher and the feedlot or rancher and slaughterhouse). Most of this change in sales would take place in the meat animals sector of the economy (Sullivan 1983:30). In this example, we assume that this is a $20,000 increase in beef sales.

To present a simple example, we assume that the $50,000 timber sale stimulates sector X_1 (wood products) to the amount of $65,000 (the $50,000 of wood plus $15,000 of extra inputs labor directly required in industry X_1 to process the $50,000 of wood). For livestock assume that $20,000 of sales to feedlots directly stimulates X_2 to the amount of $30,000. Table 7.5 presents the application of respective output multipliers from table 7.4.

The results of table 7.5 show that the direct local change of $95,000 results in $199,050 in total business volume or gross output. As mentioned earlier, what is often of interest is the gain in local income rather than the flow of business through the community, that is, the amount of this $199,050 that becomes income to local households. This depends on the percentage of local value added in the first and subsequent rounds. If nearly all the processing or manufacturing steps that translate the raw material into a finished product occur locally, then most of the change in gross output is translated into local income. If much of the processing or other inputs used is imported, then the local value added is relatively small.

In our example, let us assume from information given in the transactions table that 88% of the total gross output value in X_1 is local value added and that 92% is local value added for sector X_2. In this case the type I income multipliers would be applied to 88% of the *direct* change in output in X_1 and 92% of X_2. Continuing our example, this would be $57,200 for X_1 and $27,600 for X_2. If the income multipliers were the same as the total gross output multipliers, the total (direct plus indirect) effect of the change in income would be $128,700 for X_1 and $48,576 for X_2. These are substantial gains in income from a given amount of output because of the very high local value added assumed in the example. In remote rural economies local value added might run as low as 10% for gasoline purchases to as much as 50% for general retail purchases.

Finally, employment calculations are shown in table 7.6. This approach adopts the use of direct employment coefficients times the total (direct plus indirect) gross output calculated earlier. From table 7.6 we see that this public land management

Table 7.5
Application of Output Multipliers to Timber Sale

	Direct Change	Multiplier	Total Change
X_1: Wood products	$65,000	2.25	$146,250
X_2: Animal meat processing	30,000	1.76	52,800
Total direct and indirect sales			199,050

Table 7.6
Calculation of Employment Effects

	Total Output	Direct Labor Requirements	Total Jobs
X_1: Wood products	\$146,250	.00004	5.8500
X_2: Animal meat processing	52,800	.00001	.5280
Total direct and indirect employment			6.3780

decision will result in the equivalent of 5.85 persons being employed for one year in the wood products industry and about a half a person in the meat processing industry. Now of course a company cannot employ .528 people. Rather, these figures are full-time equivalent jobs. Thus .528 persons can be one person halftime for a year, or if the industry is seasonal it might mean two workers for three months.

Assumptions of Input-Output Models

The ability of input-output models to summarize succinctly the interrelationships between industries themselves and other segments of an economy (i.e., households, government, and so on) requires several simplifying assumptions. The term *linear additive production* refers to the first of the following three assumptions of input-output models:

1. Each industry's and firm's production function is a fixed and constant relationship between inputs used and outputs produced. For example, using 10% more inputs yields exactly 10% more output. To obtain 10% more output requires that all the inputs be increased by 10% because the production recipe or mix that translates inputs to outputs is exactly the same at 10% more output and 100% more output. In the short time period of a few years this assumption is plausible. The firms and industry have a given set of machines and other capital that limits the deviation from the existing production mix. In the long run, however, this assumption rules out several common types of long-run economic behavior of firms. The fixed relationship between inputs and outputs rules out any economies (or diseconomies) of scale when producing large increases in output to meet major increases in final demand. Again, one often observes as firms expand in the long run that they do so to take advantage of economies of scale.

2. Each commodity or group of commodities is produced by a single industry in which each firm produces only that output. More recently, however, I-O model structures do not require such an assumption and allow for multiproduct industries (Alward et al. 1985). However, all models still require that each firm use the same mix of inputs to produce that output. When

product groups or industrial sectors are broadly defined, this assumption is at best approximated. The more broadly defined the industry, the more likely it is that some firms will also produce other, related goods and will use production technologies that are somewhat different from one another.

3. There are no constraints on a firm or industry in expanding output to meet increased demand; that is, there are no bottlenecks in terms of land, labor, or capital that would interfere with industry expansion (Richardson 1972:27). This assumption may sometimes be violated in the short run when firms cannot quickly locate the additional specialized labor or capital goods necessary.

Although these assumptions may not be fully met, attempts to meet them often result in far more complex and computationally burdensome regional economic models. Most federal agencies continue to rely on simple input-output models, since the cost of additional complexity outweighs the benefits of greater precision. When possible, the analyst can qualify the results of input-output model runs when communicating these results to decision makers. For example, the analyst familiar with a local economy can indicate whether bottlenecks hindering the ability of local firms to expand might result in only part of the increase in final demands being satisfied locally and the rest being imported (even though the simple model would indicate full local expansion of production).

Development of Input-Output Models: The Survey-Based Approach

The data necessary to construct a transactions table for substate areas often involve a combination of secondary data from government agencies and primary data from the industries themselves. Often both mail questionnaires and in-person surveys are used to obtain the detailed information on what inputs an industry uses, where those inputs are obtained, and the rate of output associated with them. Since total outlay must equal total output, the various sources of information must be reconciled, so that the overall transactions table is balanced.

Development of an input-output model from primary data is a substantial undertaking and requires extremely detailed knowledge of regional economics. Generally speaking, most land management agencies do not build such input-output models themselves. Instead they use existing input-output models or contract out to regional economic specialists to develop such models.

Use of Existing Input-Output Models

Many states have developed primary data input-output models; for example, Utah has developed its own input-output model for the entire state (Bradley and Gander

1968). Other states, such as Colorado, Oregon, and Washington, have developed input-output models for various substate regions. Whenever these models match the geographic scope of interest in a public land management decision, it is wise to use them.

RIMS Multipliers

A consistent set of state-level output, income, and employment multipliers is available for each state in the United States from the U.S. Department of Commerce (1986). These multipliers are known as RIMS (for regional input-output modeling system). They are based on the U.S. national averages of production technology. However, the multipliers do reflect the presence or absence of specific industrial sectors in the state economy.

The RIMS income multipliers are slightly different from those discussed earlier. They reflect the net effect of both state value added and a type I multiplier. Palmer (1976:19) calls such a number a "response coefficient" rather than a multiplier. An income multiplier converts the direct change in income to the total change in income. The response coefficient converts the direct changes in sales volume or dollar value of output to total income. Therefore the RIMS income coefficient/multiplier can be used directly against gross output or sales volume, since the state value added has already been accounted for in determining the size of the multiplier. Of course this makes the resulting RIMS multiplier smaller. For example, most of the RIMS income multipliers are between .45 and .78. Do not be fooled by this smaller number; it contains the same full indirect effects of a more standard income multiplier (Palmer 1976:20).

Often the geographic region of interest and the available input-output models do not match perfectly. For example, if one were analyzing the effect of tourism in Bryce and Zion National Parks on the southwestern Utah economy, the state of Utah input-output model would not be appropriate. This is because the state of Utah as a whole is much more self-sufficient than the remote southwestern Utah economy. Salt Lake City is a major manufacturing, wholesaling, and distribution center for the Intermountain United States; hence it has a relatively high income and employment multipliers.

This is contrasted with the small towns of Cedar City and St. George, Utah, near Bryce and Zion National Parks. These small towns import much of their retail goods, food, and so on. Thus the leakages from these economies are also quite high. Consequently, the income and employment multipliers would be much smaller than those for the state of Utah as a whole. Thus the state of Utah input-output multipliers would greatly overstate the economic effects of Bryce and Zion National Park tourism on the southwestern Utah economy. If, however, the geographic scope of interest was the state of Utah as a whole, then the state input-output model would be appropriate.

Tailoring an Input-Output Model to a Substate Local Economy: The Forest Service IMPLAN Model

The need to develop input-output models and associated multipliers that reflected the degree of economic interdependence of small remote rural economies that surround many National Forests was also apparent to the Forest Service. It (and the taxpayers) could not afford the time or money needed to develop survey-based input-output models for over one hundred local economies influenced by National Forest management.

The Forest Service (Alward and Palmer 1983), along with Lofting, developed a modular input-output model that could be adapted down to the individual county level for any county in the United States. The IMPLAN data base consists of two major parts: (1) a national-level technology matrix and (2) estimates of sectoral activity for final demand, final payments, gross output, and employment for each county. The data base represents 1977–1990 county-level economic activity for 528 sectors or industries. The national technology matrix denotes sectoral (industry) production functions and is utilized to estimate local purchases and sales. This 528-sector, gross-domestic-based model was derived from the Commerce Department's national input-output studies.

In essence, the information on the sectors that exist in particular counties is used to construct a transactions table with only the processing sectors that are present in the local economy. Everything else is assumed to be imported. The national average technology is used to develop the direct coefficients for the sectors that are present locally. It is an added assumption of IMPLAN that local firms and industries utilize the national average production technology and are usually as efficient at turning inputs into outputs as the typical U.S. firm. Comparisons with locally derived survey-based input-output models have shown that IMPLAN may overstate the local multipliers slightly (Radtke et al. 1985).

The strong feature of IMPLAN is that the analyst can essentially tailor or customize an input-output model for any county or group of counties affected by a public land management action. Thus in our example of Bryce and Zion National Parks, a southwestern Utah regional economy can be constructed by combining surrounding counties into one input-output model.

The IMPLAN data base is updated as the Commerce Department collects new data and reestimates the national input-output model.

The structure of IMPLAN is also being updated to reflect recent advances in design of I-O models. To avoid the homogeneous, single-product industry assumption of traditional square interindustry I-O models, an industry by commodity format is being adopted in IMPLAN. As described by Alward and colleagues (1985), "In this format, industry demand for commodities is given by the 'use' matrix and industry production of commodities is provided by the 'make' matrix." The make-matrix viewpoint allows for a distinction between local commodity production that is for domestic consumption versus export. Another extension is to a social accounting matrix (SAM) approach that allows for tracing incomes received not just by location of employment but also by location of residence and by use of that income in the consumption process. More important, the SAM approach allows for explicit incorporation of nonproduction flows of income, such as governmental grants-in-aid from federal to local governments. For example, the SAM approach better allows for explicit consideration of how changes in timber production affect the 25% revenue sharing to local counties. See Alward and colleagues (1992:Appendix A) for a more detailed discussion of the SAM approach.

IMPLAN currently exists on the Forest Service mainframe computer and is also available for IBM-compatible microcomputers. The University of Minnesota provides training in the use of IMPLAN to non–Forest Service personnel. At present, IMPLAN is used not only by the Forest Service but also by the BLM. Its part of the Forest Service planning process will be highlighted in chapter 9.

Conclusion

Almost every public land management agency uses input-output models or multipliers derived from them to display the effects of public land management actions on the local economy. As discussed in this chapter, models exist or can be developed to quantify the direct and indirect change in local income and employment. This information is often quite helpful to managers in informing them whether local economic effects are or are not a real concern in choosing among public land management alternatives. This information on local distributional equity helps to complement the economic efficiency analysis, which focuses on overall or national economic analysis.

In addition, we must remember that increases in income in one county are rarely net economic gains; rather, they just transfer economic activity from one region of the country to another. Consequently, local county income *cannot* be compared to the overall project costs to arrive at a meaningful measure of the economic desirability of a project. This mistake is frequently made, since the multiplier analysis results in dollars of income. It is hoped that the arguments of chapters 6 and 7 will prevent any more people from making this mistake.

The only figures that can be compared meaningfully are those for local change in income or employment versus the local share of the costs. This would yield a "local" benefit-cost analysis. But as chapter 3's discussion of externalities and chapter 6's discussion of economic efficiency reveal, this *very partial* local analysis is likely to be quite misleading. Such a local benefit-cost analysis would ignore both benefits and costs accruing to people outside the local economy. Since these are very real benefits and costs, they should not be ignored. Thus although it is sometimes useful to display local benefits and costs separate from overall benefits and costs, it is important to remember which is the more complete measure.

8

Principles and Applications of Multiple-Use Management

What Is Multiple Use?

In many ways multiple use (MU) is as much a philosophy of land management as it is an operational guide to appropriate management activities. The evolution of the concept of multiple use was discussed in chapter 2, where the history of the Forest Service and BLM was presented. The recognized multiple uses include mining, outdoor recreation, range, timber, watershed, wildlife (and fish), and wilderness. To continue that discussion in this chapter we will start by analyzing a generic version of the often quoted definition of *multiple use* from Section 4(a) of the Multiple Use Sustained Yield Act of 1960:

> (a) The management of all the various resources so that they are utilized in the combination that will best meet the needs of people; (b) making the most judicious use of the land for some or all of these resources or related services over areas large enough to provide sufficient latitude for periodic adjustments in use to conform to the changing needs and conditions; (c) that some lands will be used for less than all of the resources; (d) harmonious and coordinated management of the various resources, each with the other, without impairment of the productivity of the land; (e) with consideration given to the relative values of the various resources; (f) but not necessarily managing for that combination of uses that will give greatest dollar return or the greatest unit output.

Several principles are worth emphasizing here. The first sentence stresses using the resources in the combination that best meets the needs of people. This provides a strong direction that the objective of managing public lands for multiple use is to provide things of value to *people*. This implies a more anthropocentric philosophy of management of multiple-use lands rather than a biocentric philosophy (which might be associated more with National Parks or Wildlife Refuges).

Note the repeated reference in (b) and (c) to the fact that multiple use does not mean every use on every acre. Also note the emphasis in (b) on flexibility to adjust the mix of uses in light of changing resource conditions and society needs. This is an important element of multiple use: just because the land had been managed with one emphasis in the 1960s does not mean the agency is locked into managing that land the same way in the 1990s or beyond. Society changes; therefore multiple-use land allocations must be dynamic, not static. A multiple-use planning process must accommodate these changes, not resist them.

This definition also stresses in (d) coordinating resource uses and hence the need for integrated management of all resources rather than considering one resource at a time. That same clause also emphasizes an ecological constraint against impairing the productivity of the land. Thus the long-term effects of land management practices on the capability of the land to provide multiple-resource services must be evaluated in all decision making.

Sections (e) and (f) may appear a bit contradictory at first. Clause (e) states that the mix of output should reflect the relative values of the various resources, but (f) indicates that the mix should not necessarily be the one that maximizes the dollar return. The appearance of contradiction is quickly eliminated if one recognizes that in 1960, when this statement first appeared in the Multiple-Use Sustained Yield Act, dollar values were synonymous with financial or cash flow. The idea that resources would have an economic value that would capture the full range of society's benefits from these resources and amenity services was only emerging in the field of economics. Today one would interpret these two clauses to mean that consideration should be based on their relative value (as measured in a comprehensive economic valuation sense, including public goods), not on narrow financial grounds. This is of course consistent with the rationale for continued public ownership. That is, economic values account for externalities and public goods, whereas financial cash flow ignores these components of value. Finally, clause (f) also suggests that maximization of physical output is not the primary criterion for multiple-use management. Thus multiple use involves maximizing neither cash flow nor physical output.

Just as it is worth noting that multiple use does not mean every use on every acre, we should also point out that the opposite is not multiple use. That is, multiple use does not mean that every acre is allocated to a different single use. For example, multiple use on 100,000 acres is *not* 5,000 acres with just mineral production, 35,000 acres with just timber harvesting, 10,000 acres with only developed recreation, 20,000 acres with just livestock grazing, and 30,000 acres with just wildlife. Although over the entire 100,000 acres several multiple uses are provided, such a strict "zoning to a single or dominant use" ignores the potential for joint production relationships and resulting economies in producing several multiple uses from the same area.

Multiple Use as Packages of Compatible or Complementary Uses

Because some lands can provide several multiple uses at one time, multiple-use land management can be viewed as a mosaic of different uses. The compatible uses can be thought of as ''bundles of uses'' that make up a multiple use. Determining what uses fit together ecologically is one of the challenges in multiple-use management. Specifically, the staff and managers must determine the complementary

YEAR ZERO

YEAR 25

YEAR 50

Figure 8.1.   Multiple Use Over Time

or compatible uses that fit together and those that are noncompatible and cannot be provided simultaneously. Although antagonistic uses cannot occur at the same time on the same piece of ground, they can occur on the same piece of ground at different times.

Thus the choice set available to managers involves determining which activities are compatible on the same piece of ground at the same time and which are compatible at future times. Figure 8.1 illustrates this diagrammatically. In the upper panel, timber harvesting, cattle grazing, and four-wheel drive recreation are compatible and to a certain extent complementary (i.e., joint production relationships exist, so that timber harvesting stimulates more forage production, and logging roads provide four-wheel driving opportunities). The area is not attractive to camping. But by year 25, with the regrowth of trees, the land will be attractive for deer hunting, fishing, and hiking. In fifty years, when the trees get much larger, the area will be attractive for backpacking and primitive camping. Thus, multiple-use choices can exist over space (different land areas) and time.

In many ways the capabilities of the land and the resource requirements of each of the multiple uses will determine which specific uses fit together in a bundle that we might call a multiple use and which do not. But good multiple-use managers are alert for ways to modify the production of one multiple-use resource so as to enhance the production of another. In some areas, a particular timber-harvesting pattern may increase net water yield downstream and production of certain grasses and brushy plants beneficial to livestock and deer. These multiple uses (timber, water yield, range, and deer) can all be jointly produced because timber production creates a by-product that is an input for the production of more water and forage. Viewed differently, the *production function* for water yield and forage is influenced positively by timber production. Of course some other multiple uses, such as trout fisheries or primitive recreation (e.g., backpacking), may be related negatively to timber production. Since the concept of a production function is so important in multiple-use management, it warrants additional discussion.

Relationship of Inputs and Outputs in a Simple Production Function

It is important to define terms clearly to avoid confusion in discussing multiple-use production functions. We will define the different multiple uses as *outputs*. Thus timber and minerals are commodity outputs, water is an output, wildlife and fish are outputs, wilderness is an output, and livestock are range outputs (e.g., beef, leather, wool, and so on). Inputs for producing these multiple-use outputs include the land (which includes soils), vegetation cover, labor (skilled and unskilled workers), and capital. Capital represents various inputs, including machinery, fertilizer, materials (e.g., fencing, steel, concrete), and knowledge about how resources interact with their environment and each other. Therefore natural resource inputs found on public land, plus private- and public-sector workers and capital goods are combined to produce multiple-use outputs. The task in planning is to determine (1) which combinations of multiple uses to produce (i.e., which combination of outputs provide the most value to people); (2) what combination of inputs is most cost effective to produce that output mix; and (3) on what specific lands or parcels specific multiple uses will be produced. Clawson (1978) suggests that three types of information must be developed to answer these three questions.

Clawson's Three Elements for Multiple-Use Decisions

Clawson (1978) suggests that three types of information are required to determine optimum multiple-use management: an inventory of available and potential resources, production interaction or trade-off functions between resources, and preference ratings or valuation functions.

Inventory of Available and Potential Resources

The first requirement is an inventory of the resources (natural resources, labor, budget, and so on) that currently exist in the area and of the multiple-use outputs they produce. This starts with a listing of what exists. The next step is determination of how much of each resource is present. For example, how many acres of each type of tree, how many acres of land by soil type, what quantity of recoverable minerals, what populations of wildlife, and what water resources (e.g., size of lake) are present in this area.

This step should also include an evaluation of the land's capability to produce each multiple use. That is, if the manager wanted to devote all the available resources (land, labor, budget, machines, and so on) to production of one use, how much could be produced? A simple two-dimensional graphical representation of the points of maximum production of wilderness and timber is illustrated in figure 8.2 as points *W* and *T*. The point *W* is the most visitor days of wilderness recreation

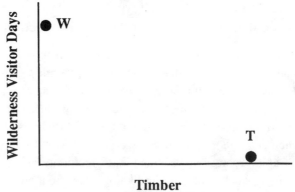

Figure 8.2.   First Step in Multiple Use: Determine Maximum Resource Potential

that could be produced with the current roadless acreage, labor, and budget (for building trails, patrolling, and so on). Point *T* is the maximum amount of timber that could be produced from these roadless areas, given the same amount of labor and budget (for laying out timber sales, surveying roads, and so on).

Production Interaction or Trade-off Functions Between Resources

Clawson's second requirement is to investigate what would happen to the output level of one multiple-use resource if an additional unit of the other multiple use was produced. This is the core of multiple-use management: determining which resources are complementary in production or which can be jointly produced (i.e., positively linked so that producing more of one makes more of another available) and determining which resources are competitive or antagonistic (i.e., linked together negatively). Alternatively, some multiple uses that do not use the same type of input resources in their respective production might have output levels totally independent of one another. Thus waterfowl production might not be related to timber production, since the basic land types used do not provide common inputs (e.g., forests versus wetlands).

The competitive production relationship between timber and wilderness (two multiple uses) that could both be produced from a roadless area containing old-growth timber is illustrated in figure 8.3. This "production possibilities frontier," or curve, illustrates several principles. First, note its general downward slope. This implies a competitive relationship between timber and wilderness outputs from this roadless area. The more wilderness visitor days one wants to produce, the less timber is produced. The concave shape means there is not a constant trade-off between board feet of timber and wilderness visitor days. For example, some of the roadless area may be heavily stocked with timber and may not be well suited to recreation (trees are too dense and too damp). At the other extreme, some of the

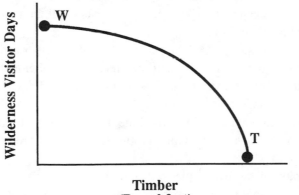

Figure 8.3. Second Step in Multiple Use: Biophysical Production Trade-offs Between Resources

land may be made up of numerous relatively isolated "flat-top peaks," which provide considerable opportunity for solitude but make road access for timber harvesting difficult. Thus as one approaches the horizontal intercept, adding the last few acres of flat-top mountains to timber harvesting adds few board feet but results in a large reduction in wilderness visitor days. If there was a constant transformation of roadless acres into either wilderness visitor days or board feet of timber, the production possibilities curve would be a straight line between points *W* and *T*.

In essence, Clawson's first two elements for multiple use inform us as to what is biologically and administratively feasible, since a given production possibilities curve is drawn holding current land, labor, and capital constant. But these first two factors do not tell us which of the many possible technically efficient points on the curve has the greatest economic and social value. For that we must consider the third requirement.

Preference Ratings or Valuation Function

A preference function portrays society's rating of the desirability of different combinations of wilderness visitor days and board feet of timber. That is, if people could barter or trade visitor days of wilderness recreation for board feet of timber like we implicitly trade one good for another in the market, what would that valuation rate be? Would we trade one visitor day for 1 board foot of timber? Or would we trade two visitor days for 1 board foot of timber?

How would we identify these relative preferences in the real world? It could be done by majority rule at public meetings in which people state the combination they would like. Or it could be done by counting the number of letters received in favor of each alternative during a public comment period. However, in both of these approaches many people who use the outputs might not attend the meetings or write letters. Individually, they just do not have a large enough gain or loss. Only loggers, commercial backcountry guides, and sport anglers might show up, and they might not adequately represent the interests of all the people affected by these resource allocation decisions.

Another way to ensure that all the people who value lumber and fish are heard, and in proportion to the intensity of their preferences for timber and fish, would be to use economic values of the two outputs. In one sense, all consumers of timber and fish vote every time they decide whether to buy the good, and if they decide to buy, how much to purchase. Comparing the market price of timber to the value of fish will often give us a more unbiased estimate of society's relative preferences for the two goods.

Thus the ratio of the two values will tell us their preference rating or value function. This function for wilderness recreation and timber is illustrated by a series of parallel lines in figure 8.4. The slope of the valuation function is given by the

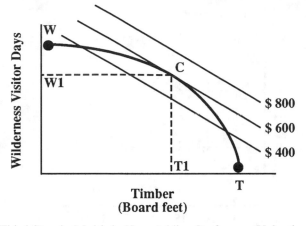

Figure 8.4. Third Step in Multiple Use: Adding Preference-Valuation Trade-offs to Determine Optimum Mix.

ratio of the value per visitor day to the value per board foot of timber. Valuation functions further to the right represent a higher total value and are preferred to those to the left (i.e., closer to zero, or the origin). But the combination of output levels in the valuation function selected cannot exceed the output levels the resource can produce. In essence, the task is to find the highest valuation attainable with the available resources.

The highest valuation attainable occurs where the highest- valuation function is tangent to the production possibilities curve. This indicates the most valuable combination of uses of this roadless area. The optimal combination is *C,* which provides W_1 visitor days of wilderness recreation and T_1 board feet of timber. This most valuable bundle occurs at the tangency because that is where no gains remain from shifting land from production of wilderness to production of timber. At the tangency the rate of transformation in production equals people's preference rating. More specifically, the rate of change of the productivity of the land in producing visitor days and timber just equals the rate of change in value for trading visitor days and timber. As will be demonstrated in the following numerical example, this point has the highest total value to society.

A Simplified Example of Multiple-Use Management Using a Production Possibilities Curve

The following example, although simplified, parallels analysis that can be found in the Loomis (1988) analysis of the Draft Siuslaw National Forest Plan and in the Loomis (1989) analysis of the Gallatin National Forest EIS on Wilderness designation of the Hyalite–Porcupine–Buffalo Head Wilderness Study Area.

Figure 8.5 illustrates a production possibilities curve calculated from six points,

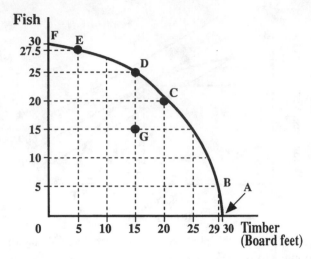

Figure 8.5.   Biophysical Production Trade-Off Between Fish and Timber

A, B, C, D, E, and *F.* These points represent alternative combinations of fish (salmon on the Siuslaw and trout on the Gallatin) and timber (thousands of board feet, or mbf) that can be produced from a fixed-size forested area that has no roads (with a given stream density per square mile of land), given the currently available amount of biologist and forester labor hours, heavy equipment time, budget, knowledge of fish-timber production and technology. Points *A* and *F* represent Clawson's first element of inventory data. Point *A* is the fisheries biologist's estimate from angler harvest surveys and habitat models of what current annual fish harvests are now in the roadless condition. Point *F* is the timber specialist's best estimate of the annual board feet of wood recoverable from the roadless area if it were logged with conventional methods.

Points *A–F* trace out the maximum combinations of fish and timber possible from these amounts of land, labor, and capital; that is, they represent technically efficient land management practices that provide the greatest possible bundle of outputs, given the available inputs. Points to the right of *A–F* (containing higher levels of fish and timber) are not attainable with the available land, labor, and capital. Point *G* lies to the inside of the production possibilities curve and would be a technically inefficient combination of land, labor, and capital. If one kept the same amount of timber production at point *G,* one could have ten more fish at the more efficient point *D.* If one wanted more timber and more fish, point *C* also would be possible. Point *G* may well represent underutilization of some inputs or inappropriate utilization (i.e., forcing marginal timberland into production) and will not be considered further. In the terminology of chapter 4's matrix approach, point *G* is dominated by points *C* and *D.*

This curve also represents Clawson's second element: knowing how production of different resources affects the other resources. The field studies and models of both the Siuslaw National Forest and the Gallatin National Forest show a competitive relationship between fish and timber.

The production transformation between fish and timber can be approximated by calculating the slope of the curve between the two points. Thus the segment of the curve from E to D has an approximate slope of one-fourth. This means in production that we can transform our land, labor, and capital into one more fish by giving up four more board feet of timber $(27.5 - 25)/(15 - 5) = .25$.

Role of the Preference or Valuation Function in Determining the Optimum

Turning to Clawson's third element, we need to know which of the technically efficient bundles $A–F$ society prefers. If people prefer one fish to 1 board foot of timber, then point C is the optimal point. Point E would not be the optimal point at this rate. That is, at point E it is possible to produce one less fish and get 4 board feet of timber. If we value one fish and 1 board foot equally, we would be much better off to shift land, labor, and capital to producing more timber. Every fish we give up results in a profit of 3 board feet of timber. We would keep shifting inputs until we reach point C, where giving up one more fish results in gaining just one more board foot of timber.

To start the example, assume an era when fish were plentiful in other rivers and streams of the Pacific Northwest. Because fish were plentiful, their relative price or value at the margin for a few additional fish was low, say a dockside price of $40 per thousand fish. At the same time there was a residential construction boom, requiring much lumber. Thus, at the margin, additional board feet of timber that could be turned into lumber was quite valuable. These forces might translate into a price per 1,000 board feet (mbf) of $200.

The resulting preference rating or valuation ratio would be $200 for timber to $40 for fish. People value an additional 1,000 board feet of timber five times as much as an additional thousand fish. Given this valuation function, the best combination in figure 8.6 would be point B. This is where the slope of the valuation function equals the slope of the production possibilities curve. To obtain a better understanding of why this point is the optimum, let us calculate benefits at point B versus those at point C. This is done in table 8.1.

Point B produces a bundle worth $6,000, whereas point C produces a bundle worth just $4,800. In essence, by shifting land, labor, and capital from point C to point B, the gain in timber (9 mbf $\times$ $200, or $1,800) more than offsets the loss of forgone fish value ($-15,000$ fish $\times$ $40 per thousand or $-$600$). Thus economic efficiency is improved by reallocating inputs from relatively low-valued (at these relative prices) production of fish to more highly valued production of timber.

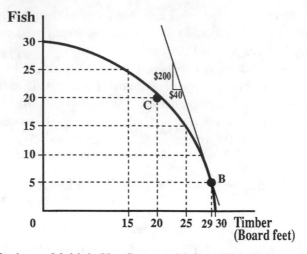

Figure 8.6. Optimum Multiple Use Output with Low Fish Values

Table 8.1
Calculating the Value of Different Resource Allocations in Decade 1

Point	Output Level	Prices	Total Value
B	29 mbf	$200	$5,800
	5 fish	40	200
	Total value of point B		6,000
C	20 mbf	200	4,000
	20 fish	40	800
	Total value of point C		4,800

Computing the Most Beneficial Output Mix in the Second Decade

Now, twenty years later, one might be ready to reevaluate the allocation of land, labor, and capital between fish and timber. Although some will argue that this allocation should stay because it has "always been that way," the principles of multiple use suggest periodic reevaluation is worthwhile if our objective is to serve people best. If people's preferences or relative values have changed, then a new multiple-use mix may be in order.

Consider the case where the source of alternative fish supplies has dropped and the demand for fish has risen because of health concerns about beef and changes in consumer tastes. Now suppose that the price of fish has risen (in real inflation-adjusted dollars) to $200 (per thousand) and that timber prices have stayed the same in real terms at $200. These consumers now equally value one more fish and one more board foot of timber. A glance at figure 8.7 suggests that point C would now be the optimum point, since there the slope of the production possibilities curve and the valuation function are equal. Table 8.2 illustrates the calculations to prove that point *C* is the optimum point.

As shown in table 8.2, point *C* is a more socially valuable mix of output than point *B*. That is, the additional fifteen thousand fish at $200 per thousand fish ($3,000) adds more to total benefits than the loss of 9 mbf of timber at $200 mbf ($-$1,800) takes away. Thus the total goes up by $1,200. Therefore point *C* is the new optimum.

Of course there will be some temporary costs of adjustment to the economy in transitioning the use of the public land from less timber to more fish. There will be a reduction in derived demand for loggers in this tract of land and a corresponding increase in the number of commercial anglers and fishing guides for the sport anglers. Unfortunately, since different people have different skills, the loggers may be unemployed until they find other logging work elsewhere or retrain themselves to the skills that have increased in demand as a result of the shift in inputs to fish.

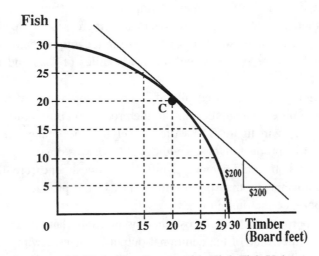

Figure 8.7.   Optimum Multiple Use Output with High Fish Values

Table 8.2
*Calculating the Value of Different Resource
Allocations in Decade 2*

Point	Output Level	Prices	Total Value
B	29 mbf	$200	$5,800
	5 fish	200	1,000
	Total value of point *B*		6,800
C	20 mbf	200	4,000
	20 fish	200	4,000
	Total value of point *C*		8,000

Observing how changes in this valuation function translate to new optimum points on the production possibilities curve illustrates how even this simple model fulfills several of the principles of multiple use identified in the introductory section of this chapter. Land, labor, and capital should be shifted as relative prices change reflecting changes in society's relative value of the two multiple uses. This agrees with two important principles of multiple use: (1) output should be produced in accordance with relative values to people, and (2) flexibility in responding to changing needs of society is necessary.

Shifts in the Production Possibilities Curve

Society would often like to have more timber and fish. Scarcity of inputs such as land, labor, and capital limit our ability to produce more of both. But society can devote more labor to provide more intense management of both the fishery and the timber stand. We could provide more capital in the form of timber stand improvements (e.g., fertilization) or fish habitat improvements (removing barriers to upstream migration, water quality improvements, increases in riparian vegetation). Technology associated with timber-harvesting techniques could change that might allow timber to be removed with far fewer miles of road and far less soil disturbance.

If the technological changes equally affected the production of both fish and timber, we would observe a parallel shift in the production possibilities curve outward to the right, as in figure 8.8a. Alternatively, if the change in technology affected only timber production or the additional labor was devoted only to timber production, figure 8.8b would depict the shift. If the investment were made only in fish habitat improvements, then the resulting shift in the production possibilities curve would look like that in figure 8.8c.

One of the purposes of benefit-cost analysis in public land management is to evaluate whether the value of the additional output associated with the expanded production possibilities curve is worth the investment cost necessary to bring it about. For example, if additional labor and capital to shift the production possibilities curve outward cost $1,000 and the benefit of the additional output is worth $2,000, then such a reallocation of scarce labor and capital from other sectors is worthwhile.

Using Linear Programming to Make the Production Possibilities Curve Operational and Add Other Criteria of Public Land Management

In the simple examples cited earlier we have explicitly dealt with two of the five criteria of public lands management reviewed in chapter 4: biological feasibility and economic efficiency. Administrative feasibility was dealt with implicitly, since

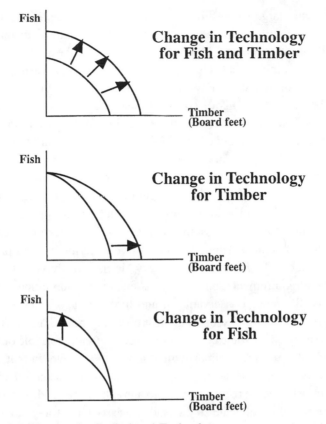

Figure 8.8a–c.   Changes in Capital and Technology

labor and budget were held constant in drawing the location of the production possibilities curve. But we have not explicitly dealt with other factors, such as cultural acceptability or distributional equity.

Linear programming is a mathematical optimization technique that allows us to make the production possibilities approach operational as well as to incorporate these other criteria into our choice of the optimal mix of multiple uses. We will start with a very simple example.

A Simple Livestock-Elk Linear Programming Example

As the name implies, *linear* programming (LP) represents all equations or functions as straight lines. There can be no curvilinear relationships in LP models. Although this is a simplification, it is often not too serious. Curvilinear relationships can be approximated by piece-wise linear segments, if necessary. If curvilinear relationships are critical, then nonlinear programming can be used. However, LP models are much easier to work with when doing real public land management problems, which involve very large models. In addition the basic principles of LP models are applicable to most optimization models. As a result we shall focus on LPs.

The first step in building LP models is similar to the first consideration in multiple use: determining what is biologically and physically feasible. Specifically, given the land, labor, and capital available, we must find how much of the various outputs can be produced. This of course is just like the production possibilities curve in figures 8.2 and 8.3. But now we are going to construct both a graphical and mathematical representation. In real LP models that are not two-dimensional, the computer finds the optimum mathematically.

In this example, a resource manager must decide how many of a rancher's cattle she will allow on a 300-acre area of public land. The manager consults with a range conservationist and wildlife biologist to determine the stocking level for cattle and how this affects elk. Assume the biologist has determined from research studies that elk and cattle are directly competitive in terms of food preference for grasses. Furthermore, both the range conservationist and biologist agree that in terms of dietary requirements, one cow consumes the same amount of forage as one elk. Lastly, the range conservationist and biologist agree that forage productivity is such that the area can support at most thirty animals. This means that either thirty cows or thirty elk or some combination is ecologically feasible on a sustained-yield basis. More than thirty animals will "mine" the forage, in that too many of the plants will be consumed, leaving the plants stressed and on a declining trend.

One can translate these requirements into a graph of the production possibilities curve in figure 8.9. If we allocate all the land and associated forage to elk, we could have at most thirty elk. So thirty elk is the vertical intercept (point *A*) of the production possibilities curve. If we allocate all the land and forage to cattle, we could have at most thirty cows. So thirty cows is the horizontal intercept of the production possibilities curve (point *D*). This corresponds to Clawson's first element in mul-

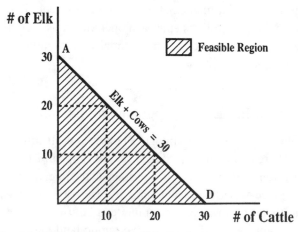

Figure 8.9.   Graphical Example of Feasible Region

tiple-use planning, a determination of what is possible on an individual resource basis.

Representing the Production Trade-off Relationship in an LP Model

The next element in Clawson's multiple-use strategy is to determine the production trade-offs. Since elk and cows consume exactly the same amount of forage, they are competitive in production. Producing more cattle means fewer elk. However, we know from the biologist and range conservationist that we can produce any combination as long as it does not exceed thirty animals. If we are interested in having as many cows and elk as possible, we want to find the combinations that exactly equal thirty. A little trial and error will show that twenty elk and ten cows and vice versa will satisfy this requirement. Connecting these points yields the production possibilities curve for all possible combinations of elk and cattle equal to thirty.

In a real LP model it is easier to represent these biological facts in an equation. The preceding description states that one cow and one elk consume the same amount of forage and that we can have no more (but we can have less) than thirty animals total. The production equation or *constraint* (i.e., production limit) that describes the line in figure 8.9 is represented by equation (8.1):

$$1 \times \text{Cow} + 1 \times \text{Elk} \leq 30 \tag{8.1}$$

Equation 8.1 states that we can trade in production one cow for one elk but that we can have no more than thirty animals total. Of course we can have less. So the striped area underneath is called the *feasible production region.* Any combination of cattle and elk in the striped area is feasible. That is, five cows and ten elk are feasible, but if society prefers more cows and elk to less, fifteen cows and fifteen elk would be a preferred bundle.

Representing the Valuation Function in the LP Model

This brings us to the third element in Clawson's multiple-use strategy: a preference or value function to determine which of the technically feasible points on the production possibilities curve is the socially most valuable. In an LP model this preference or value function is represented by an *objective function,* which depicts how the possible outputs (here elk and cattle) contribute to attaining a particular objective.

An objective could be to maximize the number of animals, or the pounds of meat, or the value to society. If we are interested in maximizing the value to society, it is likely that society receives different benefits from one cow than from one elk. We can rely on market prices for beef to derive a value for the cow. If the primary benefits of elk are for hunting and viewing, we can use the travel cost method to derive a recreation value of elk for viewing and hunting. The explicit objective is thus to maximize the sum of the beef and elk values from this land.

Suppose each cow is worth $60 to society (i.e., consumers are willing to pay $60 for a cow, for its beef and leather value). Assume that this is the early 1900s, when elk also had value primarily for its meat, but that on a per-animal basis there was less recoverable meat in an elk and people preferred the taste of beef. Consequently, an elk might be worth $30. (We shall change this to a recreational hunting and viewing value later in this example.) These values indicate that our objective of economic value goes up by $60 for each cow we produce and by $30 for each elk we produce. The objective function can be expressed as an equation of the following form:

$$\$60\text{Cow} + \$30\text{Elk} \tag{8.2}$$

This equation also describes the series of steep lines in figure 8.10. These lines are all parallel to each other because all have the same slope of two elk to one cow (i.e., relative prices of cows to elk implies we would trade two elk for one cow and not change the value of the objective function, or social value). The lines to the right have a higher value of the objective function (greater economic value to meat consumers) than the lines to the left.

Solving the LP Model for the Maximum Benefits

The overall purpose of this exercise is to use an LP framework to determine the amount of cattle to permit on this area of public land (and therefore, implicitly, the number of elk). Since the objective is to maximize economic value, we want to maximize the objective function's value subject to the production constraint between cattle and elk. Graphically, we want to select the highest objective function line as long as it does not exceed or lie to the right of the feasible production line.

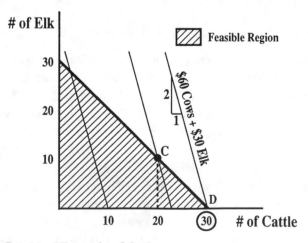

Figure 8.10.   Graphical Example of Optimum

In figure 8.10 the maximum value of the objective function occurs at point *D*, a point known as a corner solution. That all of the land is allocated to cattle in this example is not too surprising, because if cattle and elk have the same opportunity cost in terms of consuming scarce forage but cows provide twice as much benefit to meat consumers as elk, we can get the greatest value by allocating all the forage to cows.

That point *D* is in fact the maximum value of the objective function as compared to point *C* can be easily checked using the objective function. At point *D* the value of the objective function is $1,800 ($60 per cow × 30 cows + $30 per elk × 0 elk). At point *C* the value is $1,500 ($60 per cow × 20 cows + $30 per elk × 10 elk). The difference in value is a direct result of giving up $30 per animal when we shift forage from ten cows to ten elk (which are worth $30 less than cows).

Adding Other Public Lands Management Criteria via Constraints

Now let us look at a strong feature of LP models, the ability to incorporate the other criteria of public land management besides just economic efficiency and biological feasibility. Suppose that it is not socially or culturally acceptable to eliminate all elk and just have cattle on this area of public land. Alternatively, there may be a question of distributional equity: although the rancher would like to have the maximum number of cattle, this area is the only one for which a hunting guide has a permit to take nonresident hunters elk hunting. Also, the Department of Fish and Game might indicate that ten elk constitute the smallest number that would ensure, with high certainty, a viable population that would support the minimum level of hunting necessary for the hunting guide's permit.

How do we represent these other considerations in the LP model? This is done by way of imposing an additional constraint on the output combinations from which the computer model will be allowed to choose. In particular, if the consensus is that at least ten elk are needed to meet both cultural acceptability and minimum viable elk population concerns and to avoid putting the hunting guide out of business, then we will rule out any combinations of elk and cows that do not include at least ten elk.

This is illustrated in figure 8.11 by the addition of a line at elk = 10. The new feasible production region is the striped area in figure 8.11. Now any feasible solution must meet this new elk constraint. That is, any optimal solution must meet or exceed the "output target" for elk as well as meet the original production constraint (i.e., what is biologically feasible). This illustrates an important point about LP models: meeting the constraints comes first! Only after the solution space has been redefined by the constraints do the values in the objective function influence which of the newly defined combinations is the optimum. If numerous constraints are added that substantially shrink the solution space, the LP model is almost forced to find the most efficient way to meet the constraints and little else.

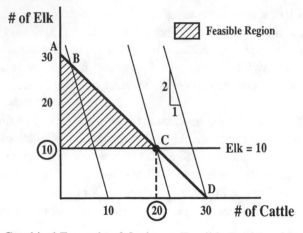

Figure 8.11. Graphical Example of Optimum Feasible Region with Added Constraint

In this example, the new elk constraint precludes us from choosing the most economically valuable combination (our original optimum) of thirty cows. That is, using the same objective function as before ($60cow + $30elk), we are precluded by this new constraint on minimum elk numbers from going to the old optimum. (This means that elk > 10 is a *binding constraint;* it changes the optimum solution.) The maximum value of the objective function (given the new constraints) is found at point *C* in figure 8.11.

The value of the objective function at point *C* is $1,500, which is higher than any other point along the production possibilities curve within the feasible region. That is, point *C* has a higher economic value than points *B* or *A*.

The Opportunity Costs of Constraints.   But notice what happened to the value of the objective function when we imposed this binding constraint. The value of the objective function fell from $1,800 in the original minimally constrained case to $1,500 when we include the elk output target. The difference in the value of the objective function when one binding constraint is tightened is a *shadow price* of that constraint. In this case, consideration of cultural values of elk and distributional equity to the hunting guide had an opportunity cost of $300, or $30 of forgone livestock benefits per elk. By itself we cannot say this is good or bad. But we can use this information to ascertain the importance of these other management criteria in the constraint: are they worth $300 in this case? For example, is the sum of the profits made by the hunting guide on his guided elk trips in this area plus the benefits to the hunter over and above their expenses (amount paid to the guide plus other trip costs) less than $300? If the profits plus hunting benefits are less than $300 (and assuming no positive externalities associated with the hunting trip), then it might be cheaper for society to compensate the guide and the hunter not to hunt here so that the forage could be allocated to cattle for more valued beef production.

However, the public land manager might conclude that it is so socially important to have ten elk in this area that it is still worth maintaining them, even if the sum of guide profits and hunter benefits is less than $300. That is fine. The key is that the results of the LP model forced the manager to recognize the opportunity cost of giving more emphasis to these other factors.

The other utility of looking at LP-generated shadow prices is that there may be situations where setting the constraint at ten elk is quite costly as compared with eight elk. For example, if the land is heterogeneous and there are other multiple-use constraints as well, the opportunity costs for having ten elk might be $300 but those for having eight elk might be only $100. That is, the last two elk produced result in a very large opportunity cost of $200 ($100 apiece), but the first eight elk have very low opportunity costs. If the Department of Fish and Game is willing to have less certainty about maintaining a viable population (i.e., is willing to accept a little more risk of not having a viable elk population) and the quality of elk hunting does not fall more than this decrease in opportunity costs, then overall social well-being (meat consumers, hunters, guides, and so on) might be improved by setting the elk constraint at eight elk rather than ten. Once again, part of the advantage of this LP framework is that it makes it possible for the resource specialists and managers to think through their perceived benefits relative to costs of what we might term "preference constraints" (i.e., constraints that are targets or limits voluntarily chosen by a manager or reflecting social/cultural concerns).

An Example of When the Targets/Constraints Dictate the Solution

What if the Department of Fish and Game and others request that minimum elk numbers be set at fifteen elk? What if the rancher says that the fewest cows that would be financially viable on that parcel of public land is fifteen. Figure 8.12 represents the addition of these two constraints to the basic feasible production curve. Note the feasible production region is now one point (*E*). The output targets have dictated the solution! It does not matter what the objective function looks like. If there is to be a feasible solution, the objective function must just pass through one point. The objective function could place zero values on elk (and hence be a

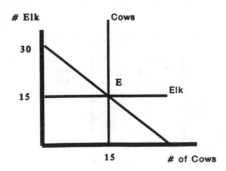

Figure 8.12.   Constraints Dictate Solution

series of vertical lines), place zero values on cows (and thus be a series of horizontal lines), or be any other combination of values of elk and cows. The point is that great care must be taken in setting preference constraints or output targets, since these can override everything else. The only thing the model might get to choose is what set of management activities to select to obtain the prespecified output levels.

The advantage of LP models in these cases is that preferences of managers and the way they have interpreted the other criteria of public land management are explicitly stated in these constraints. As the saying goes, "You can run, but you cannot hide." Preference constraints reflecting great emphasis on one factor, such as local distributional equity (e.g., local industry) or a bias toward one particular multiple use, can be "discovered" by careful examination of the components of an LP model.

This example also illustrates a pattern of optimal LP solutions. Note that in figures 8.10, 8.11, and 8.12, all the solutions occurred at corners. The computer program in an LP uses what is called the *simplex* method to compute the value of the objective function at all feasible corners. The optimal solution is the feasible corner yielding the highest value of the objective function. The proof that possible optimal solutions lie at the corners is contained in advanced texts on linear programming (Danzig 1963).

Checking to See If There Is a Feasible Solution

The other advantage of the LP format is that it will indicate when no solution is feasible. Constraints chosen by the manager, such as minimum output targets, may not be feasible. For example, if one held a public meeting and asked ranchers their desired number of cattle, they might indicate twenty cows. At the same public meeting, hunters and hunting guides might state that their desired number of elk is twenty elk. If we put these preference constraints or targets into figure 8.13, we can see that there is no feasible solution. That is, there is no solution space or point that can simultaneously meet cows = 20, elk = 20, and the production constraint of the land that 1 cow + 1 elk < 30. Those three equations do not share a common solution. Maintaining twenty elk will solve two out of the three equations and put

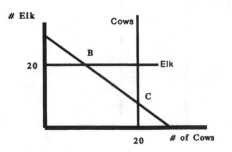

Figure 8.13.   No Feasible Solution

us at point *B,* but the cow constraint (cow > 20) will be violated, since only ten cows can be produced when there are twenty elk.

Thus the LP model provides feedback to both the manager and the public that the land will not produce as much as people desire. Something must give. In determining what it will be, the LP model can produce different values of the objective function with different targets, and the shadow prices of different targets can be computed. This will aid the decision maker directly and may help in communicating to the public why a particular solution appears optimal. The importance of this type of analysis is discussed in the last section, entitled "Another View: The Role of Analysis in Negotiation."

Continuing the LP Example with an Updated Objective Function for Elk and Cows

To illustrate how the LP's selection of an optimal mix of output changes when the relative values of output change, we will reconsider our model in figure 8.10. The basic production possibilities curve and the elk preference constraint or target of ten elk are reproduced in figure 8.14. But now the slope of the objective function is different. In particular, we have updated our economic value of elk to recognize that since the 1970s people value elk not primarily for the meat but for recreational hunting, viewing, and photography. Suppose the economist, using techniques discussed in chapter 6, determines that visitors would pay, at the margin, $120 to have one more elk to view, photograph, and hunt in this area. The objective function would be as follows:

$$\$60\text{cows} + \$120\text{elk} \tag{8.3}$$

This new series of objective functions is drawn in figure 8.14. What is the optimal

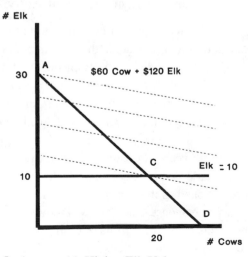

Figure 8.14.   New Optimum with Higher Elk Value

Table 8.3
*Recalculation of Total Value with New
Objective Function*

Point	Output Level	Prices	Total Value
A	30 elk	$120	$3,600
	0 cows	60	0
	Total value of point A		3,600
C	10 elk	$120	1,200
	20 cows	60	1,200
	Total value of point C		2,400

mix of cattle and elk now? It is point *A*. Table 8.3 illustrates the calculations. As shown in this table, point *A* has the highest value of the objective function (i.e., the greatest economic value). Given the new values for elk, which are twice the value of a cow, this makes sense. Once again, if it takes just as much of the scarce inputs (land, labor, and capital) to produce an elk as a cow but you get twice as much for an elk, the optimal solution is to produce elk.

Given the new value of the objective function, pictured in figure 8.14, the elk preference constraint or target of elk = 10 is no longer binding. That is, it is superfluous and has no influence on the optimal solution. In some ways, the original elk target was added to the LP model in figure 8.11 in an attempt to reflect the fact that elk had a value greater than that for meat. By incorporating the recreational hunting, viewing, and photography value of elk into the objective function, these concerns may be better reflected than through setting an absolute constraint or target.

Using LP Models to Decide What Management Activities to Select

Another strong suit of the LP model lies in its ability to help decide what combinations of management actions to use to attain the objective. For example, we might be interested in increasing the carrying capacity of our public land for cattle and elk so that we could have fifty elk, fifty cattle, or some combination (e.g., 25, 25). This would shift out the production possibilities curve and increase the feasible production region. However, several different management actions might be used. The range conservationist might suggest prescribed burning every third year or reseeding the land every fifth year. Alternatively, a stock pond might be developed to provide water so that cattle grazing is more uniform over the entire area.

Each one of these actions has its own production function, resulting in a different amount of vegetation response. Each one will require a different mix of labor and capital over time. Thus each one will have a different cost. Which treatment or combination of treatments would be the best? That depends partly on the objective function and the other constraints. Reseeding might produce the most forage, but it could also be the most costly. Prescribed burning might have the least vegetation response but could also be the most inexpensive. If the objective function is to maximize animals (cattle and elk) or maximize grass production (or minimize erosion), then reseeding might be most preferred. If economic efficiency is the objective, then the program will search over the alternative management actions to see which one gives the greatest grass production per unit of cost (i.e., the ''biggest bang for the buck'').

This ability of an LP program to compare rapidly the merits of alternative management activities to maximize the objective function and attain the constraints is a significant advantage of LP models. When there are few constraints and few alternative management actions to evaluate, a simple benefit-cost analysis performed on a microcomputer spreadsheet will likely identify the optimum alternative. But where a dozen different management actions could be undertaken in some combination to meet an objective subject to hundreds of constraints, the LP model is a useful way to solve the problem. Thus the optimal solution in an LP model will indicate not only what combinations of outputs maximize the objective function, but also what management actions on each type of land are best to produce those combinations of output.

A Few Notes on Assumptions of LP Models

Like all the modeling approaches discussed in chapter 5, LP models involve assumptions to keep the model tractable. The assumptions in LP models are as follows:

1. *The objective function and all constraints must be linear or can be approximated by a linear function.* Thus equation (8.4) is acceptable as an objective function or constraint, but equations (8.5) and (8.6) are *not* acceptable in an LP model. The following is an acceptable linear form:

$$10X + 5Y \leq 100 \qquad (8.4)$$

Unacceptable functional forms are the following:

$$10X^2 + 2Y^3 < 100 \qquad (8.5)$$

$$\log X + \log Y < 100 \qquad (8.6)$$

This linearity assumption may be inconsistent with principles of both economic theory and ecology. In economics, a linear production function means there is no diminishing marginal return or diminishing marginal productivity. Most economists believe that many production processes are governed by diminishing returns at some point. Moreover, ecologists often believe that natural systems are not linear in response to management actions. Nonlinear programs are available if departures from linearity are likely to be serious.

2. *The total, or aggregate, value is simply the sum of the individual components; this is called additivity in production.* No interactive or synergistic effects are explicitly allowed. Thus one could *not* have the following equation:

$$10X + 5Y + 2XY \qquad (8.7)$$

To reflect any interactions that might exist they would have to be built into the model as a bundle or fixed package of attributes.

3. *All variables are continuous and hence perfectly divisible.* This means that all numerical values (including fractional ones) are possible for inputs and outputs. Thus an optimal LP solution might involve allocating 55.7 acres or producing 34.5 visitor days. This is not a problem for these types of inputs or outputs, because fractions are possible. It is somewhat troublesome for such resources as elk or campgrounds, which come as integers (i.e., nonfractional) values. It is even more troublesome for ecosystem units such as roadless acres, which, if they are to have ecosystem boundaries, come in fixed sizes that cannot be continuously divided up. A Wilderness area or deer winter range needs to be of some minimum size that varies with topography or particular location. This problem varies from being a nuisance to being a serious consideration, depending on the resource issues. The availability of more advanced integer programming packages allows the analyst to avoid this assumption if many of the critical variables are integer in nature.

4. *Certainty in all coefficients is assumed.* The LP model treats the values per unit of output in the objective function or coefficients as deterministic constants. For example, technical coefficients that relate vegetation response to livestock grazing or soil erosion to timber harvests are all treated as if they were known with certainty. The LP model has no way of knowing which ones are known with more certainty than others.

At times these assumptions can be quite restrictive. Therefore public land problems may often be handled best with benefit-cost analysis (BCA), as described in chapter 6. Because BCA does not require the assumptions of LP models, it is more flexible in dealing with cases where values are not strictly additive or where variables of interest are not perfectly divisible. Uncertainty is also somewhat easier to carry through a BCA than in an LP model.

Of course, there are sophisticated mathematical optimization programs that will overcome several of the assumptions required in LP models. They are, however, much more complex, require more information, and are more difficult for computers to solve when large-scale problems are involved. The expertise necessary to assemble, run, and interpret such models is also correspondingly greater. Nonetheless there may be times when such mathematical optimization techniques are warranted.

See Danzig (1963) or Dykstra (1984) for a discussion of these more sophisticated approaches, such as nonlinear optimization, integer programming, and stochastic programming.

LP Model Tables: A Preview of Things to Come

When there are numerous multiple-use outputs (e.g., ten different commercially valuable tree species, six types of recreation, twelve types of wildlife, four different grazing seasons, and so on), coupled with land, labor, budget, and sustained-yield constraints, a graphical exposition of the LP solution process is inconceivable. Often the LP is used to determine the entire time path of use, as in figure 8.1: which multiple uses should occur on which lands and when. Adding a time element for scheduling of timber harvests makes LP models even more complex.

In real LP models all the equations representing production relationships, constraints, and objective function are represented in a matrix or table. We shall examine the structure of more realistic LP models in the next chapter when we investigate how the Forest Service implements forest planning under NFMA, because the agency relies heavily on LP models.

The Importance of Joint Production in Multiple-Use Management

The production possibilities curves shown so far have been relatively simple. We now want to discuss the concept of joint production. Joint production between two or more outputs occurs when an input is shared or when the output of one affects the cost of producing the other. For example, timber and increased water yield are joint products because removal of timber from the land increases net water runoff into streams. Another example might be timber removal and the stimulation of forage due to more sunlight reaching the ground. In both of these cases one action, timber harvesting, results in a production of two desirable outputs: wood and water in the first and wood and forage in the second. But these interactions in production are not always positive. In steep slope areas with unstable soils, removal of timber increases sediment yield, reducing the number of trout in the adjacent stream.

Because of the inherent interrelatedness of most natural systems, joint production of desirable and undesirable outputs is common. In evaluating the optimality of a particular management action, the entire bundle of positive and negative results of the action must be compared. This can easily be done in benefit-cost analysis and can be incorporated into LP models. Joint production can be viewed from either the production side (e.g., inputs to outputs) or the cost (costs to output) side (Bowes and Krutilla, 1989). Synergistic effects allow one to produce two goods jointly for less than the cost of producing each one individually. Thus a positive joint production would allow for a reduction in costs of producing the associated output. That is, water yield could be produced more cheaply by modifying the timber-cutting practices slightly in a given watershed, rather than performing a separate watershed treatment practice on a different block of land in the watershed. The reason for the lower cost relates to shared inputs: access roads and loggers to remove trees. Rather than build two road systems to different blocks of land in the watershed, if the timber-cutting pattern can be modified to enhance water runoff, then two road systems are not needed. As long as the gain in value of the additional water exceeds any loss from modifying the timber practice, this is a sensible modification.

But sometimes one management action raises the costs of production of another multiple use. Unrestricted grazing in riparian areas raises the cost of producing a given amount of fish. Cattle may consume the riparian vegetation that would normally stabilize the stream bank and shade the water. Therefore more effort will be required to produce a given amount of fish in this stream. These costly extra efforts will have to be taken to stabilize the river banks so that they do not slough off and deposit sediment into the stream.

Evaluating outputs individually when production is really joint can result in substantial errors. If there is a positive joint effect, then some (but not all!) individual activities that do not make economic sense might do so when the other benefits are included. Alternatively, if the joint effect is negative, then some of the individual activities that look economically efficient may not be when the associated negative effects on other multiple-use resources are included.

Although the need to account for both the joint benefits and costs may seem obvious, much debate ensued over the Forest Service timber sale program as a result of Barlow and colleagues' (1980) discovery that for many sales the costs exceeded the revenues. The sales were deficit sales or below cost, because the cost to the government for offering the timber (sale preparation, roads, sale administration, and so on) exceeded the revenue from the sale. The Forest Service was quick to point out that timber sales provided many beneficial joint multiple-use outputs, such as roaded recreation, better rancher access to livestock, and so on. Thus losing money on timber is not inefficient as long as the other benefits are large enough to offset the costs. This leads to the recommendation that the only valid way to evaluate the efficiency of a management action involving joint benefits and costs is to compare total benefits from all outputs affected by the management action (positive and negative) with the total cost of the management action. We shall provide an example of this type of benefit-cost analysis at the end of chapter 9.

Another View: The Role of Analysis in Negotiations over Multiple-Use and Natural Resource Management

Recently, a new approach—negotiation—has been tried to determine the optimal mix of multiple uses to produce on both federal and state lands. The objective is to bring all the competing interests together and develop a consensus about which resource management actions are acceptable. The first part of negotiation involves each interest group stating its desired position. Next, an education process begins whereby the groups share information with each other on how they arrived at those positions. The intended and unintended effects of accepting each group's position are also discussed. Then compromises and negotiation or "give and take" occurs. Each party gives up something it may want so as to attain its most important objectives. In some cases, discussion among the interest groups identifies alternative means to attain everyone's objectives. This often involves some innovative management action that departs from traditional ways of doing things.

Such a negotiation process has had some success in resolving timber, fish, and wildlife issues among American Indians, foresters, conservation groups, and public agencies in the state of Washington. The goal is consensus building and cooperation rather than conflict and litigation. By funneling the intellectual effort into a search for compromise rather than strategies to beat the other side in court, effective resource management is often improved.

For negotiation to work, all the affected parties must be represented. The people at the meeting must have the authority to make binding compromises for the groups they represent. These are difficult goals to attain, because many groups are often affected and affected differently. Large timber purchasers, small logging contractors, and established lumber mill owners versus more recent ones may not have exactly the same interests. Conservation groups and sport fishing groups often may not have the same goals and objectives. Some public land users are not well organized and may be excluded from the process. Consumers of the products from public lands may also not be represented.

Nonetheless, negotiation is a promising tool in dealing with public land management issues in which there is much disagreement over the importance of various criteria (e.g., biological, economic, and cultural) and the means of attaining them. Analysis still plays an important role in negotiating solutions, as the compromise solution still must be biologically and administratively feasible. If the resource interactions are complex, it is primarily through models that one can determine whether the negotiated compromise can actually be produced. Recall figure 8.13, where the sum of what the ranchers wanted and hunters wanted exceeded the productive capacity of the land. In addition, it is important to quantify all the environmental and economic consequences associated with the negotiated compromise. It may be that the compromise simply shifts the costs from parties represented in the negotiations to others not represented. There may be unintended environmental or economic effects that are difficult to anticipate without use of a formal model. Often the compromises are expensive to implement, so that attention still must be paid to economic efficiency.

In all these cases, it is important to mesh analysis with negotiation to ensure feasible compromises and ones that actually result in improvements in social well-being. Thus there may be a series of rounds of negotiations, followed by analysis of suggested compromises, followed by renegotiations when the results of the new analysis are presented. Together, analysis and negotiation can provide public land managers with an effective tool to augment traditional land management planning approaches, to be discussed in the next chapters.

9

Multiple-Use Planning and Management in the National Forests

Having established the basic principles of multiple-use planning in chapter 8, it is now time to see how they are actually applied in the management of National Forests. As discussed in chapter 2, the National Forest Management Act (NFMA) and its implementing regulations not only give broad direction for planning and management of the National Forests but also specify how they are to be done. Thus NFMA and its implementation in Forest Plans prepared in the 1980s will serve as our foundation for discussion in this chapter.

Forest Service Planning Process Under NFMA

Broad Direction in NFMA

NFMA requires that a plan be prepared for the management of each National Forest. This document is to be a comprehensive plan for all the lands within that National Forest and is to consider simultaneously all resource uses in an integrated, interdisciplinary fashion. The plans must reflect the constraints facing the National Forests, including physical and biological capabilities of the specific forest. Consideration should be given to administrative constraints, such as budget, personnel, past management practices and commitments, although budgets need not be treated as a fixed constant (Teeguarden 1990:26). The proposed management of the National Forest must meet changing demands and societal values of forest resources. In short, "Forest planning should find the match between capabilities and demands that best meets the needs of the American people" (Larsen et al. 1990:2).

The Forest Service NFMA planning training manual (Jameson et al. 1982:4) best summarizes the other legislative intents: "Legislative acts which govern

National Forest lands generally require these lands to be managed: (1) efficiently, (2) for the general welfare, and (3) to ensure their long-term productivity. Although these requirements characterize the past operating philosophy of National Forests, NFMA now indicates a shift toward increased emphasis on efficiency.'' NFMA calls for each National Forest to develop a plan for management of the forest every fifteen years or so, although the actual time period over which the effects of the plan are to be evaluated covers fifty years. Such a plan serves as a guide to which multiple uses will occur and when and where on the forest they will occur; it also sets forth the methods by which they will be produced. The Forest Service is to develop a range of alternative plans and then to evaluate them. The plan must also be accompanied by an Environmental Impact Statement that describes the effects of each planning alternative considered on the entire National Forest (which usually encompass a half million to a million acres). The final plan is chosen from a range of alternative ''visions'' or management emphases that best resolve the issues facing the forest.

The final plan serves as a proactive, strategic plan to move the forest from its present condition to some vision of what it will be over the next five decades. The plan guides on-the-ground management of everything from timber sales and wildlife habitat improvements to recreation facilities. Most Forest Plans usually recommend whether any existing roadless areas should be considered by Congress for Wilderness designation. The plan serves as the foundation for developing the forest's annual budget.

It also informs the public and private industry about what uses of the forest are allowable and where they may occur. For example, it often states which areas are open and which are closed to off-road vehicles. Plans determine what areas will be available for commercial timber harvesting and firewood cutting. These determinations are made as part of a comprehensive plan rather than on a piecemeal, case-by-case basis in reaction to requests by outside parties.

We now turn to the principles and techniques used in the process of multiple-use forest planning.

Overview of the Steps in USFS Planning Process

NFMA planning regulations describes ten steps in the planning process. These are

1. Identification of issues, concerns, and opportunities (ICOs).
2. Development of planning criteria.
3. Collection of data and information necessary to address ICOs.
4. Analysis of the management situation (AMS).
5. Formulation of alternatives.
6. Estimation of effects of each alternative.
7. Evaluation of alternatives.

8. Selection of a preferred alternative (proposed Forest Plan).
9. Plan implementation.
10. Monitoring and evaluation of the plan.

Implementation of these ten steps involves dozens of people from the field (ranger district) level, the supervisor's office, the regional office, and the Washington, D.C., office. Such an ambitious effort as a Forest Plan requires the work of many people for many years. In general, there are two formal groups: (1) an interdisciplinary team (called the ID team), made up of staff resource specialists (e.g., archaeologist, biologists, foresters, hydrologists, recreation planners, and so on) and (2) a management team, made up of line decision makers, such as the forest supervisor, and district rangers. The management team has the authority to make major decisions, such as the final ICOs or selection of the preferred alternative. It also provides the overall direction and supervision of the ID team, which actually implements most of the other steps in the planning process. We now examine these ten steps in more detail.

1. *Identification of issues, concerns, and opportunities (ICOs).* In this first step the agency solicits input from its own staff, other federal agencies, state and local agencies, and the public on what the current problems or conflicts are on the forest, what issues need to be resolved, and the opportunities that exist for improving the forest's management. Current use conflicts, whether those over recreational access or those arising from competition at the same time of year for the same area (e.g., wildlife versus livestock, hikers versus hunters and firewood cutters, and so on), need to be clearly identified.

Both NEPA and NFMA require that the public be notified through a variety of media of the opportunity to express concerns about general management activities (e.g., clear-cutting, road construction, road closures, and so on) and specific parcels of ground (e.g., their favorite recreation site, mining claims, and so on). Public involvement at this early stage is also crucial to make sure that the problems and issues perceived by the agency are the same as those perceived by the public. Often symbolic issues will cloud the real underlying problems. Thus continuing dialogue with users is desirable to develop sufficient understanding to make sure the "right problems" are being solved.

The focus should be not only on which existing conflicts or problems can be resolved but also on what can be done to avoid new problems and on which improvements will serve the public better. Discussion with field managers (the people who deal with the forest resources on a day-to-day basis) should help uncover the trends in resource conditions that if left uncorrected could lead to new problems. Given that many ecological systems have subtle limits or thresholds, only a perceptive field person might notice that a threshold is being approached for some resources such as soils, wildlife hiding cover, trails, and so on. One goal of the plan is to implement preventive actions.

Finally, it is important to identify which problems or issues are simply outside the authority granted to the forest supervisor in forest planning. Public concerns that motorized vehicles should be allowed in Wilderness Area X cannot be resolved at the level of the Forest Plan. These issues can be addressed only at the national level by the president and/or Congress.

Once a list of ICOs is prepared and accepted by the agency managers (forest supervisor and regional forester), this list serves as an informal agreement with the public as to what the Forest Plan will and will not address (Jameson et al. 1982:16).

2. *Development of planning criteria.* The Forest Service has developed three general types of criteria that help define or circumscribe the standards that the planning effort must meet and that help to guide ID team members on the level of analysis. In addition, these criteria guide decision makers in making judgments about suitability of management actions and in interpreting the social consequences of different physical effects. The three criteria are

a. *Public policy criteria.* These are drawn from requirements in congressional laws and acts including NFMA, RPA, NEPA and administrative direction, as reflected in Executive Orders or explicit agency policy. These criteria can be broad, such as the mandate to engage in multiple use and sustained yield (NFMA) and to avoid adverse environmental impacts as well as irreversible and irretrievable commitments of resources (NEPA). However, the legal requirements can be quite specific at times, as in the Clean Air Act.

b. *Process criteria.* These reflect agreed-upon standards for acceptable quality in data and analysis. In some sense, these are the "rules of the game" in carrying out the remaining steps in the planning process. Although everyone would like perfection, it is costly. The process criteria set up an informal agreement on the acceptable level of detail for mapping and analysis. For example, they establish how fine a level of resolution the mapping of ecological capability areas will be (e.g., whether they will be 100 acres or 500 acres). In addition, they indicate how much of the details of the data, models, and analysis will be presented in the plan. Many of these decisions are covered under "standard operating procedures" of the agency, as reflected in Forest Service manuals (Jameson et al. 1982:20).

c. *Decision criteria.* These criteria describe judgments about the desirability of different outputs or preferences toward specific management actions. In chapter 4 we referred to these as evaluation criteria, because they often reflect the factors by which decision makers will judge whether or not the outcome of some management action is desirable. The decision criteria are critical in National Forest planning, because the primary objective (net public benefits) includes many unquantified effects. Thus the determination of "what's best" may depend on how much weight a decision maker gives to various unquantified effects. For example, the weight given to preferences of local rural citizens living adjacent to the forest versus those from urban areas that come for recreation influences the determination of whether expanded livestock grazing is desirable. Another example of decision criteria relates to how the forest supervisor deals with distributional equity (discussed in chapter 4). Does the forest supervisor weigh gains and losses to workers in different industries equally (e.g., logging versus tourism)? How does the supervisor weight impacts to people who prefer roaded recreation versus those who want a more wilderness-oriented recreation? Decision criteria might influence the determination of land suitability for different uses, such as timber harvesting or mining (Jameson et al. 1982:20). For example, does land automatically qualify as suitable for timber harvesting or mining just because it has trees or minerals?

3. *Collection of data and information necessary to address ICOs.* This step begins with a comparison of data required in the remaining steps of the planning process with those currently available. If data exist, do they meet the standards set in the process criteria in step 2? Sometimes a "ground truthing" exercise may be needed to determine how well the existing inventory data match the current state of the forest. If the ground truthing shows that the data do not meet the process criteria or that critical data needed in subsequent steps of the plan are unavailable, then actions must be undertaken to fill these gaps. Thus new data collection may be necessary.

Several types of data will be required to complete a multiple-use plan. Many of these data requirements are obvious from Clawson's three elements of a multiple-use plan:

a. Current resource inventory data to determine the existing or "no-action" resource capability.
b. Data to estimate the expected change in output levels of each resource in response to alternative management actions. For example, how does timber volume respond to precommercial thinning and how does forage production respond to prescribed burning? In addition, one needs data in order to estimate the production interactions between different multiple-use resources. For example, do joint production effects exist among resources, and if so, can the magnitude of the positive or negative joint effects be quantified?
c. Data to determine the valuation function (i.e., benefits and costs of each output). The ID team often develops its own values for timber based on past sales and acquires forest-specific livestock forage values from USDA Economic Research Service. Forest-specific recreation values are sometimes estimated using the travel cost method with existing data, such as campground fee receipts and wilderness permits or occasionally, special surveys.

When time or expertise does not permit forest-specific estimates of recreation and wildlife/fisheries values, the planning team can request permission to use site-specific travel cost method or contingent valuation studies done by others. Often the ID team relies on use of regional average values for recreation and wildlife developed for the national Resources Planning Act (so-called RPA values).

4. *Analysis of the management situation.* This step involves several tasks, many of which relate to the first two of Clawson's three elements in a multiple-use plan. The common theme of all the following factors is a determination of the multiple-use supply capabilities of the forest (e.g., the production possibilities relationships). The individual resource production capability and resource production interactions are then translated into equations that can be used in linear programming (LP). Once in LP format, maximum output levels for each resource can be determined in a series of computer runs called benchmarks. This information on production capability serves as the foundation for the next step of formulating alternatives. Knowledge of the production capability of the forest helps to ensure that the alternative solutions to the ICOs can actually be attained on the forest. In terms of the graphs in chapter 8, this means the alternatives are either along the production possibilities curve or inside it (but not outside). Some of the key tasks in this step include

a. *Determination of minimum management requirements for all resources.* A minimum management requirement (MMR) for a particular resource is the legal minimum protection that NFMA requires the Forest Service to provide to ensure that the resource on the National Forest is sustained. This would be the "custodial" role in terms of protection of resources from degradation. Also included are minimum levels of water quality required under the Clean Water Act, protection of threatened or endangered species under the Endangered Species Act, and other nondiscretionary responsibilities regarding protection of resources on the forest. The MMRs set a floor of minimum levels of each resource below which no alternative or any of the subsequent benchmark analyses are allowed to fall.

b. *Classification or grouping of land into strata based on land capability.* Specifically, each stratum contains lands that are relatively homogeneous in terms of their soil, vegetation, hydrology, and so on, and more important, relatively homogeneous in terms of the land's response to such management actions as timber harvests, grazing, habitat improvements, and so on. Thus lands in each stratum have the same capability to produce a particular mix of multiple-use outputs. Along with capability is the determination of availability and suitability of land for different multiple uses. Availability relates to whether land has been legislatively or administratively withdrawn from certain multiple uses. For example, is it a Wilderness Area, which would preclude timber harvesting? Suitability for particular multiple uses may be based on the decision criteria in step 2. Suitability requires judgment about the best use of this land. It may be capable of producing both timber and recreation, yet few other areas in the forest may provide scenic opportunities for hiking or backpacking. In a sense, this land has a relative recreational advantage compared with other lands within the forest. Consequently, it might be deemed more suitable for backcountry recreation. This process of determining capability, availability, and suitability is most well developed for timber and is illustrated in figure 9.1 (next page) from Jameson and colleagues (1982:72). This three-stage approach reflects application of a screening process described in chapter 4.

c. *Development of management actions or prescriptions to be applied to different land types.* The ID team and resource specialists determine what types of management practices are appropriate for each different type of land. For example, one might develop timber-harvesting actions (e.g., one for clear-cutting, one for selection cutting), a livestock grazing system (e.g., rest-rotation), a campground building action, a big-game wildlife habitat improvement action, and so forth.

They must then decide about the land types for which these management actions would be appropriate. If a particular land stratum has no timber, then it would be nonsensical to have timber harvesting as a potential action for that land type. It may be easiest to think of assembling these data as a matrix, with land classifications as rows and potential management actions as columns. The ID team's job is to determine which combinations of management practices are relevant for each land type. That is, if there are ten land types and five possible management practices, fifty combinations are theoretically possible. Yet only a much smaller subset makes ecological and economic sense. The ID team's job is to determine which combination to make available to the LP model as choice variables.

d. *Operationalization of the ICOs.* This means translating the agreed-upon issues or concerns into well-defined performance indicators. If possible, these performance indicators should be in quantifiable units. For example, such issues as "improve water quality in Fish Creek" must be made specific to the water quality parameters that need to be improved. Measurable water quality parameters whether dissolved oxygen, turbidity, or salinity must be specified. If the water quality parameter of interest is increases in dissolved oxygen, the biologist must determine the magnitude of change that constitutes "improved water quality." If the performance indicators for the ICOs are not well defined or are unmeasurable, it will be difficult to convince anyone that a particular plan alternative "resolves" a concern. How would we know?

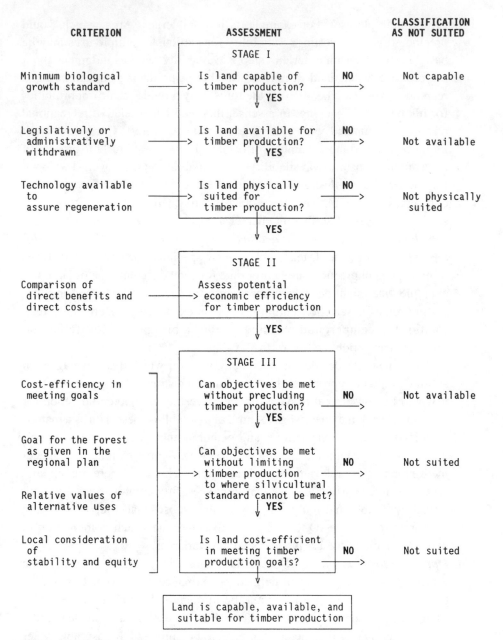

Figure 9.1. Timber Land Use Suitability Screening Process

e. *Development of ecological response effects.* This involves translating the management actions into the multiple-use output rates (e.g., board feet of timber, recreation visitor days, water yield, wildlife numbers, livestock animal unit months [AUMs], and so on) as well as environmental effects (tons of sediment, air pollution, and so on). The ID team needs to know at what rate soils will erode with logging, livestock grazing, and other management actions developed in task c. How many board feet of timber per acre logged will each land type yield (sometimes called *yield tables*)? How much dispersed recreation use will a particular land type provide? This is a major technical step and is often quite a challenge for the ID team members. Each resource specialist must draw on existing research and on his or her own field studies to answer these questions. If these methods do not yield definitive answers, the individual's own professional judgment may be required to determine how the particular lands in each land type would respond. For many resources, formal models of some portion of the biophysical process are used. For example, a hydrology model is often used to predict changes in water yield associated with different management practices. Fish habitat models and wildlife habitat relationship models are commonly developed and used to estimate the response effects of different management actions on fish and wildlife populations. (An example of such a fisheries habitat model was presented in chapter 5.)

f. *Benchmark runs.* A benchmark run is a series of LP model computer analyses that determines the maximum physical output of each resource possible, given the productive capacity of the forest and the minimum management requirements. To perform such an analysis, the information from the previous substeps is put into the LP model format (described in more detail in the next section). For each benchmark analysis a particular multiple-use output is selected to be maximized. The LP model computes the maximum amount of the resource being maximized subject to the resource capability and MMR constraints. For example, there are frequently timber, range, recreation, wildlife, and water benchmarks that compute the maximum board feet, AUMs, visitor days, deer, and acre-feet of water, respectively, that could be produced if the forest concentrated on maximizing that one resource. With the information produced in this step, the feasible production region is traced out (i.e., the area beneath the production possibilities frontier of chapter 8). One key contribution of the benchmark runs is to ensure that no one expects alternatives formulated in the next major planning step to produce multiple-use outputs in excess of the benchmark level. Besides computing the maximum physical outputs, the maximum economic benefit the forest could produce subject only to the biophysical and MMR constraints is also analyzed. With the maximum economic value (net present value) determined as one benchmark, the opportunity cost of adding preference constraints or targets to reflect cultural or distributional equity considerations can be calculated. Specifically, the reduction in NPV associated with additional constraints can be calculated by subtracting the NPV of the more highly constrained LP model runs from the NPV of the maximum economic value benchmark.

g. *Determination of how well the current management direction meets the public policy criteria identified in step 2.* During this phase the management actions currently being carried out on each existing land type are run through the LP model to quantify their biophysical effects. These results are compared to the public policy and decision criteria to evaluate how well the current management direction is doing in meeting the goals and objectives in NFMA. This will provide insights into formulating additional alternatives that better meet the spirit of NFMA and other environmental laws to which the Forest Service must adhere.

h. *Quantification of the public demand for each of the multiple-use outputs.* The amount of demand for each of the multiple-use outputs is important for two reasons: (1) if it is costly to supply more of a particular multiple use, it makes no sense to supply more than people want (at current prices); (2) if, as a by-product of producing one particular multiple use (e.g., timber), additional amounts of one or more other multiple uses are produced (e.g., water and roaded recreation), these additional units only provide benefits to society if there is a demand for them. For example, producing additional capacity for roaded recreation may be of little value if many unused roads are already available for recreation. Multiple regression models are often used to estimate the amount of public demand for wildlife and general recreation on the forest.

The result of completing these tasks is often summarized in an internal document named after this step in the planning process, the analysis of the management situation, or AMS. This document serves as the foundation for the ID team and management teams' formulation of the alternatives.

5. *Formulation of the alternatives.* This step involves specifying groups of alternative management actions that will ''resolve'' the issues raised in step 1. Of course one alternative cannot usually solve all the ICOs, since several might conflict. For example, one alternative might give more emphasis to solving environmental-quality problems on the forest (i.e., more emphasis on protecting water quality, old-growth wildlife habitat, cavity-nesting bird habitat, scenic beauty, and so on), whereas another might address an ICO related to lack of areas for off-road vehicles, firewood cutting, and other more consumptive uses of forest resources. Since allowing the public to cut more standing dead trees (e.g., snags) for firewood results in reducing cavity-nesting bird populations, one alternative cannot resolve both public demands for protecting cavity-nesting bird populations and firewood demand. This leads to the first requirement in formulating alternatives:

a. *Formulate a broad range of alternative management actions.* This range of future visions of what the forest might look and function like must be within the public policy criteria and within the production capability identified in steps 2 and 4, respectively. In practice, the ID team often makes each alternative a different theme or vision. A common alternative is an amenity alternative that might emphasize protecting scenic vistas, nonmotorized recreation, nongame wildlife, and so on. Another frequent alternative is a commodity alternative that would emphasize production of timber, minerals, forage for livestock, and other saleable multiple-use outputs.

It is important to have a wide enough range of alternatives because there may be many dimensions to amenities that are not ecologically bound together. For example, having more Wilderness might not provide a great deal more recreation capacity, since Wilderness recreation management stresses providing opportunities for solitude. Thus recreation capacity might be dealt with by building more miles of trails in semideveloped areas and adding more campgrounds in developed areas, more boat ramps at lakes, and so on. These two different features (Wilderness and developed recreation) are not physically linked, so they could be offered in three combinations: more Wilderness alone, more developed recreation alone, or more of both Wilderness and developed recreation. Since the point of alternatives is to provide choices, it is better to keep features that are physically separate uncoupled from other actions in at least one alternative. This facilitates determination of the desirability of each feature. For example, if adding more Wilderness was economical but adding more campgrounds was very costly relative to the benefits, then if forced to evaluate a bundle of the two actions together one might reject the bundle as uneconomic even though one physically unrelated component (here Wilderness) would be economical to implement. Thus a broader spectrum of alternatives is preferred as long as there are not so many that it becomes unmanageable.

b. *Be certain that each alternative represents a cost-efficient way to resolve the ICOs.* There are often several ways to resolve a particular issue, but the emphasis should be on the least expensive way, all other things (e.g., environmental effects) being equal.

c. *Ensure that there is a "no-action" alternative that is a continuation of current management direction.* This is required by NEPA for the EIS and serves as a useful reference point for comparison against the other alternatives.

d. *Be certain that the alternative is responsive to the Resources Planning Act (RPA).* There is usually an RPA alternative that reflects that forest's share of the national output goals under the Resources Planning Act national program (discussed in chapter 2). These National Forest System output goals in the program are often disaggregated down to each forest for implementation. The RPA alternative is the forest's attempt to formulate an alternative that comes as close as possible to meeting as many of these production goals as possible. This may be done using constraints or targets. This RPA alternative then reflects how this forest would be managed so as to respond to national direction for forest management.

6. *Estimation of the effects of alternatives.* In this step of the NFMA planning process the environmental, economic, and social effects of each alternative must be determined and then displayed for each performance indicator associated with the ICOs and multiple uses. This is where the hydrological, biological, and economic models are used along with the response effects developed in step 4 to calculate the changes in multiple-use outputs (board feet of timber, AUMs of grazing, visitor days of recreation, fish populations, and so on) and environmental effects (water quality, air pollution, and so on) of each alternative. The regulations require specifically that the direct and indirect benefits and costs be analyzed (Forest Service 1982:43045). Economic effects are quantified as the sum of benefits of all the outputs minus agency, other public, and private costs over time. Social effects are often quantified through performance indicators, such as local jobs and payments to local counties. An example of the estimated effects associated with the Siuslaw National Forest will be provided in the case study later in this chapter.

7. *Evaluation of the alternatives.* The alternatives are compared in terms of how well they perform on the planning criteria (specifically, public policy criteria and decision criteria) and in solving the ICOs. The planning regulations (Forest Service 1982:43046) require that, ''The evaluation shall include a comparative analysis of the aggregate effects of the management alternatives and shall compare present net value, social and economic impacts, outputs of goods and services, and overall protection and enhancement of environmental resources.''

8. *Selection of the preferred alternative (the Draft Forest Plan).* The first seven steps (especially steps 2–7) are carried out largely by the ID team staff members with guidance provided by the management team, But step 8, selection of the preferred alternative, is by far the more important prerogative of the management team (i.e., the forest supervisor), with final approval resting with the regional forester. In some cases, the ID team participates with the management team in formulating what they all believe will be a preferred alternative when performing step 5. If the management team decides to construct a preferred alternative by combining features of other alternatives, it is up to the ID team to ensure that such a preferred alternative is biophysically feasible and to predict the environmental and socioeconomic effects associated with the preferred alternative.

The forest supervisor and regional forester must document and justify the selection of the preferred alternative against ''(1) Any other alternative considered which is environmentally preferable to the selected alternative; and (2) Any other alternative considered which comes nearer to maximizing net present value'' (Forest Service 1982:43046). This documentation is made in a formal Record of Decision (ROD), which accompanies the Final Plan. The Final Plan and ROD reflect an implicit agreement between the Forest Service and the public over the management of the forest for the next fifteen or more years.

Therefore this step is more than selection of the preferred alternative. It is also a demonstration that this alternative, better than any other considered, maximizes net public benefits from that National Forest. Since the net public benefits go beyond economic efficiency (net present value) to include other social and environmental factors that may not be quantified, the rationale for selecting the preferred alternative must be defensible. If the plan is appealed and the Record of Decision is not defensible, the chief of the Forest Service or the assistant secretary of agriculture may send the plan back to the forest supervisor so that he or she can reconsider the preferred alternative. This has already happened on the San Juan National Forest (MacCleary 1985), so it is no idle threat.

9. *Plan implementation.* This step involves at least three continuing elements. For routine matters within the discretionary budget and authority of the forest supervisor, the planned management actions scheduled for that fiscal year are implemented. For other Forest Plan actions that require specific funding, budgetary requests are made as part of the normal budgetary preparation cycle. Obtaining adequate funding and personnel to implement the plan is a key step. This is especially important when conflicts have been ''solved'' in the Final Plan by increasing the intensity of management in one area to preserve some other area. If inadequate funding results in selective implementation of the plan, this jeopardizes the balanced multiple use for which the plan has striven. Finally, all uses of the forest must be brought in conformity with the plan, whether these be Forest Service actions (e.g., fire fighting, timber-harvesting practices, and so on) or private uses (e.g., use of ORVs in certain areas).

10. *Monitoring and evaluation.* This is a continual process in which the actual quantities of multiple-use outputs and associated environment effects are compared to those projected under the plan. The purpose is threefold: (1) to ascertain if the plan's management actions are being implemented; (2) to determine if the implemented actions result in the outputs and effects expected in the plan; and (3) to determine why the management actions are not resulting in the expected effects, if this is the case, so that the plan can be amended. In addition, monitoring provides the Forest Service with an opportunity to learn about ecosystem responses to management actions so as to improve the next round of Forest Plans that will start ten to fifteen years from implementation of the current NFMA plan. In many ways the first comprehensive National Forest Plans are an experiment in which the public and agency will refine their understanding of how each forest responds to different managements.

Just completing the first eight steps of this process is a challenging and time-consuming process. The first Forest Plans took as many as seven years to complete because of changes in forest planning regulations, changing direction in the secretary of agriculture's office, major forest fires that reduced timber inventories, endangered species decisions, and a great deal of "learning by doing." As the planning environment has stabilized and the agency moves down the learning curve, many plans are completed in as little as four years. But the more complex the issues facing the forest and the more valuable the resources being evaluated, the more detailed the analysis and hence the more time required.

The next part of this chapter reviews the basic principles underlying the key analysis techniques used by the Forest Service in its first round of forest planning during the 1980s.

How the Forest Service Implements the NFMA Planning Regulations: A Look at FORPLAN Linear Programming Models and IMPLAN Input-Output Models

In implementing the first round of comprehensive forest planning under NFMA the Forest Service chose to use linear programming models. The particular form of this LP model has become known as FORPLAN (Johnson et al. 1986). *FORPLAN* stands for FORest PLANning. The model is an extension of what was originally a timber-scheduling linear programming model. As such, it contains vestiges of that early model. (For an interesting history of the evolution of FORPLAN see Iverson and Alston [1986].)

The remainder of this chapter is composed of six parts: (1) a description of the basic structure of a FORPLAN LP model and interpretation of the output of an LP model; (2) a description of the IMPLAN input-output model; (3) a description of how these two models are used to implement NFMA; (4) an illustration of the output of a Forest Plan using the Siuslaw National Forest in Oregon as an example; (5) an example of performing comprehensive benefit-cost analysis for a National Forest resource allocation decision regarding a roadless area to illustrate an alternative to LP models; and (6) a presentation, in the chapter's "Another View," of the emerging perspectives on the "New Forestry" practices. Chapter 11 provides a detailed critique of this first round of forest planning.

Structure of FORPLAN LP Models

Chapter 8 presented a graphical description of LP models. The two key elements of the LP model consist of constraints and an objective function. In the multiple-use context, the basic constraints reflect scarce resources, such as different types of land. The objective function determines what combinations of outputs are most valuable. Large LP models are no more than a large series of equations that are often represented in tabular or matrix form. We now want to present a schematic overview of the tabular or matrix structure of an LP model. Although many FOR-PLAN models involve hundreds of rows and columns, this overview will provide a good idea of the basic structure. This overview draws on Jameson and colleagues (1982); Joyce and colleagues (1983), and Dyer and Bartlet (1974). It is only intended to familiarize the reader with LP models, not teach how to build them. Additional details on LP models in forest planning can be found in Davis and Johnson (1987).

Conceptual Map of an LP Matrix

Figure 9.2 is a map of the basic parts in an LP structure.

Land Type.   Starting in the upper-left-hand corner, the very first row and first column heading is *land type*. This is a grouping of land into ecologically homogeneous categories. Type 1 land groups all acres on the forest that contain, say, brushy, arid, range-type land largely devoid of commercial timber. Type 2 land is another grouping of ecologically distinct lands that might be commercial timberlands.

Management Actions.   The next row down and set of columns $(X_1–X_8)$ consist of the management actions, sometimes called "management prescriptions" or land treatments, that could be applied to that type of land. Thus only treatments X_1, X_2, X_3, and X_4 make sense to engage in on brushy type 1 rangeland. These variables might be "do nothing" (X_1), prescribed burning (X_2), reseeding (X_3), and both prescribed burning and reseeding (X_4). The model can then choose how much of type 1 land to allocate to each of these four management actions to maximize the

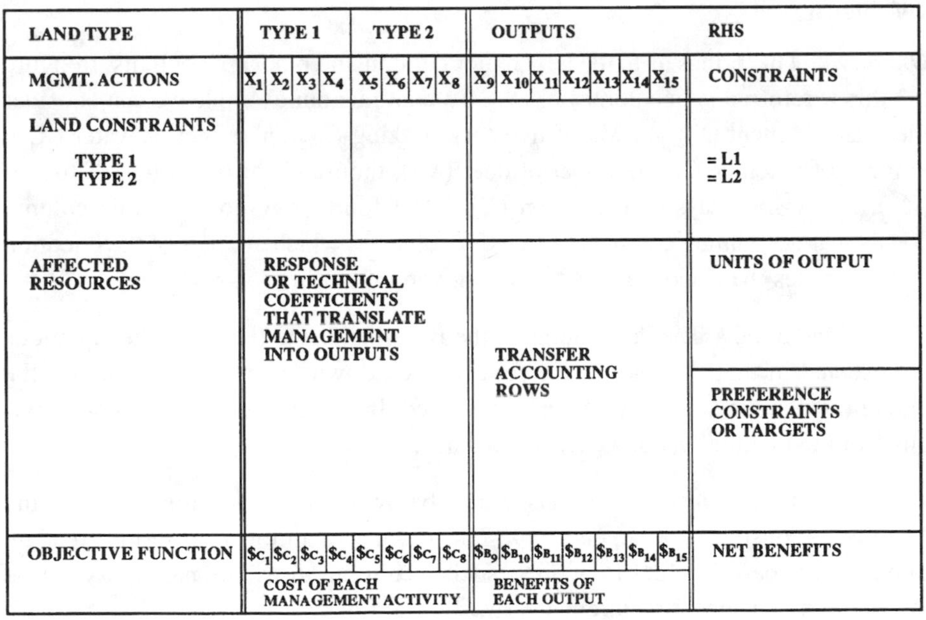

LAND TYPE	TYPE 1	TYPE 2	OUTPUTS	RHS
MGMT. ACTIONS	X_1 X_2 X_3 X_4	X_5 X_6 X_7 X_8	X_9 X_{10} X_{11} X_{12} X_{13} X_{14} X_{15}	CONSTRAINTS
LAND CONSTRAINTS TYPE 1 TYPE 2				= L1 = L2
AFFECTED RESOURCES	RESPONSE OR TECHNICAL COEFFICIENTS THAT TRANSLATE MANAGEMENT INTO OUTPUTS		TRANSFER ACCOUNTING ROWS	UNITS OF OUTPUT PREFERENCE CONSTRAINTS OR TARGETS
OBJECTIVE FUNCTION	$\$c_1$ $\$c_2$ $\$c_3$ $\$c_4$	$\$c_5$ $\$c_6$ $\$c_7$ $\$c_8$	$\$B_9$ $\$B_{10}$ $\$B_{11}$ $\$B_{12}$ $\$B_{13}$ $\$B_{14}$ $\$B_{15}$	NET BENEFITS
	COST OF EACH MANAGEMENT ACTIVITY		BENEFITS OF EACH OUTPUT	

Figure 9.2.   Conceptual Map of an LP Matrix

objective function. Since this particular model has maximizing net benefits as its objective function, the LP model will determine how many acres of type 1 land will go to each treatment based on the relative costs and returns and on other constraints in the model. Thus X_1–X_4 are choice variables for the LP model for land type 1. The same basic structure of land treatments exists for type 2 forested land, although the specific treatments are different: we have a do-nothing treatment in forestland (X_5), a clear-cut treatment (X_6), a selection-cut treatment (X_7), and a build campground action (X_8).

Land Constraints.   The next row down contains land constraints; it reflects how much of each type of land the forest contains. The total acreage of type 1 land available for management actions X_1–X_4 is equal to L1. The total acreage of type 2 land to be allocated to management actions X_5–X_8 is L2.

Affected Resources and Technical Coefficients Matrix.   The next major section of the LP table deals with the fourth block down, the affected resources and production effects. The affected natural resources (e.g., timber, water) or environmental consequences (e.g., sediment) would be listed in the first column. Below the columns labeled X_1–X_8 and in the fourth block down is the technical coefficient matrix. This block contains numbers that translate the management activities X_1–X_8 into resource output and environmental effects. These coefficients are in some sense the ecological response coefficients. That is, if one of the affected resources is AUMs of forage and X_1 is ''do nothing,'' then this would be the baseline forage production on type 1 land. For X_2 it would be the forage production (per acre) with prescribed burning.

Outputs.   This brings us to the next major column and block: the outputs. In figure 9.2 this is referred to in columns X_9–X_{15}. To continue our example we can consider these as sediment (X_9), AUMs of livestock grazing (X_{10}), acre-feet of water (X_{11}), number of antelope (X_{12}), number of deer (X_{13}), thousands of board feet of timber (X_{14}), and visitor days of recreation (X_{15}). The fourth row down in this column consists of accounting columns or transfer columns, which are part of the equation that translates the response effects into a given output in each row.

Right-Hand Side.   The last column is the right-hand side (RHS) of the equations represented in the LP table. In the fourth block down under this column are the units in which the outputs X_9–X_{15} are expressed. In our example the units are obvious from the definitions of X_9–X_{15} given earlier.

Preference Constraints and Targets.   Just below the units-of-output block is the preference-constraints or targets block. This is where additional constraints that are imposed on the LP model to address social, cultural, distributional equity or the minimum management requirements (MMRs) are placed.

Objective Function. The very last row in figure 9.2 is the objective function row. This is the equation that is maximized or minimized (depending on how the problem is structured). In figure 9.2 this row represents the costs associated with management actions in X_1–X_8 and the benefits associated with the outputs X_9–X_{15}. To attain economic efficiency in resource management this objective function would be maximized subject to all the constraints represented in the rest of figure 9.2.

LP Matrix as a Representation of Integrated Natural Resource Management

In its totality, the LP matrix is one way to represent multiple-use management. Different land types (e.g., types 1 and 2) can be managed (X_1–X_8) to produce one or more multiple uses. Joint production is evident if a single management activity (say, X_7) affects many different resources. All the multiple uses (e.g., range, recreation, wildlife, timber, water, and so on) can be represented as different affected resources and outputs. The connection of this modeling framework with multiple-use management will become more obvious as we continue to expand this example.

A Numerical Example of a Multiple-Use LP Table

Figure 9.3 fills in just the upper half of the LP matrix in our next simple example (adapted from Dyer and Bartlett 1974). The first thing to notice is that the amount of each type of land required for each management action or land treatment is now spelled out. The series of 1s indicates that actions X_1–X_7 require 1 acre of land. The campground (X_8) requires 10 acres of brushland and 15 acres of forestland, however.

Notice the block that was called "affected resources" in figure 9.2 is now filled in figure 9.3 with an affected resource (sediment) and four multiple uses (forage, water, timber, and recreation). Of course an actual LP would have all

LAND TYPE	MGMT ACTIONS TYPE 1				TYPE 2				OUTPUTS							RHS
	X_1	X_2	X_3	X_4	X_5	X_6	X_7	X_8	X_9	X_{10}	X_{11}	X_{12}	X_{13}	X_{14}	X_{15}	CONSTRAINTS
TYPE 1 BUSHLAND	1	1	1	1				10								= 22,300 ACRES
TYPE 2 FOREST					1	1	1	15								= 15,000 ACRES
SEDIMENT	.06	.05	.05	.02	.01	.05	.02	1	-1	0	0	0	0	0	0	= 0 TONS
AUM'S	.2	.4	.42	1	.01	.5	.2	0	0	-1	0	-.2	-.17	0	0	≥ 0 AUM'S
WATER	.3	.2	.2	.1	.1	.3	.2	10	0	0	-1	0	0	0	0	= 0 ACRE FEET
TIMBER	0	0	0	0	0	10	4	0	0	0	0	0	0	-1	0	≥ 0 MBF
RECREATION	.03	.025	.025	.02	.05	.03	.07	2M	0	0	0	0	0	0	-1	≥ 0 VISIT DAYS

Figure 9.3. Upper Half of Example LP Matrix

multiple-use resources and several other environmental consequences. But this is enough to start with.

The next thing to note is the technical coefficients that relate management activities X_1–X_8 to sediment, forage, water, timber, and recreation. These numbers reflect the ecological response of the two different land types to the different management actions.

Interpreting the Technical Coefficients

Reading Down the Columns in the LP Matrix Illustrates Effects of Management Actions on Affected Resources.   If one reads down a column of the matrix, one can begin to see how the management actions (the Xs) affect environmental and multiple-use resources. For example, under ''do nothing'' on land type 1 (X_1) we get .06 tons of sediment per acre, whereas when we ''do nothing'' on land type 2 (i.e., X_5), we get only .01 tons of sediment. This reflects the difference in natural erosion rates from soils and slopes in land type 1 versus land type 2. If land type 2 is forested, the presence of trees intercepts rainfall and reduces erosion rates; thus this is a reasonable difference. If one looks at the sediment rate under X_8, every campground built yields 1 ton of sediment.

Reading down a column also allows one to see the joint production of multiple-use outputs from one management action. In column 1 we see that management activity X_1 yields .06 tons of sediment per acre, .2 AUMs of forage per acre, .3 acre-feet of water per acre, no timber per acre (by definition, because it is brushland), and .03 recreational visitor days per acre. In the case of X_2 (prescribed burning) the amount of sediment is down slightly (to .05 tons), which is caused partly by the doubling of vegetation (forage) in response to burning (.4 AUMs per acre). Water yield is down slightly, because with more vegetative cover, there is more percolation of water into soil, and water uptake by plants is greater.

Turning to type 2 land and looking at X_6 (clear-cutting), one can read down the column to see the associated effects and outputs as follows: .05 tons of sediment per acre clear-cut, .5 AUMs per acre, .3 acre-feet of water, 10 mbf of timber per acre, and .03 visitor days per acre (the same as expected in the brushland). By comparing the X_6 column to the X_7 column, one can determine the changes in environmental effects and multiple-use outputs caused by clear-cutting (X_6) versus selection-cutting (X_7).

As you read down a column, you can observe the extent of complementary and antagonistic production affects of each management action. For example, in X_2 (prescribed burning), sediment, forage, water, and recreation are all joint products, because they all change from their values in the do-nothing alternative (X_1). In the case of X_2 the joint production effect is adverse for water and recreation, because the level of these outputs falls as forage increases. With clear-cutting, X_6, all the multiple-use outputs are affected. Positive joint products include forage and water yield, whereas negative joint products include sediment and recreation.

Reading down column X_8, the building campground option, sediment increases to 1 ton per acre, AUMs of forage drops to zero (because we do not allow livestock grazing in campgrounds, none of the forage will be leased as cattle AUMs), water yield increases to 10 acre-feet per acre, and timber output is zero (we do not allow timber harvesting in campgrounds). Based on recreation demand studies, we determine that 2,000 visitor days would be expected at this 25-acre (10 of brushland and 15 of forestland) campground annually.

Reading Across the Rows Displays Outputs. The other way to read the technical coefficients matrix is across the rows of multiple uses. Reading across a given row tells us how the different management actions $(X_1–X_8)$ translate into output of each resource. Reading across the water row states that acre-feet of water runoff equals .3 acre-feet of water per acre every time we apply management activity X_1 to 1 acre of land type 1, .2 acre-feet from applying X_2, and so forth, up to 10 acre-feet from every campground built (X_8). Reading the timber line shows that there is basically no timber for any actions except X_6 (clear-cutting), which yields 10 mbf per acre, and X_7 (selection-cutting), which yields 4 mbf per acre. We can rewrite this row back to an equation form and see why there is a 1 under X_{14}, the timber output column. Ignoring the zero entries, the timber row is as follows:

$$10(X_6) + 4(X_7) - 1(X_{14}) = 0 \tag{9.1}$$

The -1 is the result of rewriting the equation so that the output variable is on the left-hand side instead of on the right-hand side (RHS). Specifically, one would normally write equation (9.1) as

$$10(X_6) + 4(X_7) = X_{14} \tag{9.2}$$

By subtracting X_{14} from both sides of equation (9.2), the equation is put into LP matrix form as in equation (9.1). Thus the block of -1s on a diagonal just reflect the output variables in the extreme-right-hand column (e.g., tons, AUMs, acre-feet, mbf, and visit days). Therefore the -1s simply ensure equality between what is produced by actions $X_1–X_8$ and what is output by the LP model. This is a valuable feature of LP models an accounting system to ensure that all effects of management actions show up somewhere in the model outputs.

The competitive nature of big-game wildlife and livestock for the same forage is illustrated in this table by reviewing the forage-AUMs row. Note there is a -1 under X_{10} for AUMs. This is consistent with our discussion of timber: the -1 for X_{10} is used simply to rearrange the equation algebraically so that livestock AUMs are on the left-hand side instead of the right-hand side. But note the $-.2$ and $-.17$ in the AUM row under columns X_{12} (which represents number of antelope) and X_{13} (which represents number of deer). These coefficients reflect two things. From the forage row viewpoint, the $-.2$ and $-.17$ represent the assumed competitive nature of these two wildlife species with cattle. Given the forage produced in management activities X_1–X_8, there are three uses: .2 and .17 AUMs per acre would be consumed by antelope (X_{12}) and deer (X_{13}), respectively, leaving only .63 of the AUMs produced available for livestock. Again the accounting identity is met such that all the AUMs being produced from actions X_1–X_8 are equal to the three uses (livestock, antelope, and deer).

From the column viewpoint, the $-.2$ and $-.17$ reflect the number of antelope (X_{12}) and deer (X_{13}) produced from each AUM of forage resulting from the different management actions. The $-.2$ and $-.17$ will be translated into outputs of antelope and deer in the lower half of the expanded matrix shown in figure 9.4. Therefore let us turn to the lower half of the LP matrix in figure 9.4.

Preference Constraints and Output Targets

The first thing to note in the LP matrix in figure 9.4 is the lower right-hand side, where several rows of preference constraints have been added to the model. The first row states that sediment ($1 \times X_9$) must be less than or equal to 1,000 tons over both land types on the forest. This may reflect a minimum management requirement associated with compliance with the Clean Water Act. While sediment is a by-product of the management activities X_1–X_8 that produce social benefits (timber, livestock, recreation), in fact sediment itself reduces social benefits. We could handle this by putting in a cost per ton of sediment to downstream irrigators or municipal water users to incorporate this negative externality explicitly into our objective function. But there may be some maximum that the forest cannot exceed because of legal factors, such as the Clean Water Act, so it is represented as a constraint. As we know from chapter 8, a feasible solution requires that the constraint be met, so representing sediment as a constraint will have an effect (if that constraint is binding on the optimal solution).

The remaining rows specify minimum output targets that the managers have chosen. That is, rather than let the model completely determine the output of these multiple uses based solely on the productivity of the land, the costs of producing the output and the benefits of the outputs, the manager wants to ensure that some minimum level gets produced. As we demonstrated in chapter 8, this may have a substantial opportunity cost in terms of forgone benefits. In addition, these target levels may end up dictating the level of multiple-use outputs in the Final Plan.

Targets and Constraints to Incorporate Multiple Criteria.   Frequently, the reason for setting the minimum output targets relates to consideration of the other criteria affecting public land decisions, such as distributional equity (often jobs) and social/cultural acceptability. The preference targets also may reflect some administrative mandates, such as this forest's contribution of timber to the national RPA timber target. There may also be a concern that for some outputs, such as certain wildlife species, the benefits in the objective function (the last row in the table) may only reflect partial benefits to society. This is particularly true if, because of time and budget limits, no valuation of a particular output was made. For example, no val-

LAND TYPE	X₁	X₂	X₃	X₄	X₅	X₆	X₇	X₈	X₉	X₁₀	X₁₁	X₁₂	X₁₃	X₁₄	X₁₅	CONSTRAINTS
		MGMT ACTIONS (TYPE 1 / TYPE 2)								OUTPUTS						RHS
TYPE 1 BUSHLAND	1	1	1	1				10								= 22,300 ACRES
TYPE 2 FOREST					1	1	1	15								= 15,000 ACRES
SEDIMENT	.06	.05	.05	.02	.01	.05	.02	1	-1	0	0	0	0	0	0	= 0 TONS
AUM'S	.2	.4	.42	1	.01	.5	.2	0	0	-1	0	-.2	-.17	0	0	≥ 0 AUM'S
WATER	.3	.2	.2	.1	.1	.3	.2	10	0	0	-1	0	0	0	0	= 0 ACRE FEET
TIMBER	0	0	0	0	0	10	4	0	0	0	0	0	0	-1	0	≥ 0 MBF
RECREATION	.03	.025	.025	.02	.05	.03	.07	2M	0	0	0	0	0	0	-1	≥ 0 VISIT DAYS
									1							≤ 1000 TONS
										1						≥ 500 AUM'S
											1					≥ 1000 ACRE FEET
												1				≥ 50 ANTELOPE
													1			≥ 100 DEER
														1		≥ 1000 MBF TIMBER
															1	≥ 200 VISIT DAYS
OBJECTIVE FUNCTION	0	-6	-10	-15	0	-95	-65	-4M	0	8	60	88	78	50	17	NET BENEFITS
	COST OF EACH MANAGEMENT ACTIVITY								BENEFITS OF EACH OUTPUT							

Figure 9.4.   Complete Example LP Matrix

uation may be performed on nongame species such as spotted owls, even though they obviously have a value to some members of society. For other wildlife species the estimated benefits may be represented only by recreational hunting, although there are both nonconsumptive wildlife viewing and a significant existence value for additional animals. In these cases the measure of benefits in the objective function understates the social value of these species.

In this example in figure 9.4, the manager decides this alternative should have at least 500 AUMs of livestock grazing, 1,000 acre-feet of water, fifty antelope, one hundred deer, 1,000 mbf of timber, and two hundred visit days. These may be the minimum quantities that make the ranching operation viable (in terms of AUMs) and/or keep a timber mill operating at least one shift (in terms of mbf).

Changing Target Levels to Reflect Alternative Management Emphases.   The level of these targets may change from one plan alternative to another, reflecting the alternative's emphasis in solving the ICOs. That is, an alternative formulated to address wildlife and environmental-quality ICOs may have greater wildlife targets and lower AUM targets. An alternative reflecting an ICO relating to timber jobs and community stability might set the timber target at historic harvest levels.

Strengths and Weaknesses of Targets.   These preference targets represent both an asset of LP models as a tool (the flexibility to incorporate other factors besides what is represented in the objective function to be maximized) and a potential liability of LP models as a tool (the ability of the manager to override completely the social benefits and costs reflected in the objective function and impose specific output levels). Of course the saving grace is that the manager's value judgments as reflected in targets and preference constraints are explicitly visible in the LP model table. This aids in public review of the forest planning process.

The Objective Function: Net Benefits

The last row in figure 9.4 is the objective function row. This is the relationship or function the manager is trying to maximize. As is typical in most Forest Service LP models, this example illustrates maximization of net benefits (i.e., benefits minus costs).

Costs.   The bottom row of columns X_1–X_8 reflects the cost every time one of those management actions is undertaken. Doing nothing in brushland (X_1) and timberland (X_5) has no direct variable costs. But prescribed burning (X_2) costs $6 each time an acre is burned. It is shown as a negative 6 (-6) because costs reflect a reduction in benefits, which is what we are trying to maximize. Seeding (X_3) costs $10, and a combined burning and seeding (X_4) action cost $15 an acre (there are some economies in doing both). Clear-cutting (X_6) costs $95 an acre. This is an average of the road costs required per acre accessed, sale preparation, and administration. Selection-cutting (X_7) costs $65 per acre. Building a 25-acre campground using 10 acres of brushland and 15 acres of forestland along with associated roads, fire grates, tables, water, and so on, costs $4,000 (or $2 per visitor day).

Benefits.   The columns X_9–X_{15} reflect the benefits to society of outputs produced (X_{10} AUMs, X_{11} acre-feet of water, X_{12} number of antelope, X_{13} number of deer, X_{14} mbf of timber, and X_{15} visitor days of recreation) or disbenefits of environmental effects (X_9 tons of sediment). This example does not include a reduction in benefit (i.e., a cost) for tons of sediment produced. The concern about sediment is manifested as a threshold, which is reflected in the preference constraint that no more than 1,000 tons of sediment be permitted.

The benefits of livestock grazing (X_{10}, last row) are reflected in the $8 per AUM. This is the change in net income to the rancher per AUM, or using the terminology of chapter 6, the rancher's producer surplus or net willingness to pay for the forage. It is a net value reflecting the rancher's gain in beef sales from having more forage (greater cattle weight gain or more cattle) minus the incremental costs of raising that additional beef. This value exceeds the administratively set grazing fee (that has averaged around $1.50 per AUM) charged by the Forest Service, since the fee is not closely related to the value the rancher actually receives from being able to run his or her cattle on public land.

The next output, water (X_{11}, last row), has a value of $60 an acre-foot. This is also a net value that reflects the net willingness to pay for additional water in its most likely use: irrigation. The $60 represents the value of additional crop yield (higher yield per acre or more acres) minus the incremental costs of cultivating and harvesting the additional crop output. Thus the $60 represents the change in farm income or farmers' net willingness to pay for the water (i.e., a producer surplus).

The value of additional antelope (X_{12}) and deer (X_{13}) is the incremental value to hunters of having an additional animal on the forest. It could be calculated from either a contingent value survey or from the travel cost method demand curve (discussed in chapter 6). In this example the value is $88 per antelope and $78 per deer. These values represent hunters' willingness to pay to hunt these animals over and above their harvesting and transportation costs.

The value per thousand board feet of timber (mbf) of $50 is a forest average of accepted bids received over the last five years for similar timber species mix. These bids represent loggers' willingness to pay over and above their harvesting and transportation costs of moving the logs to the mill. As discussed in chapter 6, the stumpage value is the producer surplus per thousand board feet.

Finally, the value of visitor days to the campground (X_{15}) is $17. This value is a visitor's net willingness to pay (e.g., consumer surplus) for visiting similar campgrounds on the National Forest. It might be estimated using campground permit data and the travel cost method. Just like the value per AUM of forage, this value reflects the social benefits, not the actual payment to the Forest Service in the form of a campground fee.

Thus the benefits in the objective function represent the economic value to society from the various outputs, not necessarily the cash flow realized by the Forest Service. Given the rationale for public ownership of the forest (discussed in chapter 3), it would be inappropriate to measure benefits only as cash flow and then maximize financial profits. This would ignore the positive and negative externalities created by multiple-use management.

What Is Learned from Setting Up the LP Matrix

Even if one never solves this LP model on the computer, the LP tabular format provides us with an accounting framework to ensure that environmental and economic effects of multiple-use management options are reflected. This is quite valuable in and of itself. The framework can focus the efforts of the ID team in determining the necessary data to be collected and the legal minimums and maximums for certain resources.

Solving the LP Model

The LP software now has all the elements it needs to perform the search process we illustrated graphically in chapter 8. The model will simultaneously determine what is the net benefit maximizing (i.e., the most valuable) combinations of multiple-use outputs subject to the production constraints on type 1 and type 2 land

Table 9.1
Optimal Solution Land Allocations to
LP Problem

Management Activity	Brushland (acres)	Forestland (acres)
X_1	0	
X_2	0	
X_3	0	
X_4	17,305	
X_5		0
X_6		138.5
X_7		7,368
X_8	4,995	7,493.5
Total	22,300	15,000.0

and the preference constraints and targets. The LP model will ensure that these preference constraints and targets are satisfied with the minimum cost combination of management actions X_1–X_8.

Using any number of software packages (such as LINDO) enables one to solve the LP model for the amount of brushland to be managed using activities X_1–X_4 and the amount of timberland to be managed using activities X_5–X_8. The optimal solutions from the computer printout are displayed in table 9.1.

Interpretation of LP Output

Nearly all acres of type 1 land (brushland) went to land treatment X_4, which was the combined prescribed burning and seeding treatment. It allocates all the brushland to this action for two reasons. X_4 has slightly more than twice the AUM output per acre of X_2 and X_3, yet it costs less than twice X_3 and produces two-and-a-half times less sediment than X_2 (.02 for X_4 versus .05 for X_2). As we shall see later, the sediment constraint was a limiting factor, so management prescriptions that minimized sediment were strongly favored. Table 9.1 shows that 4,995 acres of brushland went to the campground. The LP printout actually shows that X_8 (construction of a 25-acre campground) occurs 499.567 times. Since each time X_8 is performed it requires 10 acres of brushland, the total of brushland acres is 4,995.

Clear-cutting (X_6) is performed on 138.5 acres and selection-cutting (X_7) is performed on 7,368 acres. Of course, building five hundred campgrounds requires 15 acres of forestland per campground. Therefore, 7,493.5 acres of forestland are required (15 acres per campground times 499.567 campgrounds).

At the Forest Plan stage these solutions represent the optimal number of management prescriptions that would be applied to the specified amount of acres to maximize net benefits subject to meeting the constraints. Implementation of the Forest Plan would require activity planning to determine which specific forest acres (of that suitable land type) on the ground would be scheduled for clear-cutting, which would go to selection-cutting, and which would be used for campgrounds.

Constraints That Were Binding.   Table 9.2 presents the environmental effects and multiple-use output levels of this optimal allocation of land to each management activity. This table also illustrates which constraints were binding (i.e., effectively limiting the optimal solution) and what their opportunity costs (i.e., shadow prices) were.

Relating Management Activities to Output Levels.   By using the information in tables 9.1 and 9.2 as well as figure 9.4 we can show how the computer calculated the level of outputs. For example, the 30,857 mbf of timber are obtained from 138.5 acres of clear-cuts (X_6) at 10 mbf per acre (1,385) plus 7,368 acres of selection-cutting (X_7) at 4 mbf per acre (29,472). Thus the LP accounting framework is simultaneously tracking amounts of management activities and their output levels (as well as costs, benefits, and environmental effects—here sediment, water, AUMs, and visitor days).

Binding Constraints and Opportunity Costs.   There are several things to note about table 9.2. First, the sediment, livestock AUMs, and antelope constraints had an effect on the optimal allocation of land to different management activities. In the case of sediment, the amount of sediment-producing management activities was limited by the requirement that no more than 1,000 tons of sediment could result from the management actions in this area. If we could accept one more ton of sediment above 1,000, the value of the objective function (net benefits) would rise by $13,776 because we could perform additional management actions to produce more valuable outputs. Few existing studies estimate the damage of sediment that high, and so another computer run might be made relaxing the sediment constraint some (but not so far as to eliminate fish or other beneficial uses of the water downstream).

The livestock AUM constraint is also expensive to meet. The objective function (net benefits) would rise by $458 for each reduction in AUM we have required the

Table 9.2

Optimal Solutions: Output Levels, Environmental Effects, and Constraints

Resource/Effect	Quantity	Constraint Binding?	Opportunity Cost
Sediment (tons)	1,000	Yes	$13,776
Livestock (AUMs)	500	Yes	458
Water (acre-feet)	8,241	No	0
Antelope (number)	50	Yes	3.76
Deer (number)	107,888	No	0
Timber (mbf)	30,857	No	0
Visitor days	1,000,846	No	0

forest to produce. The ranch budget analysis that determined the value per AUM in the objective function indicates the change in rancher income (i.e., ranchers' net willingness to pay [WTP]) is only $8 per AUM. Given the extremely high cost, there are at least three options. First, relax the constraint to something like 400 AUMs and see if the costs drop dramatically. (This would be evidence that it is just the cost of meeting the last few AUMs that is so costly.) If the costs do not drop dramatically (i.e., if they are still several hundred dollars per AUM even at 300 AUMs), then it may be cheaper to society to retire the grazing privileges. If the manager chooses to maintain the ranching operation because of criteria such as distributional equity, the cheapest way to do this may be to put an upper limit on deer so that more of the AUM forage produced (which is 18,850 in the optimal solution) is allocated to livestock.

Nonbinding Constraints.   There are no opportunity costs to meeting constraints when the optimal solution surpasses the minimum target. Thus the constraint requiring at least two hundred visitor days of recreation has no effect on the optimal solution, since the production relationships and net benefits (benefit minus costs) determine that much more than two hundred visitor days is optimal.

Extreme Solutions and the Need for Judgment and Revision.   The last item worth noting is the extreme solution for the mix of big-game species produced. In this solution, over 100,000 deer were produced and only fifty antelope. This unbalanced mix would likely be unacceptable to Fish and Game officials or hunters. The preference constraint or target may need to be raised on antelope and a cap or upper limit may have to be put on deer.

All this discussion on constraints and shadow prices also illustrates the learning feature of using LP models as a simulation tool. The initial optimal solution provides feedback on the costs of preference constraints and targets. Given this information, the analyst can experiment with minor changes in constraints to see if the opportunity costs of meeting those constraints can be significantly reduced.

Maximum Net Benefits in the Optimal Solution.   As shown in table 9.3, the net benefits of the maximized objective function are $24.7 million. This is obtained by multiplying the optimal output levels in table 9.2 by their coefficient value in the objective function (figure 9.4) and subtracting the costs of the management activities in figure 9.4. This process is shown in table 9.3.

Thus in some sense the LP model is simply an elaborate way of finding the particular combinations of management actions that maximize net benefits subject to the constraints imposed to meet the other criteria of biological feasibility, environmental quality, distributional equity, cultural acceptability, and administrative practicality.

We now turn to the other major analytical model that is used in conjunction

Table 9.3
Computation of Objective Function for Optimal Solution

	$/Unit	Optimal Levels	Total
OUTPUTS			
AUMs	$ 8	500	$ 4,000
Water	60	8,241	494,460
Antelope	88	50	4,400
Deer	78	107,888	8,415,264
Timber	50	30,857	1,542,850
Visitor days	17	1,000,866	17,014,722
Total benefits			27,475,696
MANAGEMENT ACTIVITIES			
X_4 (burn and seed)	15	17,305	259,575
X_6 (clear-cut)	95	138.5	13,158
X_7 (select-cut)	65	7,368	478,920
X_8 (campground)	4,000	499.567	1,998,268
Total costs			2,749,921
Net benefits (max objective function)			24,725,776

with FORPLAN optimization models to predict local employment effects associated with producing the optimal mix of multiple-use outputs.

Use of Input-Output Models in Forest Planning

To satisfy the requirements of both NEPA and NFMA and to address potential concerns from local publics, the Forest Service is often interested in the consequences of a particular Forest Plan alternative in terms of employment and personal income. Often the concern is specific to a particular industry, such as wood products or mining. As discussed in chapter 7, the concern is not only the direct change in jobs but any indirect or industry support jobs. This concern about jobs is normally focused on the multicounty area surrounding the forest (sometimes referred to as the socioeconomic impact area), not on effects for the entire state or national economy. As discussed in chapter 6, in the long run most forest planning actions simply result in a transfer of jobs between different areas in the national economy or between industries at the local level (e.g., the timber industry might lose from a reduction in harvest levels but the fishing industry might gain). Nonetheless, in the short run, specific individuals may find themselves unemployed, even though other industries (in the same area) are stimulated by a particular plan alternative and are hiring additional workers.

In some ways there is a natural linkage between the results of the LP model and the first round or direct effects on a local economy surrounding the forest. A particular LP model solution determines the mix of outputs or goods to be produced from that National Forest within the local economy. For example, if demand exists for a new campground and Alternative A builds one, some new visitors to the region will likely result. The LP model predicts the number of visitor days every time it schedules construction of a campground. Each of these visitors spends money to buy camping-related goods and services, such as extra food, film, ice, and so on. Thus there is an increase in demand for these goods and services, a portion of which may be met by businesses in the area near the campground. If the visitors live outside the multicounty impact area (the area where we want to measure gains and losses in employment), then sales by local businesses to these ''non-residents'' reflect increased exports or ''new money'' injected into this region.

This new money becomes wages, profits to businesses, and rents to landlords. As discussed in chapter 7, this increase in wages and profits represents an increase in income to the recipients. They of course will spend a portion of this added income. Some of this additional or indirect spending will be within the multicounty socioeconomic impact area and will stimulate more spending.

The total impact on the economy is thus the sum of the direct and indirect spending. This total impact is some multiple (e.g., two or three times the initial effect) and is measured by the multiplier discussed in chapter 7. Each industry will have its own multiplier, reflecting the linkages in the local economy in producing that good. Service sectors that buy many of their inputs from local suppliers will have a much larger multiplier than a specialized industry that buys most of its inputs from outside the region.

The Forest Service Input-Output Model: IMPLAN

To allow for construction of multipliers and to trace through the interindustry effects discussed in chapter 7 we saw that an input-output (I-O) model is constructed and used. Depending on how this is done, it can be expensive and time-consuming.

The Forest Service has pioneered an efficient way of customizing an I-O model to fit local economies surrounding National Forests. Known as IMPLAN, it uses the U.S. input-output model as a data base, along with county business data to allow an analyst on the forest to create a customized I-O model for any county or group of counties surrounding a National Forest.

The model is not formally linked to the FORPLAN LP model. Instead the analyst takes the inputs used in the different management prescriptions (X_1-X_8) and the outputs produced in an LP solution and determines what the direct effects will be. For example, every 1,000 board feet of timber harvested will require J number of jobs to build the roads, cut the trees, drive the trees to the mill, and mill the logs into lumber. When a recreationist visits, the model uses a ''typical bill of sale'' for a visit (i.e., how much gas, groceries, and miscellaneous retail expenditures [film, suntan lotion, souvenirs, camping supplies, and so on] is bought in that local economy).

These first-round effects are then entered into the IMPLAN software and the IMPLAN I-O model computes the total jobs (direct plus indirect) that result from those effects. The change in total jobs can be displayed by industry type or sector of the economy. For example, both recreation and timber will affect the transportation and gasoline industries.

This information on jobs by industry under different forest management alternatives can be compared to current jobs to see if any plan imposes significant job losses on particular industries. Such a finding would provide information on the distributional fairness or equity criteria used in evaluating the desirability of each alternative.

How LP and I-O Models Are Used by the Forest Service to Implement NFMA

Now that the tools the Forest Service uses in forest planning have been introduced it is important to consider the steps for determining how they are used to implement forest planning under NFMA:

1. *Identify issues, concerns, and opportunities (ICOs).* The FORPLAN and IMPLAN models are of little help in identifying ICOs. Other techniques, such as content analysis (Bailey 1987:300; Markoff et al. 1974), are more useful in organizing and synthesizing public input from letters and public meetings. However, FORPLAN is useful later in determining to what extent resolution of certain issues or concerns is within the production capability of the forest. Eventually, several of the ICOs may be translated into preference constraints or output targets in FORPLAN. In this way, these concerns will be addressed explicitly as an integral part of determining what constitutes balanced multiple use on the forest.

2. *Develop planning criteria.* A FORPLAN LP model is of little help in determining what the planning criteria should be. The model is essentially a value-neutral tool. Once the planning criteria have been determined, they can be reflected in preference constraints or output targets.

3. *Collect inventory data.* The FORPLAN LP model does not collect data or provide its own data. Rather, the LP model requires that a great deal of data be collected (e.g., data on the technical coefficients, costs of management actions, benefits of each multiple-use output, and so on). The LP structure does require more detailed data than would be required to perform a benefit-cost analysis of the alternatives, but not too much more.

The major contribution of a FORPLAN and IMPLAN model at this stage is to help guide the data collection. As discussed in chapter 5, the model structure narrows down the universe of all possible data that could be collected into a manageable subset: the data required by the model. This helps to ensure that critical data are collected and superfluous data that cannot be integrated into the decision process are not collected. Since data collection is both very expensive and time-consuming, this is a strong reason for using any formal modeling structure, including FORPLAN. The primary drawback of a FORPLAN-type LP model is that the model structure limits the relevance of qualitative (i.e., nonquantitative data). Data on qualitative concerns must be brought to bear when deciding finally on what alternative is most desirable and when translating the LP model solution into on-the-ground management.

4. *Analyze the management situation.* The Forest Service relies extensively on the FORPLAN structure in performing this step in the planning process. To some people this step is almost synonymous with building the basic FOR-PLAN model table or matrix for a specific forest. However, there are a few steps that precede the formal model building, although they are guided by the FORPLAN model structure.

   a. *Land classification.* One of the first steps is to develop the land classifications, land types, or capability areas. All the land in the forest must go into one of these land categories or groupings. The criteria for each category are set up first. Lands within each group need to have homogeneous responses with respect to the management actions. Thus similarity of soils, topography, and vegetation are often criteria. More recent versions of FORPLAN allow for geographic specificity of parcels. For example, location in the forest in terms of watershed areas may be another characteristic by which ecologically similar lands are grouped together. Thus there may be ponderosa pine forestland with slopes of zero to 20% in Indian Creek drainage in one group and the same type of land in Smith Drainage in another land class.

   b. *Translation of ICOs and MMRs into equations.* Another important step is to operationalize the ICOs and minimum management requirements (MMRs) into equation form (often to be used as constraints) in the FORPLAN model (i.e., translation of the general issues, concerns, and minimum acceptable management levels into something specific and measurable).

   c. *Development of management prescriptions.* At this stage the management actions/prescriptions to be carried out on the ground (e.g., burning, clear-cutting, and so on) are formulated within the structure of the FOR-PLAN model. Specifically, the ID team must develop management actions and then determine which of them are suitable to the specific land types developed earlier. In addition, the costs of each management action must be calculated. In our earlier example, this involves working out the details of what a prescribed burn prescription will involve in terms of burning technique, labor, fuel costs, and so on.

d. *Technical coefficients.* Once the management actions are defined, the ID team deals with the difficult task of determining the ecological response effects of each management prescription as a matrix of technical response coefficients. The structure of the LP model identifies what response effects must be quantified and the form in which this information must be developed. This is the heart of the forest production functions that will dictate the outputs that are feasible and the environmental consequences that would follow from each management action. The nature of joint production of several multiple uses arising from a given management action is reflected in this technical coefficient matrix.

e. *Determining maximum production potential: benchmark runs.* With these factors in place, the next major task is to use the FORPLAN LP model to determine the maximum production potential of each individual resource. This is often performed by maximizing the physical output variable in the objective function in a special series of runs called the benchmark runs. By performing one benchmark run for each resource, the maximum feasible amount of that particular output or multiple use is identified. This sets an upper limit on what can be expected when formulating alternatives in the next planning step. Such information is valuable in ensuring that the alternatives formulated can actually be achieved.

f. *Evaluating the Future with Current Management.* Running FORPLAN with the current mix of management activities provides baseline information about what future output levels and environmental effects will be in the absence of any significant changes in the current direction of management.

5. *Formulation of alternatives.* The FORPLAN LP model does not formulate different alternative land management actions to address particular ICOs. The ID and management teams must form a series of alternatives that do address them. As mentioned earlier, each alternative may represent a different management theme (e.g., wildlife preservation or commodity production) or may emphasize resolving certain ICOs (e.g., biological diversity). However, this emphasis must be translated into use of different types of management activities in different land classes in FORPLAN (e.g., providing the LP model with the choice of using selection-cutting as well as clear-cutting) and imposition of constraints that reflect the emphasis of par-

ticular alternatives. In some cases, this emphasis will be reflected by modifying what lands are available for different management prescriptions. Thus a more amenity-oriented alternative might preclude all lands within heavily used travel corridors from being available to the clear-cutting management prescription. A production-oriented alternative might make available all lands capable of producing timber. In essence, the ID team must fill out this portion of the LP matrix like figure 9.4 for each alternative.

6. *Estimation of effects of each alternative.* In this step the FORPLAN LP really makes a major contribution. In some sense this is the payoff for structuring the planning process to use an LP: the ability to quantify the environmental and economic effects that would occur with implementation of each alternative. As we saw from our sample output, the LP model's solution tells a lot about what will happen on the forest with that particular alternative. The LP accounting system also helps to ensure that no impacts (that were modeled) are lost, forgotten, or unaccounted for.

   As discussed briefly earlier, the output levels and labor/supplies requirements associated with each alternative can be used as input into the IMPLAN input-output model to yield estimates of many social effects: changes in jobs (by industry), county income, and population.

7. *Evaluation of alternatives.* Although the FORPLAN LP model is value neutral and hence cannot indicate that one alternative is better overall than others, the LP output does greatly facilitate the comparison of alternatives. Just like using benefit-cost analysis, each alternative can be compared on degree of attainment of the objective being maximized—usually, net present value (NPV). Since NPV summarizes all the discounted benefits (included in the objective function) minus all the discounted costs (included in the objective function), it can provide a fairly comprehensive measure of the overall net benefits to society. That is, if one has been thorough in applying economic valuation techniques, all the multiple-use outputs should be valued accurately in the objective function. Many of the environmental effects can be valued also. For example, tons of sediment can be quantified as a cost. Thus in principle, net present value may reflect a majority of the "net public benefits" the Forest Service is to maximize when selecting the Final Forest Plan.

   In addition, since this net present value was maximized subject to constraints reflecting cultural acceptability, distributional equity, and other criteria, the differences in NPV are often a useful way to compare the alternatives, knowing that at least minimum acceptable levels of consideration have been given to these other factors. Finally, reductions in NPV associated with alternatives that give greater weight (higher targets) to cultural acceptability or distributional equity can be compared to see if the opportunity cost in terms of reduced NPV is "worth it." The planning regulations requirement for comparing social impacts (e.g., jobs by industry, changes in county income, and so on) is facilitated by the results of the IMPLAN input-output model discussed earlier.

8. *Selection of the preferred alternative.* No decision aid, whether the FOR-PLAN LP model, the IMPLAN I-O model, or any other, can legitimately make the decision about what constitutes the preferred alternative. This selection is made using the combined judgment of the management team, with the final approval resting with the regional forester.

   The models help to provide the information to decision makers on the consequences of different alternatives so that an informed decision can be made. The FORPLAN LP model (like benefit-cost analysis) is a particularly useful decision aid because of its summarization of many effects into one or more summary statistics, such as net present value. Not only do these techniques "boil down" the alternatives to one or more summary statistics but the use of net present value puts all the effects in commensurate units (e.g., dollars).

   Another strong feature of a model such as FORPLAN LP models is the ability to allow the decision maker to engage in "what if" scenarios (i.e., to tinker with some feature of what he or she feels is the most preferred alternative and see what the result would be). In this sense, the FORPLAN LP can be used as a simulation model to estimate the consequences of possible refinements in an alternative, before having to make the final decision. This assists the decision maker in ensuring that a particular multiple-use combination is the best he or she can do, given what is feasible.

9. *Implementation of the plan.* Because one of the intents of FORPLAN is to schedule the outputs over the planning horizon, the LP solutions provide a blueprint on optimal timing of management activities on the different land types. This aids in scheduling personnel and requesting funding in the appropriate budget cycle.

   The earlier versions of the FORPLAN LP models were useful primarily in describing the management actions that needed to be undertaken on particular types of land to attain the expected outputs. But these earlier versions provided little guidance about where on the forest these activities should take place. If type 1 brushland is scattered over several different areas of the forest, on which areas should we implement the particular management activities called for on this type of land? Spatial detail was lacking in these early FORPLAN LP models.

   More recent versions of FORPLAN introduced in the mid-1980s provide greater ability to track management activities not only by land type but also by general location. Coupling this with the recent capability to tie FOR-

PLAN to geographic information systems (GIS) that can provide computerized mapping of FORPLAN solutions, the ability to implement LP solutions on the ground is improving. However, FORPLAN LP models will always be nothing more than models; therefore the judgment of resource specialists familiar with the land is crucial for successful implementation of the Forest Plan.

The tenth step, monitoring, is not discussed here, because FORPLAN does not apply to it. In sum, the FORPLAN LP model is so central to the Forest Service planning process that, by one count, FORPLAN influences or drives four out of the ten steps in the overall planning process (Kent 1989).

Disadvantage of Sole Reliance on FORPLAN LP Models in Planning.   Although we have documented several advantages of the FORPLAN-type LP models, there are some arguments against using them to implement National Forest planning. These arguments include technical concerns, such as the characteristics of production relationships on National Forests, which result in violations of the assumptions of LP models. (The assumptions of LP models were discussed in chapter 8.) Other arguments relate to the complexity of LP models, which results in the general public being forced to "trust the experts" (Garcia 1987) rather than understanding how the Forest Service staff estimated the environmental effects or determined the output levels. Even many Forest Service ID team specialists viewed FORPLAN as a "black box" they did not understand (Garcia 1987; Shands et al. 1990:50). Alternatives to using LP models, such as comprehensive benefit-cost analysis performed with a spreadsheet or a qualitative multiattribute simulation model called EZ-IMPACT, may be more appropriate for some non-timber-oriented forests (Behan 1992). A spreadsheet approach to benefit-cost analysis will be discussed later in this chapter. Nonetheless, at the time NFMA was to be implemented, LP models were probably one of the best tools available for dealing with both timber scheduling and integrated natural resource planning (Garcia 1987). Since the FORPLAN LP model has been the official analysis tool of nearly all the Forest Plans performed to date and it is likely that FORPLAN or similar computer models will see continued use in the next round of forest planning as well (Shands et al. 1990:51), it is worthwhile to view a case study National Forest from start to finish.

Example National Forest Plan: Siuslaw National Forest

Many excellent National Forest Plans would serve as good examples for this review. Of the many Forest Plans I have seen, those especially worth reviewing include the Eldorado in California, the Beaverhead in Montana, the Nez Perce in Idaho, and the Arapahoe Roosevelt in Colorado. Of course it is often most interesting to review a Forest Plan and an EIS for an area you are familiar with. Check the local libraries; they often have these local plans and EIS's on file.

The Forest Plan selected for this case study is one of the more "user-friendly" ones, the Draft Siuslaw National Forest Plan. Figure 9.5 shows a map of the Siuslaw National Forest in Oregon. The forest covers what is known as the Coast Range of Oregon. The Siuslaw National Forest is a coastal forest with an average stream density of 9 miles of stream for every square mile of land (Forest Service 1986).

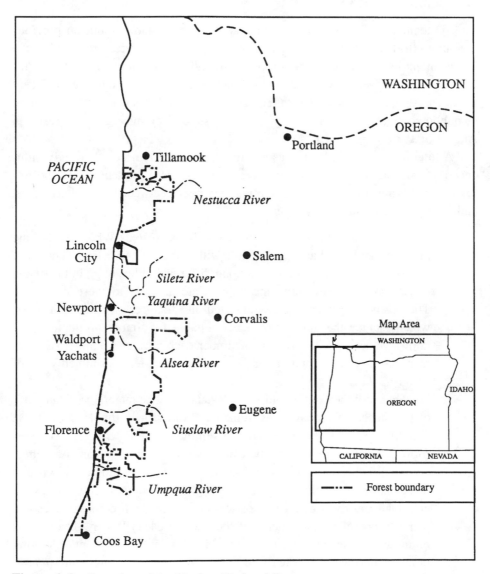

Figure 9.5.   Location of the Siuslaw National Forest

These streams contain highly prized anadromous fish. However, the forest also has one of the highest values of timber on a per acre basis of any National Forest. Given the proximity of highly valued fish and timber, the interaction of the two resources is a major concern of land management practices. We shall only touch the surface of the 200-page Draft EIS and 100-page Draft Plan. The reader who wants the details should see the full documents.

Development of ICOs

The ID team compiled a set of preliminary ICOs from public input on previous planning efforts and from Forest Service employees. These were presented to the management team (forest supervisor, his staff, and district rangers) for their comments. Then a revised list was sent to three thousand individuals, agencies, adjacent landowners, and organizations for their comments. The ICOs were revised in light of public comment and approved by the forest supervisor and regional forester in August of 1980.

A total of twenty-five ICOs was identified, of which fourteen "strongly influenced" formulation and analysis of alternatives. Of these fourteen ICOs, the following twelve are of interest in highlighting the linkage between ICOs, alternatives, and multiple-use resource allocations:

1. How much and what kind of timber will be harvested? Where, with what practices, and on what time schedule will timber be harvested? There is of course quite a range of opinion among different publics (e.g., industry and conservationists) regarding the appropriate answers to these questions. These general concerns were translated into four performance indicators, which include the number of acres suitable for timber production, the allowable sale quantity (ASQ) of timber each year, the number of acres managed at various rotation lengths, and the long-term sustained yield of wood.

2. How much of the existing old-growth stands will be maintained? Only 5% of the Siuslaw National Forest remains in old-growth timber. How these 34,000 acres are to be managed is a significant public concern. The performance indicator is the portion of existing old-growth stands retained.

3. How will lands be managed to maintain stable watershed conditions and meet state water quality standards? The plan states that timber harvesting and road construction can affect water quality for both domestic uses and fish. To meet the requirements of the Federal Clean Water Act of 1977, minimum management requirements (MMRs) limit the amount of logging in any watershed over a ten-year period, minimize logging near streams, and preclude logging land on unstable slopes. However, one alternative

will be formulated that excludes timber harvesting in municipal watersheds. The performance indicators for this ICO are estimated number of
landslides associated with timber harvesting and amount of sediment
produced.

4. What quantity and quality of anadromous fish habitat will be provided on
the forest? The public, the fishing industry, and other public resource management agencies are interested in maintaining productive fish habitat on
the five (out of the state of Oregon's seven) coastal anadromous fish
streams contained within the forest. The key concern is centers on the
effects of timber harvesting and on aquatic habitat quality (e.g., sediment,
stream temperature, and so on). The primary performance indicators are
the amount and condition of anadromous fish, such as salmon and steelhead. The actual performance indicator used is Coho Salmon Habitat Capability Index (number of smolts) developed using the Fish Habitat Index
Models reviewed in chapter 5.

5. How much habitat will be provided for wildlife and threatened and endangered species, and how and where will these habitats be managed? The
public's concern focused on elk and adequate habitat for viable populations
of other wildlife, including existing endangered species such as the bald
eagle and old-growth-dependent species such as the spotted owl. These
concerns are addressed through MMRs for all alternatives and in the
emphasis of particular alternatives. The performance indicators include
habitat capability indices for species such as elk and number of pairs of
spotted owls. (This plan was initiated before the northern spotted owl was
listed as a threatened species.)

6. What diversity of recreation opportunities will be provided? The public is
concerned primarily about Forest Service supply of semiprimitive nonmotorized (SPNM) recreational opportunities, such as hiking, backpacking,
and so on. Demand for such recreation was expected to have exceeded the
current supply by 1990. Only by reserving currently unroaded lands and
building more trails can the quantity of SPNM recreation be expected to
meet demand. The performance indicators are number of acres in unroaded
(e.g., undeveloped) condition and percentage of future SPMN recreation
demand met.

7. How much of the forest will be managed as Special Interest Areas (SIAs)?
SIAs can be designated by the regional forester to protect unusual scenic,
historic, scientific, or other special-use values. The public has expressed
an interest in having more SIAs on the forest. There are three potential
areas for SIA designation on the Siuslaw National Forest. Various alternatives will contain additional acreage for SIAs, with acreage and number
of SIAs used as performance indicators.

8. Which areas will be managed to maintain or enhance visual quality or scenic resources? Landscapes seen from highways, rivers, and developed recreation areas are called viewsheds. Protecting scenic quality (by eliminating timber harvests or modifying timber harvest layout) can contribute to an enjoyable visit to and through the forest. The performance indicator is a visual quality index (VSQ).

9. and 10.   Which areas of the forest will be managed as undeveloped areas? Which as Research Natural Areas (RNAs)? There are currently seven roadless areas, any of which could be either maintained in an undeveloped condition (to allow future consideration as Wilderness) or developed. Five more RNAs could be added to the existing two. The performance indicators for both of these factors are number of areas (undeveloped and NRAs) and acreage of both.

11. How will management of forest resources affect local communities? Forest management activities and resulting outputs influence job opportunities, incomes, and the way of life (aesthetic and recreational ties) of residents in nearby communities. Forest outputs have provided a base for lumber and wood products, commercial fishing, and tourism. In addition, a portion of receipts from the sale of forest outputs (primarily timber) goes to local counties surrounding the forest. As might be expected, those who see the forest as providing hunting, fishing, and camping and those who see it primarily as a source of commodity production do not agree on many aspects of its use. Two performance indicators are jobs and payments to local counties.

12. What economic value will the forest resources generate in the future? There is both a national and local concern about the economic value of the forest outputs. There is also a concern over the size of the forest budget and returns to the U.S. Treasury from the Siuslaw National Forest. The three performance indicators are net present value (NPV), Forest Service budget, and net receipts.

Three ICOs Determined to Be Outside the Scope of the Planning Process

Three ICOs, relating to (1) the need for more law enforcement personnel on the forest, (2) the overall number of Forest Service personnel on the forest, and (3) the need for interagency coordination were deemed to be resolvable outside the planning process. In particular, the first two issues would be addressed in normal personnel staffing decisions. The last one would have to be resolved at the national level.

The way each alternative addresses the twelve primary ICOs will be illustrated in the following sections. The performance indicators for each ICO must be estimated for each alternative (step 6 of the planning process) so as to allow comparison among alternatives (step 7 of the planning process). Therefore in determining data collection efforts, the ID team must ensure that data and models are available to quantify these performance indicators.

Data Collection

Data collection was carried out by the ID team relying on existing data, except where new data were needed "to help resolve sensitive issues or management concerns" (Forest Service 1986:B-7). Data were collected on resource capabilities, existing and future supply and demand, expected outputs, benefits, and costs.

Analysis of the Management Situation

As part of its analysis of the management situation, the Siuslaw National Forest classified lands into fourteen land type associations. For example, land type association D consists of gentle and relatively stable slopes, although soil slumps along roads are still common. The basic land classification that land type association D belongs to makes up about one-third of the forest (Forest Service 1986). Another third consists of land type association C. This type represents unstable areas with steep slopes. Timber management on this land runs a high risk of landslides and debris avalanches (Forest Service 1986). Figure 9.6 provides a map illustrating locations of the two major and similar land types on the Siuslaw National Forest. In addition to these fourteen land types, Douglas fir was modeled on three land productivity (site) classifications and ten age classes (ranging from ten-year-old trees to two-hundred-year-old trees).

Ten management actions or prescriptions were developed for such multiple-use resources as timber, wildlife, and visual quality. One of the prescriptions is minimal management; the other nine involve some form of timber harvesting in combination with other actions to protect or enhance the other multiple-use resources. For example, the visual quality prescription requires that when timber is harvested within 800 feet of roads, the loggers leave at least three trees per acre, leave some trees over 3 feet in diameter, and attain a ratio of 85% conifer to 15% hardwoods (Forest Service 1986:B-33).

The next task was to develop the technical coefficients that relate these ten management prescriptions to multiple-use outputs and environmental effects. These technical coefficients were developed by the ID team (using submodels such as Fish Habitat Index models reviewed in chapter 5). The cost of each of these ten management prescriptions as well as the benefits of the outputs were also developed at this stage in the planning process. Both the FORPLAN and IMPLAN models were constructed in this step as well. With the models constructed, a series of

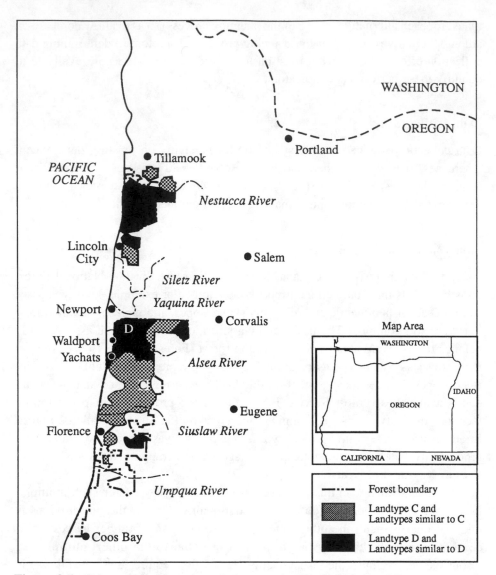

Figure 9.6.   Map of Landtype Associations on the Siuslaw National Forest

benchmarks was run to determine the maximum net present value, maximum timber output, fish output, recreation opportunities, and nongame habitat that could be provided. All these runs were subject to the minimum management requirement constraints. Finally, the FORPLAN and IMPLAN models were run with continuation of the current forest management direction to serve as a baseline for evaluating alternatives that developed to respond to the ICOs.

Development of Alternatives

Two alternatives were developed as required by NFMA: the continuation of current forest management direction (the no-action alternative) and one reflecting the Siuslaw National Forest's share of national RPA resource objectives (i.e., output targets). The Regional Office provided two more themes for alternatives: one emphasizing market/commodity outputs (e.g., timber) and another emphasizing nonmarket/amenity outputs (e.g., wildlife and dispersed recreation). Remaining alternatives were formulated to address specific ICOs. The alternatives varied in terms of how much land would be available for allocation to each management emphasis (old growth, scenic viewsheds, timber harvesting, and so on) and in terms of the management practices that could be chosen by FORPLAN for those lands.

What seems apparent in the Siuslaw National Forest case study, and is more obvious in many other Forest Plans, is that a preferred alternative is constructed at this stage in the planning process even before the effects have been estimated (step 6) and the alternatives evaluated (step 7). In essence, a package or bundle of management actions acceptable to the management team is formulated and called the preferred alternative. In some cases, the elements of the preferred alternative are drawn from the features (management prescriptions, land suitability determinations, and so on) that the management team finds attractive in the other alternatives. That is, the management team indicates it likes one feature of Alternative A, another of Alternative B, and so on, and then combines them into one preferred alternative. This alternative is checked for production feasibility; if it is feasible, it is made into the preferred alternative at this step in the planning process. This preselection of the preferred alternative can introduce several biases in the analysis. It fails to take full advantage of the information that will be generated in the next two steps of the planning process (step 6, estimating the effects, and step 7, evaluating the alternatives) and somewhat violates the neutrality of the planning regulations and hence the spirit of NFMA. In addition, it opens the door for further abuses, such as making the preferred alternative look desirable compared to the other alternatives in step 7, by building in undesirable or inefficient features (management practices, land suitabilities, or severe but unnecessary constraints on other resources) into the other alternatives.

The following is a list of the nine alternatives and the theme or intent of each.

1. Alternative A continues the current management direction, as reflected in pre-NFMA unit management plans. It emphasizes wood production, elk, and bald eagles.
2. Alternative B stresses wood production with other multiple uses at levels that would meet the minimum management requirements (MMRs) or exceed the MMRs only if compatible with timber production.

3. Alternative B—Departure from Nondeclining Even Flow. This alternative gives even greater emphasis to timber production by departing from the requirement of nondeclining even flow of timber over time. It responds to both the timber objectives of the Oregon Department of Forestry and the timber goals in the RPA. Thus it is also the RPA alternative, although RPA targets are met for some but not all other multiple uses.

4. Alternative C emphasizes economically efficient production of wood while producing big game (particularly elk) and a variety of recreational opportunities. Elk production would increase by a factor of 3 by the fifth decade of the planning period.

5. Alternative D is the market emphasis alternative. It emphasizes wood; commercial fish, such as salmon; and developed recreation activities for which a fee is paid.

6. Alternative E emphasizes economically efficient wood production and provision of a variety of recreational opportunities and wildlife habitats. The

Table 9.4a
Alternative E: Point-Counterpoint

Point	Selected Issues	Counterpoint
The ASQ* during the first decade (62.2 MMCF/year) would be 9% greater than ten-year average harvest.	1. Timber	The ASQ would be 31% lower than proposed in present plans and about 36% lower than RPA and Oregon Dept. of Forestry expectations.
75% of all suitable timber stands would be managed on long rotations (ninety years) for wildlife, which would benefit many resources.		There would be a 10% difference between the LTSY† capacity (68.4 MMCF/year) and the ASQ (62.2 MMCF/year) during the 1st decade.
Old-growth stands with high economic value would be available for harvest.	2. Old growth	Harvest of old-growth stands would reduce availability of a limited resource for aesthetic, recreational, scientific, and wildlife use.
Protection of part of the riparian zone would provide some benefits to fish habitat and watershed condition.	4. Fish	Unstable slopes outside riparian zone would be protected at MMR levels. Fish populations would be 59% of ODFW goals.
The two wildlife habitats that are in shortest supply would increase. Habitat for forty-six pairs of spotted owls in older forests and 72,000 acres of deciduous mix habitat would be provided.	5. Wildlife	No single species or group of species would be emphasized.
Elk populations would increase to 7,500 animals.		Elk populations would be lower than the 10,000 animals expected by the Oregon Dept. of Fish and Wildlife.

*ASQ = allowable sale quantity.
†LTSY = long-term sustained yield.

key feature is on longer timber rotations. This alternative eventually became the preferred alternative.

7. Alternative F would emphasize recreation and protection of scenic resources as well as habitat for fish and nongame wildlife. Only "moderate" amounts of timber would be produced.

8. Alternative G emphasizes production of nonmarketed multiple uses, such as water quality, fish, wildlife, dispersed recreation, and scenic resources. Nearly all land suitable for timber would be managed on long rotations (60% with rotations of one hundred years or more).

9. The goal of alternative H is to preserve natural systems in large areas of the forest, protect nongame wildlife habitats, and provide maximum protection to municipal watersheds.

This list of alternatives provides a wide spectrum of future ideas of what the Siuslaw National Forest could be. Each of these alternatives was compared to the key ICOs to determine what contributions it made toward addressing particular issues or resolving particular concerns. Table 9.4 shows one of the Siuslaw National Forest's "Point-Counterpoints" for Alternative E on the key ICOs.

Table 9.4b
Alternative E: Point-Counterpoint

Point	Selected Issues	Counterpoint
Scenery would be fully or partially protected along more than half of the visually important roads on the forest.	9. Visual	Scenery would not be protected along about half of the visually important roads.
Wassen Creek and Drift Creek adjacent lands would be maintained as undeveloped areas.	11. Undeveloped areas	Hebo-Nestucca and N. Fork Smith River would not be maintained as undeveloped areas; projected SPNM demand would not be met after 2006.
Three areas would be proposed as potential RNAs.	12. Research	Two potential areas would not be proposed as RNAs.
Opportunities for some personal uses of the forest—such as elk hunting, firewood gathering, fishing, viewing of scenery, and roaded recreation—would increase.	13. Communities	Some people who want the forest used for nonmarket resources would be less satisfied as the levels of old growth were reduced.
PNV would be $2 billion.	14. Economics	Long timber rotations would lower PNV.
Payments to the counties would increase.		Costs to operate would be the same as now ($26 million/year); receipts would be $64 million/year.

Estimation of Effects of Each Alternative

The effects of each alternative were estimated using FORPLAN, IMPLAN, and a host of submodels, such as a Fish Habitat Index model, that were previously described in detail.

Table 9.5 provides a detailed numerical comparison of each alternative on the performance indicators that relate to the ICOs. The comparisons involve both first-decade and fifth-decade forecasts. Although some factors are called indices, they directly translate into absolute numbers (e.g., the elk index is really number of elk).

Figure 9.7 provides a better perspective on how alternatives compare in terms of balanced multiple use and proportion of maximum possible production actually attained for each resource. This figure is adapted from the Forest Plan summary guide and does an excellent job of comparing each alternative's mix of multiple uses. The amount of the striped pie slice illustrates the proportion of the maximum potential output (e.g., benchmark) produced by that alternative. By comparing the proportions of the striped slices as one goes around the pie from resource to resource it is possible to get an idea of how balanced each alternative is. Specifically, Alternative D does not represent balanced multiple use (given the representations shown in these pie charts). Alternative F looks more like balanced multiple use, given the outputs illustrated in these pie charts.

The last technique used to give the reader a perspective of what the Siuslaw National Forest would look like under each alternative is given in figure 9.8 as bar charts. The bars trace out the percentage change from existing conditions. Since most people are familiar with the forest as it exists now, this serves as a useful reference point from which to describe changes. The public can view these to see which resources will have large changes under each alternative.

Selection of Preferred Alternative

The forest supervisor and regional forester selected Alternative E as the preferred alternative in the Draft EIS. The ID team then developed this preferred alternative into a detailed Proposed Forest Plan. Both the Draft EIS and Proposed Forest Plan were presented to the public for comment during a 120-day public comment period.

During the public comment period the forest received over 3,600 letters, although nearly two-thirds of these were form letters from a local timber operators' group. About 75% of the total comments were from Oregon, but comments were received from twenty other states. The Forest Service made several changes in the Final EIS and Forest Plan in response to both public comments and changing Forest Service direction (in response to habitat preservation for the northern spotted owl, changes to bring the Siuslaw into conformity with Region 6 unwanted vegetation

Table 9.5a
Summary of Selected Indicators by Alternative

| | | Alternatives | | | | | | | |
Variables	Today	A (Cur)*	B	C	D	E (Pref)†	F	G	H
Timber (1st decade)									
Long-term sustained yield (LTSY)		73	75	72	66	68	58	30	21
Allowable sale quantity (ASQ)		69	71	68	62	62	55	30	20
Acres harvested		6,750	6,340	6,329	5,821	5,909	4,990	2,766	1,836
Old-Growth Stands									
M acres existing maintained (fifth decade)	598	17	16	18	16	19	22	34	34
Fish									
Smolt Index (1st decade)		851	845	895	952	911	1,000	1,172	1,202
Wildlife									
Elk Index (5th decade)‡	3,000	4,880	6,000	12,560	2,430	7,470	2,050	5,500	760
Spotted owl pairs—existing maintained (1st–5th decades)	39	37/26	37/26	37/26	37/26	37/32	39/39	39/39	39/39
Spotted owl habitat areas	22	40	40	40	40	46	53	62	72
Special-Interest Areas									
Number of areas	1	2	2	3	2	4	3	4	3
Thousands of acres	2	3	2	5	1	6	7	7	5

*Current direction alternative.
†Preferred alternative.
‡Oregon Department of Fish and Wildlife (ODFW) goals are 10,000 elk.

Table 9.5b

Summary of Selected Indicators by Alternative

Variables	Today	Alternatives							
		A (Cur)*	B	C	D	E (Pref)†	F	G	H
Recreation									
% SPNM Demand met 5th decade	—	24	31	54	27	54	70	83	71
Visual Resource									
Quality index (100 is high)	—	30	7	22	9	35	40	62	71
Undeveloped Areas									
Number		0	0	2	0	2	3	4	4
Thousands of acres		0	0	8	0	12	17	21	37
Employment									
Thousands of jobs (1st decade)	8	9	10	10	9	9	8	6	5
Payments to Local Governments									
Millions of dollars (1st decade)	12	17	18	18	16	16	14	8	5
Receipts									
MM dollars (1st decade)	46	70	73	70	64	64	57	31	20
Budget									
Millions of dollars (1st decade)	26	20	28	28	26	26	22	17	14
PNV @4‡									
Billions of dollars	—	2	2	2	2	2	2	1	1
Lands suitable for timber production									
Timber M acres	—	394	414	393	354	383	317	189	144
%60–80 year harvest rotations	94	61	66	62	63	26	54	39	0
%90–100 year harvest rotations	4	35	34	36	36	62	45	61	52
%110+ year harvest rotations	2	4	0	2	1	12	1	0	48

*Current direction alternative

†Preferred alternative.

‡Monetary value of the Forest discounted to the present time at a 4% discount rate.

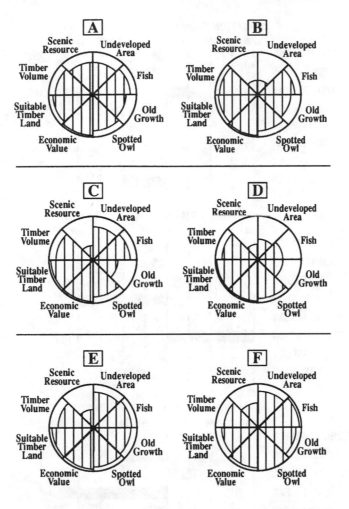

The total area of a wedge represents the maximum potential output which can be obtained. The vertical area indicates how much of the maximum potential output is produced in each alternative.

Figure 9.7. Graphic Comparison of Alternatives

managcment guidelines, and so on). Changes to the Final Preferred Alternative (hence the Final Forest Plan) included increasing the percentage of riparian zone protected by streamside buffers from 50% to 75%, increasing the amount of land on relatively short-rotation timber harvesting from 26% to 74% to maintain current timber harvest levels, and reducing roadless acreage by 4,100 acres. Further details are provided in the Land and Resource Management Plan for the Siuslaw National Forest.

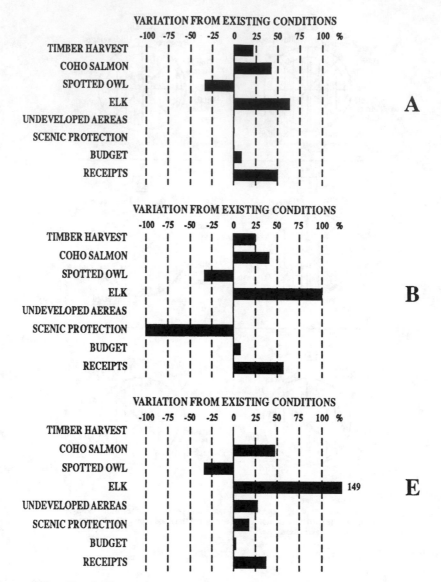

Figure 9.8.   Key Indicators-Percent Change from Existing Conditions

Implementation of Forest Plans: Nez Perce National Forest

Since the Siuslaw National Forest only recently (1990) completed its Final Plan, we cannot use it as an example of implementation of a Forest Plan. In addition, since the passage of NFMA so few Forest Plans have been implemented for a sufficient duration that it is difficult to generalize about this step. However, the Nez Perce National Forest in Idaho is an example of a careful process of implementation, monitoring, and informing the public by publishing annual reports of its progress.

Two key types of monitoring are performed: (1) implementation monitoring, to determine whether the management prescriptions, minimum management requirements (MMRs), resource improvement projects, and so on, have actually been carried out on the "ground" as scheduled; and (2) effectiveness monitoring, to determine whether the management practices and MMRs resulted in the types and amounts of multiple-use outputs expected with the acceptable environmental effects predicted by the plan. Money and personnel required to implement each planned resource activity are compared with what was actually received from Congress and from a variety of special funds.

Impressions of the First Round of Forest Planning

The early stages of NFMA forest planning were subject to high expectations both inside and outside the Forest Service (Larsen et al. 1990). Although comprehensive planning was frustrating for both the agency and the public, many within and outside the agency believe the Forest Service has made a reasonable attempt to comply with NFMA (Larsen et al. 1990:3). Much was learned in the process, and the Forest Service has made great efforts to critique this first round of forest planning (Baltic et al. 1989). Several Forest Service task forces interviewed academics and Forest Service personnel and read the many critiques of forest planning. The objective has been to learn from this experience so that the next round of forest planning (which will begin in the early 1990s) will benefit from this first experience. The results of this critique, the costs of forest planning, and the proposed planning regulations for the second round of forest planning will be reviewed in chapter 11, after we discuss the other federal multiple-use agency, the Bureau of Land Management. However, among the many conclusions of the forest planning critique was the need to provide simpler and more direct analytical methods than FOR-PLAN-type models. It is with this critique in mind that we provide an example of National Forest management planning that can be done with a simple microcomputer spreadsheet.

A Comprehensive BCA Approach to Evaluating Multiple-Use Trade-offs

Several National Forests have little or no commercial timber. Hence reliance on an extended timber-scheduling model such as FORPLAN may be an unnecessarily complicated approach to National Forest planning. In addition, some forests are dominated by such issues as roadless area suitability for Wilderness designation or free-flowing-river suitability for Wild and Scenic River designation. These actions involve large or indivisible areas for which decisions must be made between mutually exclusive alternatives instead of developing a plan that blends a little of one multiple use with a little of another. Lastly, joint production that characterizes multiple use may be handled more simply in a benefit-cost analysis, where the total benefits are compared to total costs for alternative bundles of joint products.

In these situations the agency could apply a comprehensive benefit-cost analysis to multiple use as follows: (1) Have the ID determine which bundles of multiple uses go together as joint products. Different bundles could represent different management emphases. These different themes would become alternatives in the plan. (2) Determine the expected levels of outputs from each bundle or combination of management actions using whatever submodels are required. Hydrology models, fisheries models, wildlife models, and so on, would be used sequentially to predict resource effects. For example, a timber-scheduling model might be run, with harvest levels and number of acres harvested as outputs. The number of acres harvested would feed into a "timber-sediment" model that would predict changes in water yield and sediment. These outputs would be input into a fisheries model that relates water flows and sedimentation to fish production. The output of that model would be of interest as a biological indicator but also could be translated into recreation fishing days and economic value by inputting change in fish populations into a recreational fishing demand function.

The next case study illustrates use of a comprehensive benefit-cost analysis (BCA) approach to one particular forest resource allocation question: whether to recommend designation of one roadless area as Wilderness or to allow timber harvesting. Unlike the Siuslaw National Forest case study, this analysis does not represent an entire Forest Plan. Therefore it does not strictly follow the steps in the forest planning process. Instead the case study illustrates how to compare the economic value of two different bundles of multiple uses that would flow from the two alternative management prescriptions. At the same time, the issue of below-cost timber sales is evaluated. This case study also illustrates the integration of nonmarket valuation techniques, such as travel cost and contingent valuation methods, into multiple-use planning.

Gallatin National Forest Case Study

Study Area*

Within the Gallatin National Forest is the "Hyalite-Porcupine Buffalo Horn" Wilderness Study Area (hereafter referred to as HPBH WSA). This roadless area was designated a Wilderness Study Area (WSA) in 1977, when Congress passed the Montana Wilderness Study Act (PL 95–150). This act required the Forest Service to evaluate the HPBH for suitability as Wilderness. This evaluation was done as

---

*Portions of this section are reprinted from the Journal of Forestry (Vol 87, no. 3) published by the Society of American Foresters, 5400 Grosevenor Lane, Bethesda, MD. Not for further reproduction.

part of the Gallatin National Forest Plan, although the analysis is presented in a separate report.

The HPBH WSA contains approximately 155,000 acres in the Gallatin Range in southwestern Montana. This land includes the mountain divide between the Gallatin and Yellowstone rivers. The WSA provides habitat for important trout fisheries in the Yellowstone and Gallatin rivers and trophy elk. The sensitivity of the watershed to development activities was identified during public workshops and written comments as one of the fourteen ICOs to be addressed in the Forest Service evaluation of this WSA. Figure 9.9 provides a general location map of the WSA relative to southwestern Montana.

For the purposes of this case study, the two alternatives to be evaluated are (1) the complete-wilderness alternative (formally, Alternative 5 in the Forest Ser-

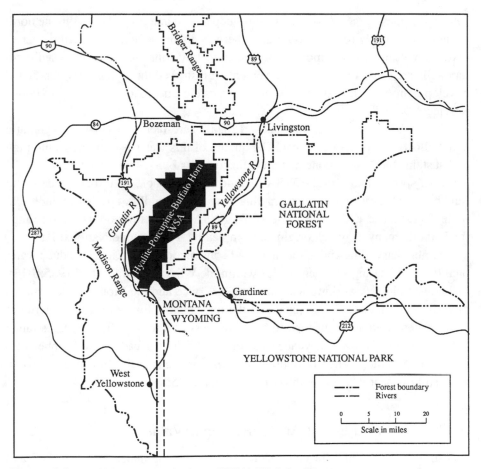

Figure 9.9. Wilderness Study Area (WSA) Vicinity Map

vice Analysis) and a development alternative (Alternative 2). Alternative 5 protects 145,000 acres out of the 155,000 acres of public and private land within the WSA as Wilderness. It provides for no timber harvesting, no road construction, and no motorized access but maintains water quality, fishing, trophy elk hunting, and primitive recreation. In contrast, Alternative 2 would not recommend any acres as formal wilderness, although approximately 60% to 65% of the Forest Service land, or nearly 70,000 acres, would be left in a "near natural" state (Forest Service 1985:II-53). Under Alternative 2, 21,354 acres of national forestland are deemed suitable for timber production and 5,383 acres of national forestland are deemed suitable for intensive grazing and an expansion of motorized recreation.

Description of the Environmental Setting and Resources

The topography of the WSA varies from steep terrain and rugged peaks in the north to more moderately rolling terrain elsewhere. Soils in the area range from coarse-textured volcanic soils to more erosive sedimentary soils. Some of these sedimentary soils are prone to mass soil movements if disturbed through poorly conducted development activities, such as timber harvesting and road construction (Forest Service 1985:S-8).

The WSA provides approximately 126,000 acre-feet of water to the Gallatin and Yellowstone rivers each year. The quality of this water is currently quite high and sustains the Yellowstone and Gallatin rivers as blue-ribbon trout streams.

Fish species found in the WSA include several brook, cutthroat, golden, and rainbow trout as well as arctic grayling. Big-game wildlife species include elk, mule deer, moose, and bighorn sheep. The WSA provides important winter range for 240 elk, many of which originate from nearby Yellowstone National Park.

Timber harvesting and associated road construction can seriously reduce water quality by increasing sediment yields within the WSA (Forest Service 1985:S-13). Fisheries biologists working with the Forest Service and the Montana Department of Fish, Wildlife, and Parks quantified the unavoidable losses in trout fisheries if development were to occur instead of Wilderness designation. These losses range from a low of 130 fish per year in the first and fourth decade, when timber harvesting is minimal, to nearly five hundred fish per year in the fifth decade, when harvesting is near its maximum (Forest Service 1985:IV-3).

Analysis of Fisheries and Angler Economic Value

To estimate the demand for and benefits of trout fishing under Alternatives 2 and 5, the travel cost method (TCM) will be used. Briefly, this method uses angler visits as a measure of quantity and travel costs (including travel time) as a measure of price to trace out the demand curve for fishing at a particular site. From the demand curve, anglers' net willingness to pay for fishing at that site can be estimated. To estimate changes in the value of fishing with changes in trout populations, angler visitation data across several rivers are pooled together and one multisite or "regional" TCM demand equation is estimated using multiple regression. This method was discussed in detail in chapter 6. The details of this particular application of the TCM follow.

The data used with the travel cost method to estimate willingness to pay for recreational fishing were collected from two surveys of Montana anglers in 1985 (Duffield et al. 1987).

To measure the net economic values of fishing, different TCM demand equations were used for the general streams within a watershed (tributaries to the upper Yellowstone study section 31 and the Gallatin study section 32) and the mainstem rivers themselves (Gallatin and upper Yellowstone). The models reflect pooled zonal travel cost demand equations. The equation for general watershed tributary streams was

$$LTRIPCAP = 2.471 - 2.619(LRTDIST) + 0.246(LSTROUTC) - 0.885(LYRSFISH)$$
$$(t\text{-statistics}) \ (2.810) \quad (-53.387) \qquad (3.897) \qquad (-8.530)$$

$$(9.3)$$

$$+ 1.492(LEDUC) - 0.017(LSUBTRTC) + 0.079(LSOTHRC)$$
$$(4.180) \qquad (-3.221) \qquad (1.955)$$

where

> *LTRIPCAP* $=$ log of trips per capita from county of angler origin i to river j
> *LRTDIST* $=$ log of round-trip distance from county i to river j
> *LSTROUTC* $=$ log of sum of trout catch at river j
> *LYRSFISH* $=$ log of average years fished of anglers in county i
> *LEDUC* $=$ log of average years of education of anglers in county i
> *LSBUSTRTC* $=$ log of substitute site fishing index. Index is based on trout catch per mile at alternative rivers k with higher fish catch per mile than the study river j
> *LSOTHRC* $=$ log of sum of other sport fish catch at river j (mostly whitefish)

This demand equation has an adjusted $R^2 = .819$. The individual coefficients are significant at the 95% level or greater. The coefficient on distance (the price variable) is highly significant. The small standard error on this coefficient indicates it is precisely estimated. The R^2 is quite high, indicating that nearly 82% of the variation in trips per capita is explained by the set of independent variables. The equation also contains statistically significant variables for the influence of substitute rivers and fish catch.

The demand equation developed by Duffield and colleagues (1987) for the mainstem rivers is as follows:

LTRIPCAP = 1.855 − 2.753(LRTDIST) + 0.314(LSTROUTC) − 1.072(LYRSFISH)
(*t*-statistics) (1.508)   (− 36.742)   (3.886)   (− 7.622)

$$(9.4)$$

+ 2.052(LEDUC) + 0.328(LNOSITER) − 0.015(LSOTHRC)
       (4.330)   (2.691)   (− 2.170)

where LNOSITER is the log of the number of developed recreational sites (other variables as previously noted).

This equation has an adjusted R^2 of .808. All the coefficients are significant at the 95% level or greater. The coefficient on trout catch is statistically significant. The very high *t* value of distance implies that our price variable is highly significant.

Process for Using TCM Equations to Calculate Angler Benefits

Consumer surplus estimates for the mainstem rivers (upper Yellowstone and Gallatin) and general tributary rivers (31 and 32) were calculated for each of two alternative expected trout populations associated with the two Wilderness study alternatives. This was done by changing the fish catch variable in the TCM demand equations estimated earlier. A river's total consumer surplus under the Wilderness alternative is estimated with the site's existing trout catch; then catch is reduced by the loss in catchable trout expected under the timber alternative. The reduction in catchable trout variable in the demand equation shifts the travel cost method demand curve inward to D_2 as illustrated in figure 9.10. The striped area can be thought of as either the loss in fishing benefits with timber development or the gain in benefits (WTP) with maintaining the area as Wilderness.

This shifting demand curve process is repeated for each mainstem river and tributary rivers for each decade. The present value of the change in fishing benefits is calculated as the present value of the striped area in figure 9.10 over the affected rivers over fifty years. The change in recreational fishing benefits so calculated is $2.073 million in 1978 dollars.

Effects of Timber Harvest on Elk Populations

In addition to increasing sediment in streams, timber harvests reduce the effectiveness of habitat for elk. An interagency research project concluded that elk will generally not use habitat within a half mile of a road open to traffic (Lyon et al. 1985:6). With spacing of logging roads as close as every quarter mile, large areas of habitat are effectively lost to elk when timber harvesting occurs.

The additional human access afforded by roads results in greater hunting pres-

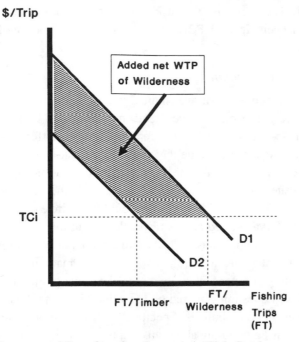

Figure 9.10.   Incremental Benefits of an Increase in Fish Catch

sure. When combined with the effect of timber harvesting, which reduces cover for elk, the net effect appears to be a change in the structure of animals harvested. Specifically, greater access and less cover result in a higher harvest rate, particularly of younger animals. Over time this results in the harvest of fewer large bull elk (six-point or better antlers) and more young bulls (two points or less). Some hunters identify opportunities to bag a trophy bull elk as a higher-quality elk-hunting experience.

Elk also tend to avoid cattle. Elk are only half as likely to use a tract of land as habitat if cattle are present (Lyon et al. 1985:13–14). Since the Gallatin National Forest proposed in Alternative 2 (the timber-development alternative) to increase the amount of cattle grazing by nearly 2,000% (from current 290 AUMs to 5,420 AUMs), there will be an additional loss in habitat over and above the road-related effect in some areas of elk habitat.

One approach to estimate the losses in recreational elk hunting from the reduction in elk habitat is to evaluate the change in quality of elk hunting associated with the effect of timber harvesting and cattle grazing. In particular, the timber harvesting, roads, and increased cattle grazing reduce the elk habitat and shift the distribution of harvest away from trophy elk toward younger, smaller bull elk. To measure the change in value of elk hunting with harvest of fewer trophy bull elk a series of willingness-to-pay questions was asked using the contingent valuation method (CVM). The questions were asked of Montana elk hunters visiting the two hunt districts that contain the HPBH WSA. Details of this portion of the case study follow.

Data Sources for Elk Hunting

A questionnaire in booklet form was mailed to a sample of elk hunters. Details of the survey and response rate can be found in Loomis and colleagues 1987. The CVM (willingness-to-pay) questions were asked for two different scenarios. First, the elk hunters were asked to value their most recent elk-hunting trip. The dichotomous-choice CVM question for the most recent elk-hunting trip to the hunt area most recently visited was, ''Would you still have made the trip if your share of the expenses had been $X more?'' The hunter would then circle either *Yes* or *No.* The dollar amount (X) was varied across respondents, but the maximum amount any elk hunter was asked to pay was $1,100 more than his or her current costs.

The context of the question in the overall survey made it very specific to the value of elk hunting at a particular site, not to the value of elk hunting in general.

The next CVM question centered on the value of having double the chance to harvest a six-point or better bull elk. Specifically, this dichotomous-choice question asked, ''Imagine that everything about this last trip were the same, except that your chance of getting a 6-point or better bull elk was twice as great and that your trip costs were $X more than your actual costs. Would you still have made the trip under these circumstances?'' The elk hunter was required to check *Yes* or *No.* Once again, different hunters received different dollar amounts (X).

Logit Equations for the CVM Questions

The answers—''Yes, I would pay'' or ''No, I would not pay''—are analyzed using a statistical technique called logistic regression. The name is derived from the ''logistic distribution'' that the error term is assumed to follow in the utility difference model. (See Hanemann 1984 for more details.)

The candidate-independent variables that are suggested by economic theory include trips (measure of quantity), income, and amount the respondent was asked to pay (X). In addition, certain other variables would be expected to influence the probability of stating that the hunter would pay. For example, in asking about willingness to pay for a current trip, variables reflecting the quality of a current trip, such as number of elk seen, number of other hunters seen, and so on, would be expected to influence the probability an elk hunter would say yes to a given dollar amount.

Equation (9.5) provides the initial specification of the logit equation that relates the log of the odds ratio to our candidate-independent variables.

$$\ln[P(Y))/(1 - P(Y))] = B_0 - B_1(\text{BID}) + B_2(\text{INC}) - B_3(\text{TRIPS})$$
$$+ B_4(\text{ELKSEEN}) - B_5(\text{HTRSEEN}) + B_6(\text{HTRYS}) \quad (9.5)$$

where

> $P(Y)$ = probability that the hunter would pay
> BID = dollar amount of increased trip cost the hunter was asked to pay
> INC = hunter's household income
> $TRIPS$ = number of elk-hunting trips to this area
> $ELKSEEN$ = number of elk seen while hunting in this area
> $HTRSEEN$ = number of hunters not in your party that were seen while hunting in this area (a measure of congestion)
> $HTYRS$ = number of years hunting elk in this area

Basically, these same factors would be expected to affect willingness to pay for double chances of bagging a six-point bull elk or better. However, in this scenario, willingness to pay to increase chances of bagging a bull elk might not be affected by variables such as number of other hunters seen.

Estimation of the Logit Equation

Sellar and colleagues (1986) have demonstrated the relationship between equation (9.5) and standard demand function. That is, a demand equation often relates quantity demanded to price and other variables, such as income. From equation (9.5) it is possible to derive an inverse demand function that relates price or value to quantity demanded, income, and so on. The estimated logit equations are used to predict the probability a hunter will pay higher and higher dollar amounts. In essence, a demand curve is traced out that relates the probability a hunter will pay a given increase and the dollar amount of the increase. Figure 9.11 displays the logit curve for current conditions and double chances for trophy elk for hunt area 40. The average or mean willingness to pay is the expected value of willingness to pay (the sum of the probability times dollar amount across the range of probabilities from zero to 1). In terms of figure 9.11, mean WTP is the area under the respective logit curve. As can be seen in figure 9.11, the WTP for double chances to harvest a trophy elk is greater than that for current conditions.

The area under the current conditions logit curve has a value of $317 and $376 per trip for hunt areas 40 and 41, respectively. These values increase $108 per trip for double chances of harvesting a trophy elk.

To apply these relationships to the HPBH WSA, it was necessary to estimate how the harvest of six-point-or-better bull elk changed relative to that of a five-point-or-less bull elk, as the hunt areas would be roaded for timber harvesting. The current harvest mix is roughly 26% for six points or better and 74% for five points or less. Using this distribution, the value of this mix of hunting could be quantified. To forecast how this mix would change, it was necessary to perform a paired comparison between two hunt districts that were generally similar except that one had an extensive amount of roads and the other did not. The Montana Department

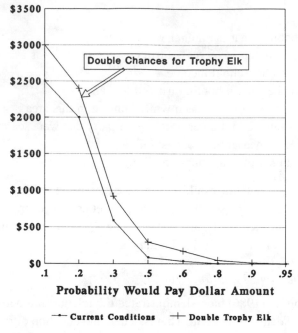

Figure 9.11.   Elk Hunting Logit Curve and WTP

of Fish, Wildlife, and Parks suggested Hunt Districts 332 and 319 would make an adequate comparison. Based on this comparison, the percentage of six-point-or-better bull elk was expected to drop by approximately 5% a decade for each of four decades as roading increased in the HPBH WSA if it was logged. Of course the percentage of five-point-or-less bull elk harvested was expected to increase by 5% a decade over the four decades. Thus, holding hunter days constant, the mix of hunters having a higher-valued experience associated with bagging a six-point-or-better bull elk would decrease. The mix of hunters having a lower-valued hunting experience bagging a five-point bull elk or smaller would increase. Although this approach may be somewhat simplistic, it illustrates an important point: even with the same number of hunters visiting an area, the value can fall over time if the quality of the hunting experience decreases.

Comparison of Benefits and Costs

To allow comparison of benefits and costs of wilderness preservation, all benefits and costs must be measured using a set of assumptions that are internally consistent between alternatives and resources being valued. Unfortunately, the original HPBH WSA report contains a few questionable assumptions. The key factors that must be reconciled involve treatment of wilderness recreation, timber values, livestock AUMs, and wildlife/fisheries recreation.

Table 9.6 displays the change in present value of benefits and costs. Under the benefit categories, negative values mean that timber results in reduced benefits. For the cost category, wilderness results in less costs (e.g., fewer roads to be built) and thus cost savings to wilderness. Costs are treated as negative benefits in computing NPV. Overall, the analysis shows that Wilderness designation has positive net economic benefits of nearly $8 million over a fifty-year planning period.

The joint benefits of timber-related multiple uses (range and motorized recreation) are less than the value of the forgone wilderness-related multiple uses. The fact that the timber-related multiple-use management carries with it higher costs makes timber management in this area particularly uneconomic.

This example also illustrates how a relatively simple spreadsheet-based benefit-cost analysis can account for multiple-use resources and their interactions without the use of a FORPLAN-type LP model. The spreadsheet and other microcomputer-based analysis/simulation approaches, such as GIS, may prove a cost-effective alternative for some forests in the future.

Another View: the New Forestry and Sustainable Forestry

For many commercially valuable tree species in many geographic areas, the traditional view of sustained-yield forestry has been the clear-cutting of "decadent" old-growth trees and their replacement with fast-growing strains of new trees. The emphasis has been on rapid and successful reforestation, coupled with maximizing wood volume. Although this practice has resulted in large amounts of wood, it has

Table 9.6
Comparison of Benefits and Costs of Wilderness Designation

Output	Timber Alternative	Wilderness Alternative	Change with Timber Alternative
	(in thousands of 1978 dollars)		
Timber	2,152	0	2,152
Range	902	70	832
Recreation			
Nonmotorized dispersed	2,105	0	2,105
Roadless/wilderness	1,216	7,794	−6,578
Motorized	860	0	860
Elk hunting	11,938	12,343	−405
Fishing	*	*	−2,073
Present value of benefits			−3,107
Present value of cost	9,472	5,091	4,381
NPV of timber/grazing/roaded recreation			−7,488

*Not reported, because benefits change over the entire watershed affected by land use in the wilderness area, not just visitation to the area. Only the difference is applicable here.

also eliminated complex ecosystems and their associated multilayered canopies. These intricate ecosystems have been replaced by relatively homogeneous ecosystems of even age and species. Often brushy vegetation that might compete with the new seedlings is sprayed with herbicides. Thus one multiple-use benefit of timber harvests, the opening of the canopy for browse species beneficial to many big-game wildlife, is temporarily lost.

There has been growing recognition that both these cutting practices and re-forestation practices are dramatically altering the structure of forested ecosystems. The terms *New Forestry* and *sustainable forestry* reflect the notion that public multiple-use agencies must change their current cutting and reforestation practices. To some people current practices are unsustainable over the long term. Rather than face complete prohibition on timber harvesting because of the ecological unac-ceptability of current forestry practices, these practices of public agencies can be modified. The modifications require that timber-cutting practices simulate natural forces rather than work against them, as current forestry practices now do. For example, cutting patterns would simulate the mosaic created by wildfires or wind-storms. Thus selection-cutting, leaving both dead snags and live trees for wildlife, would be practiced. The layout and borders of timber sales would be changed to give more recognition to relationships with uncut areas. For example, trees would be left in migration corridors to connect remaining large stands. The trend could move away from scattering numerous small clear-cuts that some ecologists believe have resulted in fragmentation of habitat.

Reforestation of both previously clear-cut and new areas would again attempt to simulate nature: there would be diversity in age and species of vegetation. Rather than focusing solely on maximizing wood volume, the function and structure of forest ecosystems would be emphasized in reforestation efforts.

USFS New Perspectives on Forestry

The Shasta Costa Integrated Resource Project on the Siskiyou National Forest in Oregon is serving as a pilot test for what the Forest Service is calling its New Perspectives on Forestry. The Shasta Costa is a 23,000-acre roadless area that is bounded by the Wild Rogue Wilderness Area to the north and that is within a few miles of the Kalmiopsis Wilderness Area to the south. This area represents an important wildlife corridor for many species that move between the two formal Wilderness Areas. There are three overall components to the Shasta Costa effort: New Thinking; New Technologies, and New Alliances.

As the Forest Service states, the New Thinking is in some sense going back to nature. Specifically, it attempts to emulate natural changes and mimic nature rather than work against it (Forest Service 1990b). The term New Technologies refers not specifically to new tools but to how existing tools are combined and used. One of those technologies is greater reliance on Geographic Information Systems for computerized mapping to visualize at the landscape level what the forest will look like in the future under different alternatives. The goal of New Alliances is to bring the greatest possible variety of interested publics together to work on common solutions at every step in the process.

The planning team developed thirteen rules of thumb to guide it in this new forestry:

1. *Follow nature's lead.* Mimic the natural disturbance patterns and recovery stages in the area.
2. *Think big.* Manage for landscape diversity as well as within-stand diversity.
3. *Don't throw out any of the pieces.* Maintain a diverse mix of genes, species, biological communities, and regional ecosystems.
4. *Side with the underdogs.* Prioritize in favor of the species, communities, or processes that are endangered or otherwise warrant special attention, such as the spotted owl, significant old growth, and riparian areas.
5. *Try a different tool.* Diversify silvicultural approaches. Reduce emphasis on clear-cuts.
6. *Keep your options open.* Use existing roads whenever possible.
7. *No forest should be an island.* Minimize fragmentation of continuous forest by cutting adjacent to existing clear-cuts and nibble away at the edge of a stand instead of creating a new hole.
8. *Encourage free travel.* Create a web of connected habitats. Leave broad travel connectors for plants and animals, especially along streams and ridges.
9. *Leave biological legacies.* Select what you leave behind as carefully as what you take out—specifically, standing live and dead trees and fallen logs.
10. *Leave it as nature would.* Leave a mixture of tree sizes and species on the site. Restore diverse forests after harvest.
11. *Be an information hound.* Use the latest studies and state-of-the-art technology to design, monitor, and evaluate new approaches.
12. *Be a critical thinker.* Use only the scientific studies that make sense for your region and social setting.
13. *Monitor, monitor, monitor.* It is the only sure way to tell if you really are conserving biological diversity.

The Forest Service has developed several alternative management strategies for the Shasta Costa area. Many of them incorporate these thirteen principles. For example, timber is harvested so as to emulate such natural disturbances as natural wildfires. In addition, less total timber is removed from the forest. The Shasta Costa Integrated Resource Project will be interesting to watch. Only time will tell if it is a precursor to what citizens can expect in the next round of forest planning or whether something new will evolve from it.

However, it is worth noting that these changes in cutting and reforestation practices will in most cases reduce the amount of wood produced from National Forests. If attention continues to be focused on timber targets, then some believe the new forestry will result in disturbance of far more acres to reach a given level of timber output. If the "New Forestry" is broad enough in vision to recognize that sustainability in forestry is consistent with a more sustainable economy that recycles more of its paper products, then timber targets too can be made more sustainable. A growing chorus of forest supervisors on several National Forests is suggesting that the next round of forest planning (1995–2000) should significantly change the type of forestry practiced on the National Forests.

10

Multiple-Use Planning in the Bureau of Land Management

Background on BLM and Federal Land Policy and Management Act

As we saw in chapter 2, the Bureau of Land Management (BLM) is the other major federal multiple-use land-managing agency. It administers about one and a half times more land than the Forest Service, and this land is by some measures more heterogeneous. Although BLM administers much more land than the Forest Service, its budget is only about one-third as large. On a per acre basis, it receives about one-fourth what the Forest Service does for planning and management. Although the Forest Service has lands throughout the United States, nearly all BLM-administered lands lie within the eleven western states of the continental United States and Alaska. BLM lands vary from sandstone canyons and mesas to large expanses of sagebrush to valuable timberlands in western Oregon. We now provide a short review of some major BLM programs before turning to this agency's planning process.

> *Mining.* BLM lands contain many valuable minerals, including coal, oil, gas, uranium, gold, silver, and much of the nation's reserves of tar sands and oil shale. To a large extent BLM is the federal government's minerals manager. As discussed in the last section of this chapter, BLM lands have been closely associated with prospecting for ''locatable'' minerals, such as gold, silver, and uranium, for many decades under the Mining Law of 1872. Coal leasing, another major BLM program, was given great emphasis during the U.S. energy crises of 1973–1980. In support of energy distribution, BLM's extensive land holdings often provide rights of way for electrical transmission lines and natural gas pipelines that move energy from producing areas (often on or adjacent to BLM lands) to urban consuming areas.

Livestock grazing. BLM lands are widely used for livestock grazing. As we saw from the history of its evolution, grazing management has been a central focus of BLM (or its predecessor, the Grazing Service) since the amendments to the Taylor Grazing Act in 1939. Much of our following discussion of planning and management focuses on the need to modify historic grazing patterns to accommodate environmental concerns (as reflected in the National Environmental Policy Act of 1969) and to adapt to the modern multiple-use mandate provided by the Federal Land Policy and Management Act.

Off-road vehicles. The heavy use of BLM lands by off-road-vehicle (ORV) enthusiasts has presented many challenges to BLM administrators. BLM has responded in three ways. First, through its planning process (described later), BLM has designated its lands as open, closed, or limited (i.e., limited to existing roads or seasonal closures of specific areas). Second, it has designated Off-Highway Vehicle Areas, which are used intensively by such vehicles. Such areas can be found in many states, including California and Utah. Third, BLM has responded by giving or denying permission for ORV use in certain situations. The most famous case is the Barstow (California) to Las Vegas (Nevada) motorcycle races. These races were initially permitted by BLM, and the agency worked with motorcycle clubs to lay out a route that would minimize the environmental damage caused by hundreds of motorcycles. When the damage to resources became increasingly evident, the race was no longer permitted (although a few renegade motorcyclists continued to race).

Wilderness. The Federal Land Policy and Management Act (FLPMA) of 1976 directed BLM to inventory and evaluate all roadless areas greater than 5,000 acres for suitability as Wilderness. While Congress gave BLM fifteen years to do this job, BLM undertook a fast-track program, completing its inventory within just a few years. Roadless areas dropped from consideration, as Wilderness could then be released for such multiple uses as mining, ORV use, timber harvesting, and so on. The roadless areas retained as Wilderness Study Areas have to be protected from loss of their Wilderness characteristics while they are further evaluated. Incompatible multiple uses are precluded until BLM determines through a NEPA EIS process that such areas are not sufficiently outstanding or do not add significantly to the diversity of the National Wilderness Preservation System. Those areas that BLM does recommend as Wilderness must be acted upon by Congress, a process that itself can take many years.

Given the potentially pervasive influence of the Wilderness designation on other multiple uses (both positive and negative), in several states the BLM used the formal planning processes to perform their evaluations. This makes a great deal of sense, because the BLM's evaluations focused on multiple-use trade-offs. In other states, such as Utah, the BLM attempted to fast-track the evaluations by performing separate statewide Wilderness EIS's. We will comment more on the shortsightedness of that approach later. At present, millions of acres of Wilderness Study Areas are before Congress, awaiting final determination.

The BLM has essentially two levels of resource planning. The field-level over-all-management plans, called *resource management plans* (RMPs), address all multiple-use resources within a given administrative unit, called a *Resource Area*. Once these are completed, more detailed, site-specific *activity plans* are developed to provide the blueprints for designing and implementing individual projects or management actions. This chapter first reviews resource management plans. After this we briefly discuss activity plans for range and wildlife.

Resource Management Planning

Origin of Initial Planning Regulations

Since the Federal Land Policy and Management Act (FLPMA) emerged from Congress within a few months of the Forest Service's National Forest Management Act, it is not surprising that many similarities in their planning processes exist. For example, they follow nearly the same steps in the planning process and in some instances use nearly identical terminology. Their respective definitions of *multiple use* are quite similar. Like the Forest Service, BLM uses an interdisciplinary approach to planning. When FLPMA's planning regulations were originally published in 1979, BLM was moving toward a comprehensive (all resources on all lands) planning process resembling that of the Forest Service. This was expected to be a major effort, as BLM manages 162 Resource Areas covering about one and a half times as much land as the Forest Service.

Director Burford's Revised Planning Regulations

The Reagan administration required both BLM and the Forest Service to revise their planning regulations, but, unlike the Forest Service, BLM's regulations were drastically altered. As described by its chief of planning (Williams 1987:33), "BLM Director Robert Burford established a task force, carefully excluding anyone associated with planning, and asked them to report back in nine working days with shorter regulations. They did." But the hastily revised regulations took another two years to be approved.

Under their initial implementation, BLM would no longer start from scratch in its planning process as the Forest Service was doing. Instead, the Reagan-appointed director would receive a "streamlined" plan "focusing" only on resource issues that needed to be solved and that were not covered adequately by other plans. The details of implementing this approach will be discussed later. However, this approach is closer to the spirit of incremental planning than to that of comprehensive planning (Garcia 1987).

BLM's revised regulations stress that its plans must be consistent with those for adjacent land use developed by other federal agencies, states, local governments, and Indian tribes. Although this could enhance coordination among landowners within an ecosystem, the requirement can reduce BLM's discretion in directing desirable land uses. That is, if all surrounding landowners have decided that mining is suitable, it may be difficult for BLM to preserve an area as Wilderness. Further limiting BLM is the requirement that its proposed RMPs be reviewed, prior to public release, by the governor of the affected state to obtain an opinion on whether it is consistent with state and local government policies regarding intermingled state lands and adjacent private lands. The governor can recommend how the plan should be changed to be more consistent with state and local land use intentions. The BLM District Office must adopt these recommendations or defend to the director of BLM why such changes would compromise the intent of FLPMA or related federal resource laws (the Clean Water Act, Clean Air Act, Endangered Species Act, and so on). This "consistency clause" sometimes makes it more difficult for BLM to defend balanced multiple use in accordance with national values for these federal lands. In addition, it gives much implied power to local governments to influence federal land management through its own local land use plans. We shall, however, return in the concluding chapter to discuss how this consistency clause can be used as a proactive vehicle for planning along ecosystem rather than administrative boundaries.

Fine Tuning the RMP Process: Supplemental Program Guidance for Planning

In 1986 BLM's resource management planning effort took a significant step forward with the issuance of supplemental program guidance (SPGs). The SPGs are noteworthy for at least three reasons. First, they provide detailed planning guidance for each of the multiple-use resources. This has stimulated more thorough consideration of all resources in RMPs. Second, they require a formal analysis to document or substantiate the basis for selecting a preferred alternative. Thus, if the rationale for selecting a particular alternative is based on "cost-effectiveness, demand, efficiency, equity or public preference," "supporting analysis" must be presented (BLM 1986:1612-.2). Although our discussion in chapter 4 indicates that all these criteria should be analyzed (not just the one used to justify plan selection), at least it explicitly recognized that a BLM manager cannot use a criterion as a decision factor without a thorough analysis. This is a step forward, and one that we shall see the U.S. Fish and Wildlife Service violate in the case study of the Charles M. Russel National Wildlife Refuge.

Third, and this element is more implicit than explicit in the SPG, is the trend back to discussing all resources on all BLM lands. That is, the narrow streamlining procedures of the early RMPs under Director Burford, which omitted consideration of selected resources, is largely being discontinued. Instead each multiple-use resource is to be addressed as both a proactive resource to be managed and a resource to be protected from the impacts of other resource management actions. However, the SPGs still stress flexibility in choosing the level of analysis and primarily offer suggested factors to be addressed in planning rather than mandatory factors that must be evaluated. In 1989 additional guidance from individual state directors of BLM required that the RMP must provide complete multiple-use plans for the entire Resource Area. Thus, the RMP process is slowly moving back to being more comprehensive, as it was originally envisioned by President Carter's director of BLM, Frank Gregg.

Overview of Resource Management Planning in BLM

BLM's planning process is driven by specific land management problems or issues. Different ways of resolving these problems serve as alternatives in the EIS. The selected alternative becomes the RMP, and it is used by BLM managers to allocate resources and select appropriate uses for the public lands within a Resource Area (an area between 500,000 and 1 million acres). "An approved plan provides a basis for making day-to-day management decisions and for considering or denying actions" (BLM Planning Manual 1984b:1617-.3). In addition, "Budget and action proposals to higher levels in the Bureau and the Department . . . must conform to the approved plan" (BLM Planning Manual 1984b:1617-.3). Thus in principle the RMP has the potential to shape nearly all of BLM's action and what will actually happen.

Steps in BLM's Planning Process

There are nine steps in the BLM planning process:

1. Identification of issues, concerns, and opportunities.
2. Development of planning criteria.
3. Inventory data and information collection.
4. Analysis of the management situation.
5. Formulation of alternatives.
6. Estimation of effects of alternatives.
7. Selection of preferred alternative.
8. Selection of Resource Management Plan.
9. Implementation, monitoring, and evaluation.

There is one less step than the Forest Service process, because BLM does not have a separate step for evaluating alternatives before selecting a preferred alternative.

Of course, this step is implicit in selecting alternatives. Although on the surface there is a striking similarity between the planning processes of the BLM and those of the Forest Service, the description of what the BLM does at each step will highlight the major differences between the two.

Step 1: Identification of Issues, Concerns, and Opportunities

Issues are defined as the ''concerns and controversies surrounding uses of the public lands in the particular Resource Area.'' These issues can be put forward to BLM by the general public, industry, other federal agencies, and state/local governments as well as by employees within the agency itself. These issues are then screened to ensure that they are appropriate to be resolved in the RMP or at the BLM administrative level. Issues that require changes in national legislation or, at the other extreme, that are site-specific and more appropriately addressed at the plan implementation stage are not included.

Since BLM planning is issue driven, this first step is critical. Which resource uses are analyzed in detail and which are given only cursory attention are largely set by the key issues identified here. For example, if current management directions for certain natural resources are deemed adequate, they will not be reevaluated but will be carried forward as a common feature in all RMP alternatives. The recent supplemental guidance in 1986 does require that the environmental and economic effects of continuing these common management directions at least be displayed in the EIS. Thus the identification of the key resource issues to be resolved in a particular RMP dictates the actions that will require real choices rather than merely continuing current management.

Step 2: Development of Planning Criteria

The planning criteria guide development of the RMP and ensure that the planning effort is tailored to the issues previously identified. In addition, they help avoid unnecessary data collection and analysis (BLM 1983b:20372). BLM's Planning Guide summarizes this planning step as ''stating what will or will not be done or considered during the planning process. Planning criteria streamline the plan's preparation and put it into focus. They may also help set the scope of inventory and data collection'' (BLM 1983a:12).

Some of the planning criteria are mandated by FLPMA, including requirements that the RMPs follow the principles of multiple use and sustained yield, that an interdisciplinary approach be used, that present and potential uses of public lands be considered, that the relative scarcity of the values found on these lands be taken into account, and that long-term benefits to the public be weighed against short-term benefits. The planning criteria may develop factors that all alternatives must address and factors that none of the alternatives will address (which are sometimes quite important, as we shall see later in this chapter). For example, all alternatives must address such mandatory factors as compliance with other federal laws, such as the Endangered Species Act or Clean Water Act. It may include discretionary items that are nonetheless of critical importance in this Resource Area. For example, all alternatives may have to address how existing nondegraded riparian areas will be preserved. At the other extreme, certain resource allocation decisions are taken out of the planning process and not evaluated by any alternative. For example, suitability of lands for coal leasing or Wilderness recommendations may be handled in a separate evaluation process.

The planning criteria also establish conventions for what the analysis will do and will not do in estimating the effects of the alternatives in step 6. This aspect of the planning criteria guides the District Office planners in determining how much detail is needed in modeling resource effects. Some BLM offices have generated screening criteria to determine lands that are suitable or not suitable for off-road-vehicle use, mining, and timber production.

Step 3: Inventory Data and Information Collection

BLM recognizes that ''facts are a critical component in sound resource management planning'' (BLM 1983a:13). Field offices are instructed to compile information on public land resources, the environment, and social and economic activity.

If data are inadequate, BLM does allow for new data collection to address ''significant issues and decisions with the greatest potential impact'' (BLM 1983b:20372) or to fill ''critical information gaps'' (BLM 1984a:1616-.3). But BLM stresses that, ''Many times it is possible to draw all information necessary from existing inventories and other sources of information developed by BLM or other agencies or scientific organizations'' (BLM Planning Guide 1983a:13). BLM's new planning system ''avoids new, costly and time-consuming inventories or data-gathering'' (BLM 1983a:4).

Step 4: Analysis of the Management Situation

The key factors involved here are quite similar in spirit to the ones considered by the Forest Service. In particular, the critical feature of this step is the determination of land/resource capability to resolve the planning issues and the demand for resources within the Resource Area. The planning regulations for determining resource capability require BLM to determine the ''estimated sustained levels of the various goods, services and uses that may be attained under existing biological and physical conditions and under differing management practices and degrees of management intensity which are considered economically viable under benefit cost or cost effectiveness standards prescribed in national or State Director guidance'' (BLM 1983b:20372). Another common step includes describing the present land management activity (the no-action alternative) to establish a baseline for judging the other alternatives. This information is compiled into an internal BLM planning document called the Management Situation Analysis (MSA).

Overlay Mapping.   The basic approach to determining capability is made using a series of resource maps. When BLM first started multiple-use planning, this was done using clear plastic overlays on a base map of the Resource Area. Each overlay would display one or more resource capabilities. For example, there would be a soils map, a vegetation map, a livestock grazing allotment map, a wildlife map (noting species distribution, critical habitats, and so on), a minerals map, a recreation map, a forestry map, and so on. Each map would delineate the presence of a resource and where possible a resource condition classification for the resources present (e.g., seasonal or critical habitat or low-productivity ponderosa pine forests). The current resource management direction in terms of land use zoning for each land area would be displayed on maps as the no-action alternative. The management prescriptions proposed by each resource specialist in a given RMP alternative would be shown on additional map overlays as well. For example, areas proposed for leasing, clear-cutting, and off-road-vehicle closure would be represented on respective overlays. When the maps were overlayed, it was possible to determine incompatible multiple uses proposed on a given area of land. That is, a minerals specialist might have proposed oil leasing in an area that the wildlife biologists had designated as critical deer winter range and had consequently proposed a reduction in livestock use coupled with wildlife habitat improvements. Either the staff or the area manager would then choose which conflicting use had priority in a particular management alternative.

Information on production relationships and value trade-offs is critical if resources are to be allocated efficiently in these cases. In this example, one should ask how much recoverable oil there is. What is the net (not gross) economic value of the oil recovered versus the net economic value of wildlife supported by the winter range? And how would the amount of wildlife and associated recreation change with various leasing arrangements (i.e., either no surface occupancy via off-site slant drilling or special stipulations that minimize drilling activity during the winter months when wildlife is present)?

More recently, BLM has adopted "computerized mapping" using a tool known as Geographic Information Systems, or GIS. The GIS approach is gaining considerable popularity among all natural resource management agencies for maintaining geographic databases on distribution of resources and their capabilities. Each parcel of land (a cell) can have all the resource descriptors previously contained on the series of maps. That is, tied to each cell is information on soil type, erosivity, vegetation types, recreational suitability, minerals, wildlife, timber species and productivity class, livestock grazing, presence of archaeological sites, and so on. A series of custom maps showing resource relationships (either compatibilities or incompatibilities) can be generated by utilizing the GIS data base and mapping capability. The advantage of relying on any type of mapping system, such as GIS, is that potential management practices and multiple uses are tied to easily identifiable land areas.

Key relationships between resource uses and the environment as well as thresholds in these relationships are also identified during the Management Situation Analysis (BLM 1984a:1616-.43A). Issues involving substantial investments are assessed through benefit-cost analysis or other standards prescribed in the planning criteria as a basis for judging their economic reasonableness (BLM 1984a:1616-.43A). If applied properly, this benefit-cost test should ensure that only management actions that are both biological and economically feasible are carried forward in the planning process. This is an important step forward as past multiple-use activities in BLM (and the Forest Service for that matter) have usually focused on what is biologically feasible without considering the cost or benefits of those actions.

In the Management Situation Analysis BLM also describes the degree of local dependence on resources from BLM lands. For example, it sets forth the amount of dependence (in terms of percentage of their production or output) of local ranchers or timber mills on BLM land in the Resource Area.

A key difference between the BLM's analysis and the Forest Service Analysis of the Management Situation relates to rigor of the analytical procedures used. As illustrated more fully in the BLM case study that follows, the BLM's planning manual requires that key relationships between resources be identified, but only a "preliminary evaluation" of the "quantitative" nature of these relationships, outputs, and effects must be made. There is little in the way of formal quantitative benchmark analyses for assessing the absolute magnitude of resource capability or resource interactions. Resource capability is often described simply in terms of what existing uses are provided (but with only minimal information given on output rates or value of the output). Current resource use is displayed on a series of land use capability maps (and often not quantified). Although such capability classifications are important, they often fail to provide the information the BLM planning regulations require on production potentials of those lands in a way that can be used to estimate the output responses from different land management prescriptions. Specifically, the range of output levels likely with different management intensities is neglected. In addition, it is important to know the maximum output level possible. Without this information, alternatives might be formulated in the next step that cannot be attained.

Nonetheless, whatever capability analysis is done, it sets the stage for what is feasible when formulating alternatives for the RMP in the next step. But before this step is taken, BLM screens the issues identified in step 1 to determine which of them must be addressed through the EIS, which can be resolved administratively without being part of the EIS, and which current resource management practices and land uses are adequate. These "adequate" management practices and land uses are considered not to need further review and may go into the Final RMP without additional analysis. This process is illustrated in figure 10.1. This additional screening process further streamlines BLM's planning process by eliminating the need to evaluate many current resource uses or management actions. However, this by-

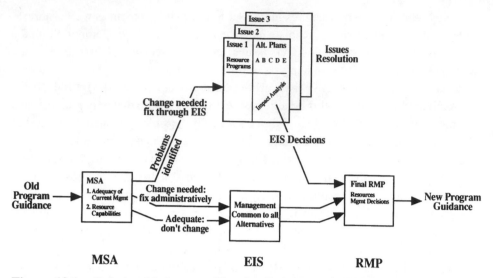

Figure 10.1.   Relationship between Planning Documents in BLM

passing of selected resources and management actions presumes that in solving the resource issues raised by the public, these bypassed resource and current management activities will not need to be changed. Unfortunately, such independence is not frequent in ecological or economic systems and raises some concern about such a process. That is, a more efficient way to respond to the same management issues may be to change existing management practices that are acceptable by themselves. The more recent Supplemental Planning Guidance suggests that the environmental effects of continued management should at least be displayed in the EIS.

Step 5: Formulation of Alternatives

Given the resource capabilities described in the Management Situation Analysis, a range of alternatives that represent different combinations of management actions and associated multiple uses is developed to resolve the issues raised in step 1. The set of alternatives must include one for continuing existing management (the no-action alternative), one that favors resource protection, and one that is more commodity/production oriented (BLM 1984a:1616-.5). Issues that can be resolved by similar management actions might be addressed in one alternative that uses those similar management actions as a theme. For example, several recreation-related issues might be addressed in one recreation alternative. The formulation of the alternatives is an iterative process performed by the interdisciplinary team in a cooperative environment. To keep the planning process manageable, BLM generally develops no more than six total alternatives for detailed analysis in the next step of estimating effects of alternatives.

The mapping or screening approach can be applied to generate different alternatives as follows: a protection alternative might zone land to the most ecologically sensitive use and not allow incompatible uses. By comparison, a commodity alternative might allocate all lands containing minerals or timber to production of those resources and allow other multiple uses only on lands without commodity potential.

Importance of Separable Alternatives. As desirable as the goal of keeping the number of alternatives manageable may seem, a broad enough range of alternatives must be constructed so that the manager and the public have real choices. There is no need to group together resource management actions under a common theme if they are not physically related or dependent on one another. For example, an amenity alternative might recommend more roadless areas as Wilderness while developing more water sources for wildlife and closing some other areas (not potential wilderness or wildlife habitat areas but different lands with highly erodible soils) to off-road vehicles (ORVs). Although these all paint the same picture of resource protection and environmental enhancement, they are not physically related. Several alternatives could be developed that implement just one of these feature even though they all belong to the same theme. The concern with linking them together is that, unlike having to vote for one candidate based on his or her stand on all the issues, there is no need to reject a desirable feature because it is linked artificially in a particular alternative to an undesirable feature. For example, ORV groups might be strongly opposed to closing erosive soils to their use but mildly supportive of the wildlife habitat improvements (perhaps many ORV users are also hunters). Since ORV groups strongly oppose ORV closures, they may write letters or speak at public meetings against the entire alternative. Thus it would make more sense to develop at least three separate alternatives and two or more alternatives that combine two of these features.

Areas of Critical Environmental Concern as a Management Tool. One management prescription that FLPMA added to BLM's menu of problem-solving tools is designation of Areas of Critical Environmental Concern (ACECs). ACECs provide special management attention to protect some unique environmental attribute of regional or national significance on a particular parcel of BLM land. The special characteristic could be a high density of cultural or archaeological sites, critical habitat for a particular wildlife or plant species, unusual geologic formations, and so on, that would be threatened if incompatible uses were allowed (or allowed to continue). The designation of ACECs is a powerful land use control tool, but one BLM has used sparingly (some feel too sparingly) in RMPs.

Finally, the alternatives are described by a series of statements and on maps of what management actions and associated multiple uses would take place on particular lands in the Resource Area. Examples of pre-GIS maps will be shown later in the chapter. Current GIS maps are color-coded by resource and activity.

Step 6: Estimate Effects of Alternatives

The physical, biological, social, and economic effects of each alternative are to be estimated and displayed in step 6. The procedures used to estimate these effects are considered sufficient if they meet the planning criteria for adequate analysis procedures. In addition, because these estimated effects are part of the EIS, impact procedures developed by the Council on Environmental Quality to implement the National Environmental Policy Act (NEPA) must be met as well. Both FLPMA and NEPA require a systematic and interdisciplinary approach to estimating the effects.

The primary purpose of step 6 is to provide the estimated results (both multiple-use outputs realized and environmental effects) that would occur if that alternative was implemented. The intent is to highlight the trade-offs within and between alternatives and the degree to which each alternative resolves the issues identified in step 1. The BLM Planning Manual (1984a:1616-.6) states that this step ''includes estimates for each alternative of the social and economic effects of resource values foregone, decreased, retained or increased, in quantified terms when possible.''

Matrix Approach to Display Effects.   Step 6 results in a set of tables or a large matrix that lists all the resources and environmental categories down the left-hand side and all the alternatives as column headings. An example table will be shown in the following case study and is similar to matrix tables presented in chapter 4. Ideally, each resource effect or output level (in the cases where they are actually quantified) is shown in the absolute units relevant for that resource. Thus coal should be represented in tons, livestock grazing in AUMs, water yield in acre-feet, and so on. In principle, most of these effects could be put into commensurate units, since the Planning Manual quoted earlier states that economic effects of resource uses are to be quantified. This is not usually done, however, nor is any attempt made to develop some other summary statistic of the net effects of each alternative. Thus BLM keeps a fairly disaggregated matrix, much like the one discussed in chapter 4, except that these individual effects are not related to any explicit evaluation criteria, such as equity, efficiency, or administrative feasibility. Consequently, the way each effect contributes toward alternative objectives of public land management or to resolving specific issues is often not clearly identified. Instead, all the effects are simply listed. In many cases, these summary tables only list acres available to different multiple uses (e.g., acres to be leased for coal, acres of wildlife habitat, and so on). In some cases the summary tables provide only an indicator of direction of change and whether the change is low, moderate, or high. The implications of neither making the decision criteria explicit nor developing a summary index of each alternative will become apparent in the next step, selection of the preferred alternative.

Finally, it is during this step in the planning process that inconsistencies of any alternatives with plans, programs, or policies of other federal agencies, states, and local governments and Indian tribes are determined.

Step 7: Selection of Preferred Alternative in Draft EIS

Step 7 combines the Forest Service steps of evaluating the alternatives in terms of how well they resolve the planning issues and selection of the preferred alternative. The preferred alternative is identified in the draft EIS and developed into the Draft/ Proposed Resource Management Plan in a separate document.

As in the Forest Service, it appears that early on in the planning process (after step 3) a preferred alternative has been already built from suggestions by managers and staff and carried through step 6 to this stage. Specifically, the prechosen preferred alternative is constructed by combining the desirable features drawn from the other alternatives, such as resource protection and resource development. This is different from the more neutral suggestions in the planning regulations that a set of alternatives be developed. Then, after the estimation of effects, managers realize which is the best one, for only after the estimation of effects does one have sufficiently complete information on the results of implementing the preferred alternative.

The chain of command in this selection of the preferred alternative starts with the area manager, with concurrence required by the district manager, and final approval being given by the state director of BLM.

How does the area manager decide on the preferred alternative? The BLM Planning Manual section on choosing the preferred alternative states, ''The Area Manager selects or develops, from among those analyzed, a preferred alternative. This alternative in the Area Manager's judgment addresses the issues and management requirements of the resource area'' (BLM 1984:1616-.7). Since the estimated effects are not put into commensurate units or in the form of a summary indicator and the evaluation criteria are not made explicit, it is hard to determine how the judgment about which alternative best meets the planning issues and reflects balanced multiple use is made. The recent requirement in the Supplemental Program Guidance on Planning that analysis be done to document the basis for selecting the preferred alternative is at least one small step toward remedying part of this problem.

The draft EIS (with the draft preferred alternative identified) and the draft RMP are mailed to the governor of the state, local governments, Indian tribes, and all interested individuals for their comments during a ninety-day public comment period. Members of the public may comment on their preferences for specific alternatives and question the estimated effects and any other feature of the EIS or plan.

Step 8: Select Final Resource Management Plan

The area manager, district manager, and the interdisciplinary planning members review the comments received and determine what modifications (if any) are needed in the preferred alternative. That is, the main thrust of public comment is in changing the mix of multiple uses or the location of some of those multiple uses as reflected in the preferred alternative. If changes to the preferred alternative occur, the estimated effects of the revised preferred alternative must be made and presented in the Final EIS.

Although the modifications to the preferred alternative are usually minor, they can sometimes be major. If major changes are made or a different alternative is adopted as the preferred alternative, then a new draft EIS and draft RMP are needed.

The modified preferred alternative is developed into the Final Resource Management Plan. Both the Final EIS and Final Plan are filed with the US Environmental Protection Agency (the agency with final authority over all federal EIS's). A notice of availability of Final EIS's is published. This announcement gives the public thirty days to protest the Final EIS and selection or composition of the Final RMP. One basis for protest by the governor is inconsistencies between the Final RMP and state or local government plans or programs related to lands adjacent to or in this Resource Area.

Step 9: Monitoring and Evaluation

Step 9 occurs along with implementation of the RMP. Therefore before we discuss monitoring, we should mention what is involved in plan implementation. Some aspects of the RMP can be directly implemented without budgetary changes or development of site-specific activity plans. For example, land use policy changes in the RMP such as mineral leasing stipulations, Wilderness area visitor day quotas, and so on can be phased in with new leases or when the next recreation season begins. Some implementation may take slight changes in personnel or more funding and must be integrated into BLM's Annual Work Planning and budgeting process that serves as the foundation of its annual appropriations request.

BLM has recently developed a computerized project management software system that will allow tracking of RMP action items. These include the RMP action, who is responsible (area manager, district manager, and so on), when it is to be implemented, budget requirements, any additional analyses required, and so on. This system will link the RMP action items to BLM's Annual Work Planning and budgeting system so that funds can be requested.

Implementing management prescriptions that require significant investment of federal dollars, such as range or habitat improvement projects, requires the preparation of detailed activity plans. These are called Allotment Management Plans (AMPs) and Habitat Management Plans (HMPs) and are used for range and wildlife projects, respectively. These site-specific activity plans are the most detailed type of planning carried out by BLM. Because a request for significant funding of improvement projects often accompanies the AMP or HMP, the Federal Office of Management and Budget requires a benefit-cost analysis of the project. BLM has developed a software package (called SAGERAM) to facilitate and standardize these benefit-cost analyses.

Once the plan is implemented, key monitoring check points include assessing whether the multiple-use prescriptions are generating the expected results and determining if any environmental thresholds are being exceeded and whether any new information alters the desirability of implementing the plan.

Much like the rest of BLM's planning process, the monitoring program is streamlined and focused. Monitoring is not continuous but occurs at regular intervals, which are not supposed to exceed every five years. In addition, some resources are not even monitored periodically; this is reserved for only the most environmentally sensitive resources.

BLM updates or revises RMPs as the need becomes evident. The RMP will be amended formally to incorporate changes in the direction of resource management and to respond to unanticipated situations. Unlike the Forest Service, which has a formal ten- to fifteen-year cycle for revising Forest Plans, BLM has not committed itself to major revision of RMP on a regular basis. Such a cycle was part of the original 1979 planning regulations but was dropped in the Burford revision (Williams 1987:33).

To obtain a better understanding of how these steps in the planning process are performed we now turn to a case study of BLM Resource Management Planning in Utah.

Case Study of BLM's Implementation of FLPMA Planning Through RMPs

To illustrate the application of the RMP planning regulations and guidance the remainder of this chapter uses the San Juan Resource Area (SJRA) as a case study. The SJRA is typical in size for Utah's Resource Areas, being 1.7 million acres, nearly twice the size of most National Forests. Figure 10.2 shows a general land ownership map of the SJRA in southeastern Utah and connections with other federal lands, such as National Forests, National Parks, and Indian reservations. The SJRA is the striped area of figure 10.2. As is evident from this figure, with its many artificially straight boundaries for Canyonlands National Park and Manti-Lasal National Forest, BLM planning should logically take into account ecological interfaces between its lands and those of these other two agencies. Of course, the converse is true as well: the Forest Service should recognize the ecological interrelationships of its lands with those of BLM. We shall return in our concluding chapter to recommendations for planning in cases where there are multiple federal ownerships that cross ecological boundaries.

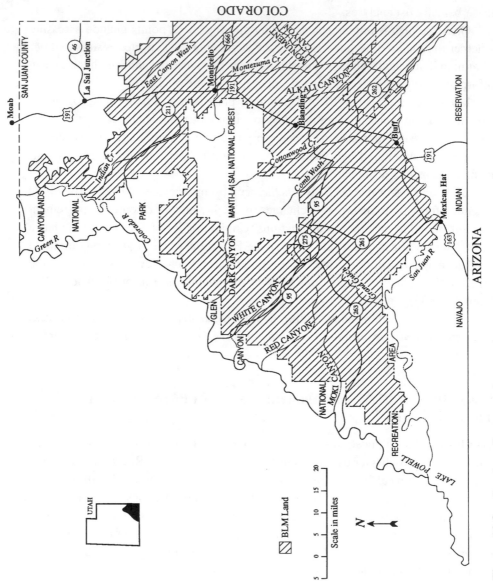

Figure 10.2. San Juan Resource Area

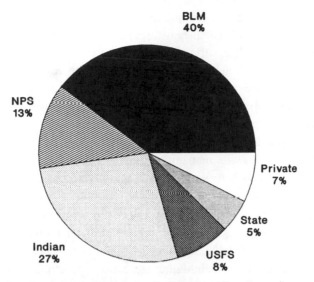

Figure 10.3. Landownership Pattern in the San Juan Resource Area

Figure 10.3 illustrates the overwhelming amount of land that is federally or state owned (93%) in the Resource Area and shows that BLM is the major land-owner, with 40% of the land. Thus activities of federal agencies in general and of BLM in particular significantly influence not only land use in southeastern Utah but also the opportunities for those who live on the 7% of private land in the Resource Area.

In terms of present multiple uses, all but 2% of BLM's land has livestock grazing on it. There is a significant amount of mining and oil/gas development as well in the SJRA. This land also contains one of the highest concentrations of archaeological sites in the United States. In addition, it is well known for its high desert canyon country, river rafting, and desert vistas.

We shall go through the steps in the RMP process to see how each one was implemented in the SJRA. Much of what follows summarizes the BLM's listing of issues, analysis, and results.

Step 1: Identification of Issues

Five issues to be addressed by the planning process were identified (Moab District, May 1986). The first was livestock management. Livestock grazing has resulted in loss of forage productivity (i.e., from grazing in excess of the plants' sustained yield) and conflicts with wildlife (specifically bighorn sheep, antelope, and mule deer). Since it was BLM's intent that the RMP and its accompanying EIS satisfy the federal court–ordered site-specific grazing EIS, the RMP must address stocking levels of livestock (i.e., number of animals) and seasons of use (i.e., spring, fall, winter, and "on and off dates" of cattle within the respective season).

The second major issue is vegetation management. In particular, the RMP must identify which areas of land can sustain sales of wood products, such as firewood and Christmas trees, as well as surface disturbance from mineral exploration. BLM must also determine which areas should be targeted for vegetative improvement efforts. Finally, watersheds with critically sensitive soils must be identified.

The third major issue relates to wildlife habitat management. This issue emphasizes protection of special wildlife habitat areas from adverse effects caused by other land uses, extraction, and production of other resources.

The fourth issue deals with managing recreation conflicts. These involve competition for the same areas between people seeking nonmotorized recreation, such as hiking and backpacking, and those seeking to use their off-road vehicles (ORVs). In addition, the impact on other resources associated with ORV use must be considered. Thus BLM must determine the optimal mix of recreation activities on specific parcels of land within the San Juan Resource Area. In addition, the balance between giving precedence to maintaining recreation opportunities and restrictions of development uses must be addressed.

The fifth planning issue deals with managing Wilderness Study Areas (WSA) if they are not designated as Wilderness by Congress. Specifically, if they are not formally protected as Wilderness, what previous incompatible multiple uses (if any), such as oil and gas leasing or roaded recreation, should be allowed? Or should these areas be managed for nonmotorized recreation and for wildlife and watershed protection? Unfortunately, the larger issue of whether they should be recommended for designation is not part of the SJRA RMP but part of a separate but parallel state-wide Wilderness EIS. If the analyses were integrated it would be easier to analyze the desirability of alternative management of these lands.

As is apparent, these resource issues are quite general. Unfortunately, no quantitative performance indicators for determining to what degree the issue has been resolved are provided. Developing observable performance indicators would help better define the extent of the issues and how the degree of resolution would be measured when comparing alternatives. However, BLM does develop more specific, operational management "opportunities" (i.e., potential management actions or decisions) in the fourth step (analysis of the management situation). The criteria BLM follows to identify these form part of the next step, the development of planning criteria.

Step 2: Development of Planning Criteria

As discussed earlier, the planning criteria tailor the remaining steps in the RMP to resolving the specific planning issues, thereby avoiding "unnecessary data and analyses" as well as defining the categories of effects that must be estimated under step 6.

The SJRA has five main types of criteria:

1. *Formal planning requirements from Section 202(c) of FLPMA, discussed earlier in this chapter.* These require planning to follow principles of multiple-use sustained yield and must be interdisciplinary, compare short-term to long-term benefits, consider relative scarcity of resource values, and so on.

2. *Criteria for problem identification.* In the SJRA these relate to three situations: where there is documented public controversy regarding management of a specific resource, where current land use allocations or management significantly curtails another multiple use, and when current uses conflict with policies or plans of other federal agencies.

3. *Criteria for identification of management opportunities.* These criteria are used to determine which changes from current management are required, which ones may be accomplished through administrative actions, and which ones can be done through the RMP process.

4. *Criteria for formulation of alternatives.* These criteria set standards all alternatives must meet such as legal minimums for specific resources or environmental effects. In addition, they prescribe the admissible range of management actions and budgets that are deemed feasible. For example, in the case of the SJRA, the feasibility criteria require that alternatives not involve appreciable changes in amount of facilities, services and scale or scope of management. However, this appears to prejudge that current facilities, services and scope of management are optimal in scale, when in fact this is part of what one should be trying to determine in the planning process.

   The criteria for formulation of alternatives also state what should be different in the range of alternatives. In the SJRA, seven themes are presented must be represented in at least one alternative. These are (a) continue present management; (b) maximize production of resources; (c) maximize recreation; (d) minimize livestock grazing; (e) protect sensitive ecological or visual environments; (f) design and protect of Areas of Critical Environmental Concern (ACESs); and (g) protect those values are relatively scarce on public lands. However, no recommendations of lands for Wilderness will be addressed in any alternative as this is being handled in a separate tatewide EIS.

5. *Criteria for estimation of effects.* These criteria specify what BLM will evaluate as effects of each alternative. In the SJRA this includes (a) impacts of BLM management upon adjacent Federal, Indian or private lands; (b) consistency with formal land use plans of state and local governments; (c) evaluation of shortterm impacts (five years or less) and longterm impacts (years); (d) evaluation of all local economic and social changes; and (e) determination of the cost to BLM to implement each alternative.

Step 3: Inventory Data and Information Collection

Existing data were assembled on climate, topography, geology, soils, vegetation, water, wildlife, existing facilities, and access. This information was summarized on a series of clear plastic sheets that literally are overlaid on the base map of the SJRA.

Step 4: Analysis of the Management Situation (AMS)

The AMS examined three major topics: the current management situation, resource capability analysis, and management opportunities. It is worthwhile to look at each of these in more detail to understand what is done by BLM at this stage.

1. *Current management situation.* This section describes current resource allocations, current management practices, socioeconomic considerations (e.g., employment, personal income, and local taxes generated), consistency with other governments' plans, and data gaps.

2. *Resource capability analysis.* This section discusses what output or use levels the SJRA is capable of providing and the current and future demand for those resources. A discussion is provided of what management practices can be undertaken to increase the resource capability to meet demand. Critical thresholds are discussed for each resource. However, much of this is qualitative, with no quantitative estimates made because of lack of baseline data and an inability to formulate production relationships. This carries over to the inability to quantify critical thresholds for wildlife, vegetation, and livestock grazing.

3. *Management opportunities and limitations.* These are specific actions or decisions that BLM can make as part of the RMP process. For example, a management opportunity exists in the recreation program for BLM to determine which lands in the SJRA will be classified as open to ORVs, closed to ORVs, or open to ORVs only on existing paths or jeep roads. Another opportunity is to deal with forage allocations between livestock and wildlife as required by the federal court order. However, the guidance in the SJRA as well as in many other RMPs reviewed is to postpone making forage reallocations and instead monitor range conditions for five years and then make forage allocation decisions. The rationale is that range condition cannot be determined prior to this monitoring (i.e., a belief that existing data are inadequate to determine the amount of forage reallocation necessary). One key management opportunity that cuts across all resource programs is designation of ACECs to protect specific resources on specific areas of land from conflicting multiple uses.

Once step 4 was completed, the framework and direction for the remainder of the RMP was largely determined. Given the planning criteria in step 2, alternatives are formulated and their effects are estimated.

Step 5: Formulate Alternatives

As established in the planning criteria and required by NEPA, one alternative (denoted *A*) is the continuation of present management (considered the no-action alternative). Using the seven themes listed in the planning criteria, "The interdisciplinary planning team developed Alternatives B, C, and D to suggest different ways of managing the public lands and resources. BLM managers selected ideas from A, B, C, and D to develop alternative E (the preferred alternative)" (Moab District, May 1986:2–1).

As discussed earlier, and with regard to the Forest Service, it is clear in this quote that a particular plan of action was built as the "preselected" preferred alternative without taking advantage of the additional information that would be generated from estimating the effects of the alternatives and comparing alternatives. This appears to go against the spirit of FLPMA's view of the planning process.

Nonetheless this was the process used in the SJRA. To facilitate understanding what management actions are represented by each of these alternatives, they are summarized here, based on the Final EIS (Moab District, September 1987:1–4).

> *Alternative A* continues current management.
>
> *Alternative B* gives priority to mineral production and livestock grazing (and therefore represents themes 2 and 4 from the planning criteria).
>
> *Alternative C* gives priority to providing a wide range of recreation opportunities and protection of wildlife habitats and watersheds.
>
> *Alternative D* preserves natural succession of plant communities by minimizing surface disturbance, protects cultural resources beyond the minimum required by law, and increases area available for primitive (nonmotorized) recreation.
>
> *Alternative E*— the preferred alternative—provides opportunities for primitive and semiprimitive recreation in selected areas, protects scenic values, protects cultural resources beyond the minimum required by law in certain areas, protects certain wildlife habitats, preserves watershed values, continues livestock grazing at current levels in areas where no conflicts with other resource values occur, and otherwise makes public lands available for production of mineral resources.

How Alternatives Influence Land Management

The key differences between the alternatives can be summarized by comparing the proactive resources to the reacting resources. For example, in Alternative B, mineral production and grazing are the driving forces (proactive resources) and the other multiple uses are produced only in amounts consistent with or residual to maximum mining and livestock-grazing uses. This contrasts with Alternative C, where recreation and wildlife are the proactive resources and minerals are produced and livestock are allowed to graze only to the extent that they do not conflict with recreation and wildlife.

The mixture of multiple uses in the preferred alternative is a result of an implicit screening approach that matches multiple uses to environmental capabilities of the land. Using map overlays, a screening approach (similar to that discussed in chapter 4) is used to filter out multiple uses that would be inappropriate on certain lands because of the soils, critical wildlife, or special recreation or scenic qualities. The lands remaining after this screening are available for other multiple uses, such as mineral production. This results in multiple use over the entire SJRA by allocating lands with special resource capabilities to those uses and allowing other consistent multiple uses but not other incompatible multiple uses. For example, scenic lands might be ''zoned'' for special recreation management and only compatible multiple uses, such as wildlife, would be allowed. In addition, screening of lands by soil type, vegetation productivity, and so on, limits specific multiple uses to only those lands capable of supporting those uses on a sustained-yield basis.

Figures 10.4 and 10.5 illustrate differences in recreation emphasis between Alternatives A and E. Figure 10.4 reflects the current situation (Alternative A), and because no expansion in recreation or cultural resource protection is proposed under Alternative B, this is also the map for Alternative B. Figure 10.5 illustrates the recreation and cultural resource protection under the preferred alternative (Alternative E). This alternative adds a recreation management emphasis in areas surrounding Canyonlands National Park, a logical step in blending the management objectives of the two agencies (i.e., recreation as one of the dual purposes of a National Park and recreation as one of BLM's multiple uses).

Figures 10.6 and 10.7 illustrate differences between the recreation-oriented alternative (C) and the preferred alternative (E) in terms of overall land uses. Nearly two times the amount of land is placed in no surface occupancy in Alternative C, and much more land around Canyonlands National Park is put in limited surface use. Note the similarities between Alternative E areas, which are no grazing use or no surface occupancy for mineral production, and the special recreation management area boundaries in the preferred alternative shown in figure 10.5. Figure 10.7 is somewhat difficult to read, because of the map-overlaying technique used by BLM (using the black and gray shading rather than color, as is now done in BLM's computerized Geographic Information System maps).

Matching Land Use Capability to Multiple Uses

The maps in figures 10.4–10.7 illustrate how evaluating land for special recreation and cultural resource capabilities screens areas out for mineral leasing and other development. This is done by designating these special recreation and cultural re-

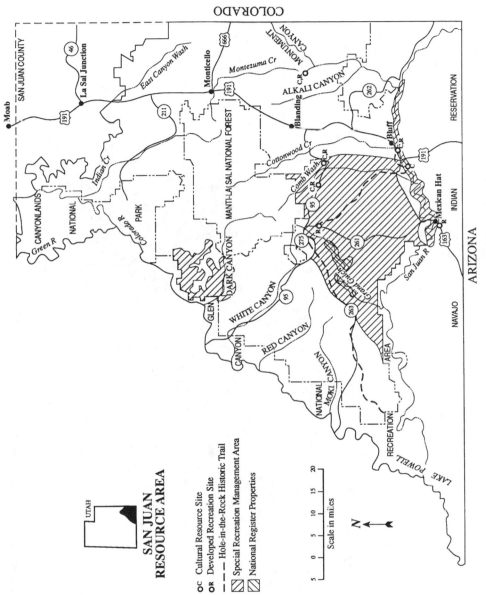

Figure 10.4. Cultural and Recreation Resource Management-Alternative A

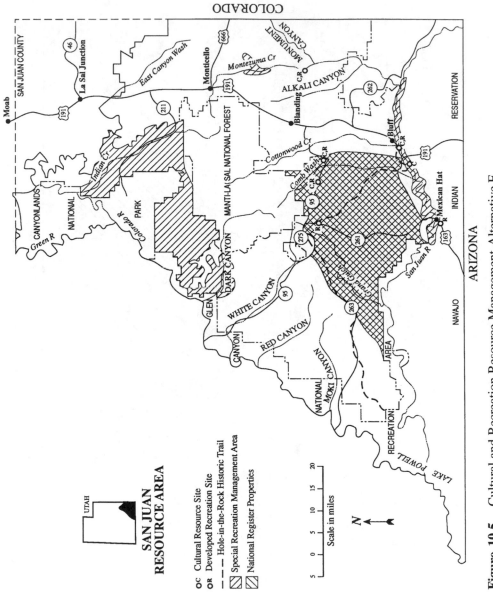

Figure 10.5. Cultural and Recreation Resource Management-Alternative E

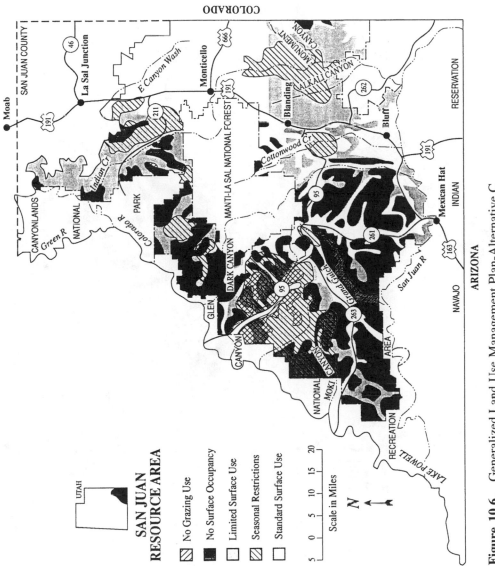

SAN JUAN RESOURCE AREA

▨	No Grazing Use
■	No Surface Occupancy
☐	Limited Surface Use
▨	Seasonal Restrictions
☐	Standard Surface Use

Scale in Miles

5   0   5   10   15   20

N

Figure 10.6. Generalized Land Use Management Plan-Alternative C

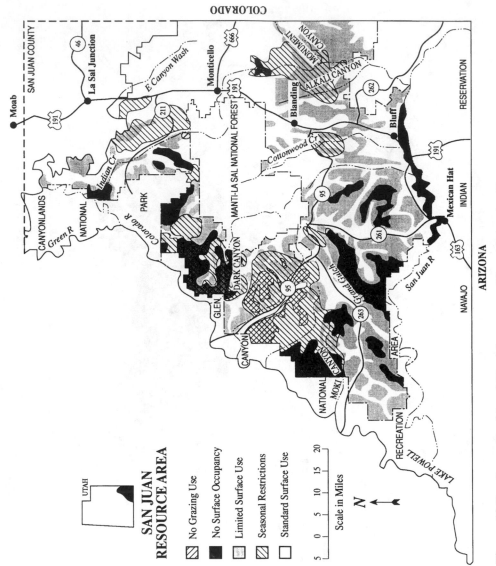

Figure 10.7. Generalized Land Use Management Plan-Alternative E

source as lands on which no surface occupancy, no grazing, and no off-road vehicles (ORVs) are permitted. Special characteristics of multiple-use resources are protected by designating some of the lands where these resources occur as Areas of Critical Environmental Concern (ACECs). The remaining lands that do not have special recreational or cultural possibilities and that do not have environmentally sensitive soils, plants, or wildlife habitats are available to other multiple uses. These other, more-development-oriented uses are referred to as "standard surface uses" in figures 10.6 and 10.7. Thus standard surface uses are lands that pass through all the other restrictive screens and that are open to ORVs with no restrictions, to mineral leasing, to livestock grazing, and to such forestry practices as firewood and Christmas tree cutting.

However, the bulk of the BLM land in Alternative E is in limited surface uses. This land use classification restricts surface disturbance on lands with sensitive soils as well as most riparian areas and nonmotorized recreation use areas. In addition, seasonal restrictions are placed on other multiple uses to protect bighorn sheep lambing and rutting areas, antelope fawning areas, and deer winter range. These restrictions are good examples of resolving multiple-use conflicts: in the first case the conflict is resolved by segregating incompatible uses away from land areas that will not support particular intensive multiple uses. In the second case, separating the incompatible uses by time reduces the potential for conflict. Thus oil and gas development or ORV use can occur during the time of the year that the migratory wildlife species of concern either are absent or are not at a critical season.

How Alternatives Address the Planning Issues

The SJRA RMP also details how each alternative addresses each of the five major planning issues. The grazing issue is addressed in all alternatives, but the response varies from alternative to alternative. The preferred alternative (E) and the no-action alternative (A) continue livestock grazing at levels equal to those of the past five-year average. Only Alternatives C and D make substantial reductions to accommodate other multiple-use values, such as primitive recreation (Alternative C) or riparian vegetation (Alternative D).

All the alternatives address the issue of how other multiple uses should be allowed to affect vegetation. There is little difference between alternatives on this issue. Alternatives A (existing management) and B (maximum production) meet the minimum requirements of law and use a mixture of native and exotic plant species for revegetation of disturbed sites (primarily associated with mining). Alternatives C, D, and E (preferred alternative) go beyond the legal minimum in protecting riparian vegetation and emphasize (but do not require) the use of native plants in ACECs and areas where primitive recreation takes place.

The wildlife habitat issue is addressed in the same general terms by all five alternatives (Moab District, September 1987:chapter 1–69). Specifically, the location and extent of livestock exclusions and ORV restrictions are similar. The primary difference is found in wildlife population goals. In the production alternative (B), wildlife population numbers are limited by the forage remaining after intensive livestock grazing (i.e., livestock get first priority in the forage allocation). In Alternative C, the wildlife goal is to approach the Utah Division of Wildlife Resource's goal of stable wildlife numbers. In the preferred alternative (E), wildlife population goals are subordinate to recreation management goals. Although none of these goals is quantified or disaggregated by wildlife species, the estimated effects of each alternative (to be discussed next) do present numbers of big-game animals by species.

All alternatives also deal with the wilderness study area (WSA) management issue of how WSAs not designated as Wilderness by Congress would be managed. The maximum production alternative (B) would make these WSAs available for mineral leasing if any companies were interested and would open them to ORV use. Alternatives C, D, and E would designate some of the WSAs as ACECs and would include others as parts of special recreation management areas. The primary differences between Alternatives C and D and the preferred alternative (E) relates to grazing and ORV use. Alternatives C and D would keep the areas closed to ORVs and reduce livestock grazing by 25% to 50% of past grazing use. The preferred alternative (E) allows some WSAs to be open to ORVs and allows the current level of livestock grazing use.

Step 6: Estimation of Effects

Using a variety of techniques (to be discussed later), the BLM staff estimated the effects of implementing the five different RMP alternatives. These effects are shown in table 10.1, which draws from the SJRA's table 2.10 in the Final EIS (Moab District, September 1987).

The first major category is minerals, which deals with oil and gas, coal, locatable minerals (in this case gold), and nonenergy minerals, such as potash. Most of these are measured in terms of acres available for the activity instead of in terms of expected outputs. Although oil and gas production are stated as barrels per year and million cubic feet (MCF), they are not quantified. What is curious is how BLM determined that changes were positive and significant or negative and significant when output was not quantified. The next major category is what BLM calls *biotic*. This includes air, soils, water, vegetation, and wildlife. In tables 10.1c–10.1e, these resources are measured in terms of tons per year of soil loss and acre-feet per year of sediment and tons of salt lost. Vegetation effects include acres disturbed and harvest of forest products. Unfortunately, what is described is the number of acres of each type of vegetation affected but not how much output results from those acres. For example, it would be useful to have an estimate of how many cords of firewood would be produced by each alternative.

The estimated effects of each alternative on wildlife are quantified in terms of three big-game species (desert bighorn sheep, antelope, and deer) and of riparian and threatened/endangered species habitat. For big-game species BLM provides information on both habitat acres (inputs) and number of animals (outputs).

The next major section is referred to by BLM as human uses of the SJRA. In tables 10.1f and 10.1g the human uses displayed include livestock grazing (quantified as acres and AUMs), archaeological sites, acreage by Recreation Opportunity Spectrum class (where P denotes primitive; SPNM, semiprimitive nonmotorized; SPM, semiprimitive motorized; RN, roaded natural), and R, rural and lands available for ORV use. It is interesting to note in Alternative E that acres available for grazing goes down by 100,000 acres, yet AUMs actually go up slightly. Evidently, additional vegetation from range improvements is expected to more than compensate for the reduction in acres.

Tables 10.1h–10j present economic and social considerations. Economic factors displayed relate primarily to direct and indirect effects on the local economy, such as personal income (wages and business income), local employment, and local tax revenues (sales, use, and property taxes). More detail is provided on the economics of livestock with returns to labor and investment as well as wealth of ranchers calculated using an analysis of ranch budget data performed by USDA Economic Research Service. Note, however, that for the recreation program income, employment, and taxes are calculated only for the baseline alternative (A) but not for any of the others. The effects of the plan budget (which are costs to BLM of implementing that particular alternative) are quantified as changes in local income and employment. The only costs explicitly quantified are those resulting from the sediment and salinity of each alternative. Social considerations are tracked through life-style effects, which, not surprisingly, are not quantified.

Procedures Used to Estimate Effects

In general, BLM uses a mixture of professional judgment, existing data, and, when available and applicable, formal models to estimate the effects of alternatives on each resource. For example, soil loss is estimated using a universal soil loss equation (USLE). Although it was not developed for rangelands, comparison trials between USLE and actual measurements indicate that USLE provides ''good estimates for soil loss from rangeland'' (Moab District 1985:4340–50). Analysis of wildlife habitat capability and production relationships relied on big-game and livestock forage diet competition studies (done on contract to biologists outside BLM) and population-monitoring work of the Utah Division of Wildlife Resources. In addition, limiting habitat factors were identified for major big-game species.

The primary economic model used is the Forest Service Input-Output model, called IMPLAN. This model, which was discussed in chapters 7 and 9, converts changes in direct expenditures or employment into total changes in employment (by commercial sector) in a given county or group of counties. Thus wildlife is

Table 10.1a

Summary Comparison of Impacts in Year 2000
(San Juan Resource Area—RMP)

Environmental Component/ Specific Indicator	Unit	Alternative A Total Baseline Quantity	Alternative B Total Quantity (Change)	Alternative C Total Quantity (Change)	Alternative D Total Quantity (Change)	Alternative E Total Quantity (Change)
Mineral Components						
Oil and Gas						
Area available for lease						
Open	acres	891,310	1,768,740	383,560	0	481,150
	(change)		(+877,430)	(−507,750)	(−891,310)	(−410,160)
Open with restriction	acres	617,170	6,540	683,040	461,670	923,450
	(change)		(−610,630)	(+65,870)	(−155,500)	(+306,280)
No surface occupancy	acres	114,120	2,550	711,230	241,770	373,230
	(change)		(−111,570)	(+597,110)	(+127,150)	(+259,110)
No lease	acres	155,230	0	0	1,074,890	0
	(change)		(−155,230)	(−155,230)	(+919,660)	(−155,230)
Oil production	bbls/year	unquantif.	unquantif.	unquantif.	unquantif.	unquantif.
	(change)		(+signif.)	(−signif.)	(−signif.)	(+signif.)
Geophysical operations	miles/year	750	750	725	725	750
	(change)		(no change)	(−25)	(−25)	(no change)

Table 10.1b
Summary Comparison of Impacts in Year 2000
(San Juan Resource Area—RMP)

Environmental Component/ Specific Indicator	Unit	Alternative A Total Baseline Quantity	Alternative B Total Quantity (Change)	Alternative C Total Quantity (Change)	Alternative D Total Quantity (Change)	Alternative E Total Quantity (Change)
Mineral Components						
Coal						
Area available for lease	acres (change)	0	212,000 (+212,000)	0 (no change)	0 (no change)	0 (no change)
Production	tons per year (change)	0	unquantif. (+unknown)	0 (no change)	0 (no change)	0 (no change)
Locatable Minerals						
Area available for location	acres (change)	1,674,480	1,776,190 (+101,710)	1,538,430 (−136,050)	710,260 (−964,220)	1,497,610 (−176,870)
Gold production	ounces (change)	50	50 (no change)	unquantif. (−insignif.)	unquantif. (−insignif.)	unquantif. (−insignif.)

Table 10.1c
Summary Comparison of Impacts in Year 2000
(San Juan Resource Area—RMP)

Environmental Component/Specific Indicator	Unit	Alternative A Total Baseline Quantity	Alternative B Total Quantity (Change)	Alternative C Total Quantity (Change)	Alternative D Total Quantity (Change)	Alternative E Total Quantity (Change)
Air						
Air quality	(change)	high	high (−insignif.)	high (−insignif.)	high (−insignif.)	high (−insignif.)
Soils						
Soils loss	tons per year (change)	643,720	834,820 (+191,100)	564,000 (−76,420)	557,910 (−83,420)	581,975 (−61,745)
Water						
Surface water quality						
Sediment yield	acre-feet (change)	160	200 (+40)	140 (−20)	137 (−23)	130 (−30)
Salt yield	tons/year (change)	630	800 (+170)	560 (−70)	550 (−80)	540 (−90)
Ground water quality						
Total dissolved solids	mg per liter (change)	unquantif.	unquantif. (+500 to 2,000)	unquantif. (no change)	unquantif. (no change)	unquantif. (no change)

Table 10.1d
Summary Comparison of Impacts in Year 2000
(San Juan Resource Area—RMP)

Environmental Component/ Specific Indicator	Unit	Alternative A Total Baseline Quantity	Alternative B Total Quantity (Change)	Alternative C Total Quantity (Change)	Alternative D Total Quantity (Change)	Alternative E Total Quantity (Change)
Vegetation						
Disturbance						
Short-term loss	acres	39,400	176,050	40,370	23,655	44,800
	(change)		(+136,650)	(+970)	(−15,745)	(+5,400)
Residual loss	acres	5,130	6,740	8,150	4,340	8,550
	(change)		(+1,340)	(+3,020)	(−790)	(+3,420)
Area available for forest product use:						
Private dead wood harvest	acres	476,160	449,900	243,520	93,690	317,970
	(change)		(−26,260)	(−232,640)	(−382,470)	(−158,190)
Commercial fuelwood harvest	acres	476,160	449,900	142,270	93,690	317,970
	(change)		(−26,260)	(−333,890)	(−362,470)	(−158,190)
Other forest product harvest	acres	536,810	449,900	142,270	93,690	317,970
	(change)		(−86,910)	(−307,630)	(−443,120)	(−218,840)

Table 10.1e
Summary Comparison of Impacts in Year 2000
(San Juan Resource Area—RMP)

Environmental Component/ Specific Indicator	Unit	Alternative A Total Baseline Quantity	Alternative B Total Quantity (Change)	Alternative C Total Quantity (Change)	Alternative D Total Quantity (Change)	Alternative E Total Quantity (Change)
Wildlife						
Desert bighorn sheep	animals (change)	1,200	930 (−270)	2,000 (+800)	1,500 (+300)	1,410 (+210)
Crucial bighorn sheep habitat	acres (change)	329,750	306,240 (−23,410)	329,850 (+100)	349,750 (+20,000)	328,750 (−1,000)
Antelope	animals (change)	50	27 (−23)	100 (+50)	75 (+25)	85 (+35)
Crucial antelope habitat	acres (change)	12,930	12,930 (no change)	12,960 (+30)	12,930 (no change)	12,930 (no change)
Deer	animals (change)	7,357	3,760 (−3,597)	10,000 (+2,643)	9,162 (+1,805)	8,000 (+643)
Crucial deer habitat	acres (change)	191,920	175,540 (−16,380)	195,000 (+3,080)	192,150 (+230)	186,966 (−4,965)
Riparian/aquatic and T/E species habitat	acres (change)	6,080	6,000 (−80)	7,880 (+1,800)	7,880 (+1,800)	6,680 (+600)

Table 10.1f

Summary Comparison of Impacts in Year 2000

(San Juan Resource Area—RMP)

Environmental Component/ Specific Indicator	Unit	Alternative A Total Baseline Quantity	Alternative B Total Quantity (Change)	Alternative C Total Quantity (Change)	Alternative D Total Quantity (Change)	Alternative E Total Quantity (Change)
Grazing						
Area available for grazing	acres (change)	1,720,970	1,776,640 (+55,670)	1,678,630 (−42,340)	1,742,430 (+21,460)	1,620,610 (−100,360)
Livestock forage	AUMs (change)	56,735	97,504 (+40,769)	43,345 (−13,390)	37,671 (−19,064)	57,076 (+341)
Cultural Resources						
Archaeologic/historic sites damaged	sites (change)	15,764	17,154 (+1,390)	15,030 (−734)	14,289 (−1,475)	14,914 (−764)
Archaeologic/historic sites protected	sites (change)	25,380	25,360 (−20)	42,940 (+17,560)	45,120 (+19,740)	28,225 (+2,845)

Table 10.1g
Summary Comparison of Impacts in Year 2000
(San Juan Resource Area—RMP)

Environmental Component/ Specific Indicator	Unit	Alternative A Total Baseline Quantity	Alternative B Total Quantity (Change)	Alternative C Total Quantity (Change)	Alternative D Total Quantity (Change)	Alternative E Total Quantity (Change)
*Recreation**						
P	acres	61,190	38,840	198,520	198,520	195,810
	(change)		(−22,350)	(+137,330)	(+137,330)	(+134,690)
SPNM	acres	561,750	522,110	512,360	512,360	421,040
	(change)		(−39,640)	(−49,390)	(−49,390)	(−140,710)
SPM	acres	393,330	353,400	326,630	324,810	289,020
	(change)		(−39,930)	(−66,700)	(−68,520)	(−104,310)
RN	acres	747,880	849,800	726,640	728,460	858,280
	(change)		(+101,920)	(−21,240)	(−19,420)	(+110,400)
R	acres	14,720	14,720	14,720	14,720	14,720
	(change)		(no change)	(no change)	(no change)	(no change)
Area available for ORV use:						
Open	acres	1,679,340	1,776,640	484,320	367,420	611,310
	(change)		(+97,300)	(−1,195,020)	(−1,311,920)	(−1,068,030)
Limited	acres	0	150	542,390	336,880	813,060
	(change)		(+150)	(+542,390)	(+336,880)	(+813,060)
Closed	acres	99,850	2,400	752,480	1,074,890	354,820
	(change)		(−97,450)	(+652,630)	(+975,040)	(+254,970)

*In the Recreation Opportunity Spectrum (ROS), P is primitive, SPNM is semiprimitive, nonmotorized; SPM is semiprimitive, motorized; RN is roaded natural; and R is rural.

Table 10.1h

Summary Comparison of Impacts in Year 2000

(San Juan Resource Area—RMP)

Environmental Component/ Specific Indicator	Unit	Alternative A Total Baseline Quantity	Alternative B Total Quantity (Change)	Alternative C Total Quantity (Change)	Alternative D Total Quantity (Change)	Alternative E Total Quantity (Change)
Economic Considerations						
Livestock						
Returns to labor and investment	dollars (change)	403,300	682,600 (+279,300)	171,800 (−231,500)	35,600 (−367,700)	384,000 (−19,300)
Wealth	dollars (change)	26,753,000	27,821,000 (+1,068,000)	24,536,000 (−2,217,000)	24,166,000 (−2,587,000)	25,280,000 (−1,473,000)
Income	dollars (change)	1,013,000	1,133,000 (+120,000)	740,000 (−273,000)	560,000 (−453,000)	868,500 (−144,500)
Employment	jobs (change)	176	199 (+23)	158 (−18)	146 (−30)	175 (−1)
Tax revenues	dollars (change)	62,000	74,000 (+12,000)	54,900 (−7,100)	48,000 (−14,000)	61,800 (−200)
Recreation						
Income	dollars (change)	307,000	unquantif. (unknown)	unquantif. (+insignif.)	unquantif. (unknown)	unquantif. (+insignif.)
Employment	jobs (change)	23	unquantif. (unknown)	unquantif. (+insignif.)	unquantif. (unknown)	unquantif. (+insignif.)
Tax revenues	dollars (change)	10,600	unquantif. (unknown)	unquantif. (+insignif.)	unquantif. (unknown)	unquantif. (+insignif.)

Table 10.1i

Summary Comparison of Impacts in Year 2000
(San Juan Resource Area—RMP)

Environmental Component/ Specific Indicator	Unit	Alternative A Total Baseline Quantity	Alternative B Total Quantity (Change)	Alternative C Total Quantity (Change)	Alternative D Total Quantity (Change)	Alternative E Total Quantity (Change)
Wildlife						
Income	dollars (change)	59,100	41,100 (−18,000)	73,700 (+14,600)	68,500 (+9,400)	62,500 (+3,400)
Employment	jobs (change)	4	2 (−2)	5 (+1)	5 (+1)	4 (no change)
Tax revenues	dollars (change)	3,000	2,000 (−1,000)	3,800 (+800)	3,500 (+500)	3,200 (+200)
Minerals						
Income	dollars (change)	7,216,000	8,726,000 (+1,510,000)	7,128,000 (−88,000)	4,133,000 (−3,083,000)	unquantif. (insignif.)
Employment	jobs (change)	250	311 (+61)	246 (−4)	103 (−147)	unquantif. (insignif.)
Tax revenues	dollars (change)	4,322,000	4,837,000 (+515,000)	4,264,000 (−58,000)	2,588,000 (−1,734,000)	unquantif. (insignif.)
Soil and Water						
Sediment cost	dollars (change)	17,500	22,000 (+4,500)	15,500 (−2,000)	15,200 (−2,300)	14,900 (−2,600)
Salinity cost	dollars (change)	36,500	46,400 (+9,900)	32,500 (−4,000)	31,900 (−4,600)	31,300 (−5,200)

Table 10.1j
Summary Comparison of Impacts in Year 2000
(San Juan Resource Area—RMP)

Environmental Component/ Specific Indicator	Unit	Alternative A Total Baseline Quantity	Alternative B Total Quantity (Change)	Alternative C Total Quantity (Change)	Alternative D Total Quantity (Change)	Alternative E Total Quantity (Change)
Plan Budget						
Income	dollars (change)	494,000	583,000 (+89,000)	623,000 (+129,000)	558,000 (−64,000)	600,000 (+106,000)
Employment	jobs (change)	25	30 (+5)	32 (+7)	28 (+3)	30 (+5)
Social Considerations						
Community	life-style (change)	unquantif.	unquantif. (no change)	unquantif. (no change)	unquantif. (unknown)	unquantif. (no change)
Individuals	life-style (change)	unquantif.	unquantif. (insignif.)	unquantif. (unknown)	unquantif. (unknown)	unquantif. (insignif.)

reflected through the conversion of hunter expenditures to income and employment in San Juan County. Grazing is reflected by translating changes in rancher expenditures and rancher returns into income and employment in San Juan County. The same is true for minerals.

Some Observations on BLM's Estimated Effects

One major criticism of the effects displayed in table 10.1 is that table 10.1 often displays acres of land affected but not the effects these acres have on production of actual multiple uses or outputs. This violates a fundamental rule in policy analysis: the importance of focusing on outputs (i.e., results), not just inputs. To know which RMP alternative makes society better off, we need to know not just how inputs change by alternative (land, labor, BLM budget), but what society receives for this use of inputs. This failure to track outputs for most resources other than livestock grazing and wildlife results in an incomplete picture of the local income-employment effects. As table 10.1 illustrates, local income, employment, and tax revenues from minerals, wildlife, and livestock are quantified, but none of these categories are reported for a major multiple use such as recreation. This is largely a result of not linking recreation land allocations to projected recreation use.

In addition, representing human uses by focusing solely on gains and losses in income and employment in San Juan County misses many dimensions of human and economic benefits SJRA provides to people in Utah and adjoining states of Colorado, Arizona, and the rest of the United States. That is, the SJRA is federal land, owned by all citizens of the United States, not just those residing in San Juan County. Much of the recreation use in the SJRA is by people residing in other counties in Utah. Unfortunately, the only aspect of this human use accounted for in table 10.1 is spending in San Juan County. The rationale for providing recreation on public lands is not primarily to provide a tourism industry for local counties, any more than the primary reason to have oil and gas leasing is to provide jobs in a particular area. Since in both cases changes in the SJRA reflect simply a transfer of economic activity from elsewhere, there is not a net national gain in employment. It makes sense to have both mineral leasing and recreation on public lands if the value of the output to consumers exceeds the costs of producing that output. Thus if public lands can supply oil for less than the costs of importing it, this saving is the human benefit and economic effect of mineral leasing. If the visitors to the public lands are better off by their visit, this benefit should be counted even if these people do not live in the SJRA or do not generate employment for those who do live in San Juan County. In the interest of knowing not only net benefits but also the distribution of those benefits between locals and nonlocals, both types of information should be developed. Distributional equity is an evaluation criterion that requires that all major categories of beneficiaries be spelled out, not just locals.

Thus a major omission from table 10.1 is any comparison of benefits and costs. The costs to BLM to implement each alternative are developed in BLM's Appendix

K but not reported in its display table. More important, no comparable benefit information is provided. As discussed in chapter 6, it is incorrect to compare local income generated with costs, since the local income and employment gained are merely a transfer and not a net gain. BLM blurs this distinction when it computes the local income gained from the costs to BLM to implement the RMP. With this analysis federal costs have been translated into local gains, without any consideration for comparing the costs of the BLM program with the benefits to society, not just residents of San Juan County.

Step 7: Selection of the Preferred Alternative

According to the formal steps in FLPMA's planning process for BLM, it is at this point in the planning process that the preferred alternative is to be identified. However, as discussed earlier, the preferred alternative was already preidentifed in step 5, when the alternatives were formulated. This results in its being tentatively identified before the estimated effects of implementing it can be compared with the other alternatives. Although the initial preferred alternative might represent a desirable set of management actions (from the view of the Resource Area manager and his or her staff), the effects of those management actions should have some influence on whether those management actions are really preferred over others in different alternatives. Of course, there is an opportunity for BLM when it moves from its preferred alternative in the Draft EIS to the preferred alternative in the Final EIS to incorporate not only the public comments but also what was learned from step 6 on estimated effects. In this way, the "final" preferred alternative can be modified from the initial preferred alternative to minimize the undesirable effects that may have been discovered in step 6.

Step 8: Selection of Final RMP

The Draft EIS and proposed RMP were put forward by BLM in May of 1986. The public originally had ninety days to submit comments. At the request of the public, BLM later provided an additional sixty days. A public meeting was also held to allow the public to discuss the draft. However, it was held in the capital of San Juan County (Monticello, Utah). The location of the meeting provided county residents with a convenient opportunity to discuss the RMP with BLM officials. Unfortunately, other public land users who live in Salt Lake City or Colorado were not given such an opportunity. Public meetings would normally be held in major urban areas, where the majority of recreation visitors to the Resource Area would originate.

A total of 112 responses was received on the Draft EIS and RMP (several of which were from the same persons). BLM grouped the comments as follows: special interest groups (conservation organizations), industry (mining companies), federal government agencies (EPA, USFS, NPS, USFWS, Bureau of Indian Affairs), state and local agencies, universities, and individuals. The geographic distribution of these public comments indicated that people in Colorado, Arizona, and urban areas in northern Utah (Salt Lake City and Logan) submitted more comments than those living in southeast Utah. This is proof that there is widespread interest in the management of the public lands in the SJRA. This is due in part to the archaeological resources and relatively unique canyon country desert environment. Nonetheless, this reinforces the importance of holding public meetings where the public land users live, not just where the Resource Area is located.

Differences Between Draft Preferred Alternative and the Final Proposed RMP

There are some major differences between the Draft Preferred Alternative and the Final Proposed RMP. The most dramatic is the substantial increase in acreage designated for special management considerations as Areas of Critical Environmental Concern. Three areas were added to ACECs to protect their scenic value, and the 50,000-acre Grand Gulch ACEC was expanded to the entire Cedar Mesa, with over 300,000 acres. The differences in number of ACECs and the extent of these areas is illustrated by comparing figure 10.8, which came from the Draft EIS, with figure 10.9, which came from the Final EIS. As can be seen by comparing these two figures, the Final EIS and RMP are more protective of recreational and cultural resources and nicely blend BLM's management with adjacent Canyonlands National Park. This Final RMP was accepted by the governor of Utah with only minor modifications.

Step 9: Implementation, Monitoring, and Evaluation

The Final EIS presents BLM's implementation schedule and monitoring objectives by resource category. Certain actions are scheduled to occur immediately upon approval of the RMP. These actions include designation of areas as ACECs and special recreation management areas (SRMAs) and new mineral leasing stipulations. Other actions are phased in over a one- or two-year period. These include implementing ORV closures and excluding grazing from particular areas identified in the RMP or changing season of livestock grazing. Other activities, such as writing site-specific management plans or nominating cultural resources to the National Register of Historic Places, occur at a rate per fiscal year specified in the Final RMP.

In the case of the SJRA, environmental groups protested to the BLM that the Final RMP did not go far enough to protect natural resources. After more than six months the protest was denied, but that delay now makes it difficult to evaluate how implementation of the SJRA RMP has proceeded. However, the Grand Resource Area to the north of the SJRA has a Final RMP that it has begun imple-

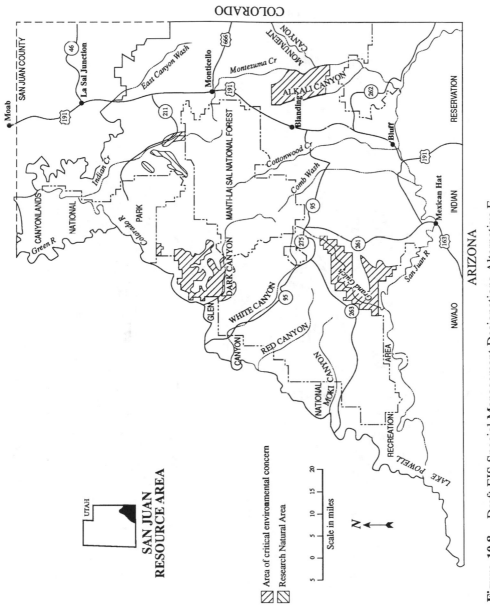

Figure 10.8. Draft EIS Special Management Designations–Alternative E

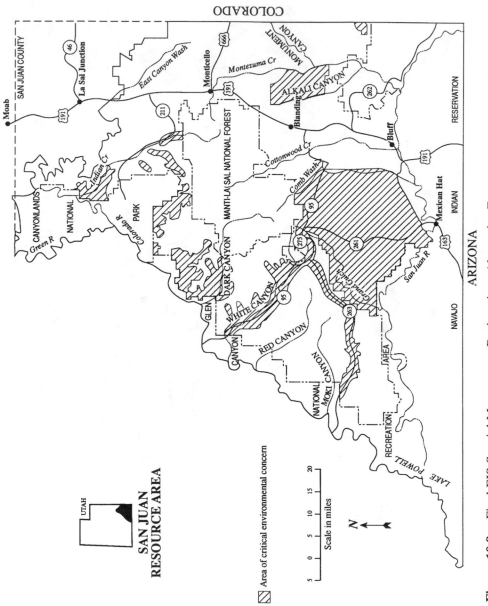

Figure 10.9. Final EIS Special Management Designations–Alternative E

menting over the last two years. Therefore it will be used to illustrate the pace of implementation.

Implementation of Existing RMPs in Grand Resource Area

The first action to be implemented involved applying the new surface occupancy restrictions to new mineral leases that were granted since the issuance of the Final RMP. In addition, the areas identified for grazing exclusions have had cattle grazing eliminated through retirement of the grazing privileges. Specifically, as the ranchers have taken "nonuse" (i.e., not put livestock out in that allotment), BLM has retired those unused AUMs from the grazing privileges. This means that no other rancher can apply for those AUMs in that area. The designation of one area as an ACEC has also proceeded as planned within the Grand Resource Area.

Importance of Funding for Full RMP Implementation. The Grand Resource Area Final RMP also recommended closing certain areas to ORVs. The detailed master maps designating areas as closed, open, and limited have been prepared. They must be printed and distributed to the public so that people are informed of the new rules. In addition, closed areas must be signed to inform ORV users that these areas are now closed. Additional rangers are needed to warn and, if necessary, issue citations for repeat offenders. Unfortunately, the District Office has not been given the funding in its budget to print the maps, sign the closed areas, or hire additional rangers. Consequently, this significant provision of the Final RMP is not implemented.

An important lesson from the Grand Resource Area is the importance of funding for implementation of the RMP. The mineral leasing stipulations, grazing, and ACEC actions were readily implemented, in part because little additional funding was required. Thus even the most balanced multiple-use plan will not result in balanced multiple use on the ground if all provisions are not funded. Given the many levels through which an agency's budget must pass, implementation can be voided at many levels. For example, it can be thwarted if the State Office of BLM fails to forward the Resource Area's request for funds to the Washington office of BLM. Or the Washington office can thwart implementation by its unwillingness to forward complete funding requests up to the Department of Interior or the Office of Management and Budget (OMB). At the Washington level (particularly in OMB), the overall federal budget constraints associated with budget deficit reduction often preclude full funding of every BLM Resource Area's request. Sometimes such lack of federal funds is used as a convenient excuse to systematically not fund management actions in the Final RMP that are politically unpopular with the Executive Branch. For example, OMB might not provide money to implement grazing exclusions or cuts in livestock grazing. This political orientation of budgeting often continues at the congressional level, where appropriation committees choose to fund some programs at full agency budget request and provide only a small fraction of what the agency requested for others. In principle, agency officials could use planning documents in testimony before Congress to argue that full funding for implementation of the Final RMP is needed if the intent of Congress' own FLPMA legislation is to be met. In the end, the agency needs to be constantly alert to ensure that its RMP is funded for implementation. Nonetheless, there have certainly been cases where the agency, frustrated in its attempts to get balanced funding for its multiple-use programs, has taken funds targeted by Congress for one politically popular multiple use (in BLM, energy production and livestock grazing) and used some of those funds to implement related multiple-use management actions.

BLM's "Second-Tier" Planning: Activity Planning

Writing specific management plans for ACECs and Special Recreation Management Areas is site-specific. The issues analyzed at this level of detail involve specific recreation use guidelines, such as no camping areas, group size limits, and so on. This second tier of planning involves activity planning. By *activity planning* BLM means the detailed, site-specific plan for recreation, range, or wildlife improvement projects. For example, if water was identified in the RMP as a limiting factor in a particular habitat, the activity plan would lay out exactly what type of water structure (pond, guzzler) would be built, would specify where it would be built, and would perform a benefit-cost analysis on the project. It is through the activity planning that the multiple-use management objectives for each area are carried out.

Recent National Strategic Plans

Recently, BLM has been developing what can best be called national "strategic plans" that set out goals for wildlife and recreation on BLM land. Both the recreation document, called "Recreation 2000," and the wildlife document, called "Fish and Wildlife 2000," are new BLM initiatives to take a proactive role in enhancing these two resources on public lands. Former President Bush's director of BLM, Cy Jamieson, had pushed for increased funding to implement these new initiatives. In many respects this is an exciting development in BLM: national direction backed by a commitment from the Washington office of BLM to secure the necessary funding. National direction can help BLM balance the clout currently wielded by local ranchers and state and local governments.

The only drawback to these new national plans is that they reflect a return to independent, resource-by-resource planning. They give only limited consideration to the multiple-use trade-offs associated with meeting the new recreation or wildlife goals. Unfortunately, the only current example of a national integrated multiple-use plan is the Forest Service Resource Planning Act (RPA) Program, which is performed every five years. Such an intensive and expensive RPA process may not be worthwhile for BLM relative to what the contribution might be in terms of less local domination of multiple use on BLM lands. Something in between these two extremes might promise to provide realistic multiple-use direction without requiring the sizable effort involved in performing an RPA-type assessment and program.

A Critique of BLM'S First Round of RMPs

As with any new endeavor, the first few steps are often the most daunting. This was certainly true for BLM. As mentioned earlier, just as the agency was getting started with its first FLPMA planning process in 1980, a radical change in presidential administrations resulted in a new director (Burford), who ordered major revisions in the planning regulations that had just been adopted. In addition, while BLM was attempting to get its new planning process started, other de facto planning efforts that focused on specific resources were either continuing or beginning (Leman 1984). Having been forced to move from custodial manager to active manager of public lands by court suits brought by Natural Resources Defense Council over coal and livestock grazing, at first BLM could not synchronize all the resource concerns. This was partly because of the court orders and partly because of a desire not to hold up such resource decisions as coal leasing until comprehensive Resource Management Plans were completed. Eventually, decisions over suitability of lands for coal mining and livestock grazing management were being made in the RMP process rather than in separate EIS efforts.

Unfortunately, such programs as the Wilderness review were often not integrated into RMPs. For example, the BLM in Utah, Wyoming, and several other states was fast-tracking the wilderness determinations required under FLPMA well ahead of the legislative deadline. As a result, states like Utah and Wyoming ran parallel Wilderness Study Area EIS's without integrating these into the planning process as Colorado did. In some states there seemed to be an effort to clear as much land from wilderness consideration as fast as possible so that it could be "released to multiple use" (even though Wilderness is a multiple use and supports several other multiple uses).

If adequate funding and personnel were available, these parallel planning efforts would at best duplicate resource planning. However, with funding for planning at half the level of the late 1970s, running three parallel planning processes—one for coal, one for wilderness, and one for other multiple uses—resulted in piecemeal partial plans that inadequately addressed all but the key resources that were the focus of the particular single-resource plans. From discussions about multiple-use planning in chapters 8 and 9, it should be obvious that similar information would be needed by all these resource allocation decisions. Consequently, it would have been far more economical to perform a comprehensive RMP on all resources rather than to rush through one abbreviated planning process after another.

Another criticism of the RMP planning process under Burford, made by a former BLM planning official, was that RMPs were being used not to make the resource allocation decisions but to postpone making them (Crawford 1986:407)! Specifically, the Final RMPs often suggested continuation of current livestock grazing levels with monitoring to determine if changes are needed. The failure to rely on existing data on range condition to allocate forage is contrary to the overall emphasis in BLM's planning process of relying on existing data. This ability to postpone making the forage allocation decisions is also inconsistent with two additional facts: (1) livestock grazing has an impact on soils, so that vegetative communities and wildlife are nearly always one of the planning issues to be resolved; (b) many RMPs are intended to meet the court-ordered requirement of a site-specific grazing EIS under the NRDC suit. Given these two factors, one would have expected that if the existing data were inadequate, the planning effort would have provided the opportunity to collect the necessary data and perform the required analyses needed to allocate forage among livestock, wildlife, and wild horses.

Crawford (a former BLM planning official) states one of the key ailments in BLM's current planning process: "Lack of solid economic, analytical procedures and hard data continually handicaps planning by failing to portray objectively trade-off values to be gained or lost through managerial decisions" (1986:409).

In addition, there continue to be managers "willing to strike a deal" with public land users even if those deals are inconsistent with what the RMP recommends for that area (Crawford 1986). Lastly, BLM's plans have been poorly linked with its budgeting process.

Adoption of a comprehensive benefit-cost analysis approach in its RMP planning process could remedy many of these criticisms. BLM, with its inadequate funding and personnel, is understandably reluctant to adopt FORPLAN-type LP models to perform BCA. However, as shown in chapter 9's case study of the Gallatin National Forest, these models are not necessary to perform comprehensive multiple-use planning and trade-off analyses. In many cases the same production information (e.g., output levels) used by BLM in its input-output models to predict changes in employment could be used to perform a BCA of each alternative plan. BLM already does BCA of its activity plans for range and wildlife habitat improvements. The agency has growing experience with BCA. There is little reason not to use BCA to address the major resource allocation decisions in its planning process. With benefit-cost analyses in hand, managers would have not only the "solid economic, analytical procedures and hard data" to make objective trade-off analyses, but the data to justify funding requests to implement the RMP. BCA economic values could shield managers from those that focus solely on employment effects to local residents at the expense of other public land users. An example of using BCA to deal with one of BLM's most central issues, wildlife versus livestock grazing, is presented in an additional case study at the end of this chapter.

Conclusion

As these comments on how BLM is implementing these planning steps indicate, BLM is taking a somewhat different approach from that of the Forest Service to the same basic multiple-use planning mandate. In some ways this does not make sense, because both agencies have basically the same multiple-use mandate and both have valuable resources. However, in other ways this different approach is reasonable, because the BLM lands may be more variable than those of the Forest Service. BLM has some lands that are as valuable as the most valuable National Forest lands, but it also has lands whose multiple-use values per acre are low. Thus tailoring planning to the values at stake may make some sense. BLM has been able to implement its multiple-use planning process quickly and relatively inexpensively. In those areas with low values per acre and no outstanding resource values or conflicts this cursory planning approach is appropriate. But in many other areas, where minerals, wilderness, or wildlife make BLM lands extremely valuable, a more careful analysis may be well worth it. Where serious resource degradation is resulting from livestock overgrazing, a cursory analysis because of limited data and a wait-and-see approach to management may be politically expedient, but it certainly is not wise multiple-use management. Chapter 11 will provide a framework to determine the appropriate intensity of planning on multiple-use lands. Focusing on the value of additional information, this chapter shows that the intensity of planning should be in proportion to the resource values at risk. If inaccurate information could lead to suboptimal resource allocations that involve forgoing millions in economic benefits, spending an additional $100,000 on data collection is wise. First, however, the following case study provides an example of how wildlife-livestock forage allocations can be optimized with a relatively straightforward application of benefit-cost analysis.

Case Study of Economic Value of Forage on BLM Lands for Big Game and Livestock*

One of the continuing issues in BLM management has been the conflicts between cattle stocking rates and their effects on fish and wildlife resources. Economically efficient use of BLM rangelands requires adjusting the numbers of livestock and wildlife so that the mixture is roughly proportional to the relative values these different animals provide to society. In addition, more wildlife and livestock can be accommodated by boosting range productivity through such investments as water developments and manipulation of the vegetation. This is an example of shifting the production possibilities curve outward. However, the U.S. Office of Management and Budget and some economists (Stroup and Baden 1983) need to be con-

*Portions of this section are reprinted with permission from the *Journal of Range Management*, 42 (2), published by the Society for Range Management, Denver, Colorado.

vinced that these investments are competitive with other competing uses of government funds.

To evaluate the economic efficiency of livestock investment projects, BLM developed a model called SAGERAM (Bureau of Land Management 1985). Although this model framework is useful, its analytical capabilities have been limited by difficulty in estimating marginal values of wildlife and forage used by wildlife in a manner commensurate with livestock forage values (see Godfrey 1982; Bartlett 1982, 1984; Dyer 1984).

However, estimation of the marginal values of elk and deer on public lands has become more frequent in recent years (Cory and Martin 1985; Keith and Lyon 1985). This case study illustrates the travel cost method (TCM), introduced in chapter 6, to estimate marginal values of two big-game species (elk and deer) and calculate the marginal-value product of an animal unit month (AUM) of forage for these species. These values are then compared to the marginal value of forage to cattle.

Marginal Valuation of Wildlife with Travel Cost Method

As discussed in chapter 6, the economic value of any good or service is defined as consumers' and producers' net willingness to pay (WTP). Measuring consumers' net WTP involves measuring the area under their demand curve but above their current price. Because TCM estimates the demand curve for recreation at a specific site, the net willingness to pay for recreation under existing conditions at a site can be calculated. For example, consider a demand equation for a recreation site of the following:

$$V = h(P, Q, Y) \tag{10.1}$$

where:

V = visits
P = price
Q = wildlife quality
Y = income

Since wildlife quality (e.g., success rate if the activity is hunting) is a demand-shift variable, the demand curve will shift outward with increases in quality. As discussed in chapter 6, the area between the existing and improved-quality demand curve is a measure of the WTP for the improvement. The marginal or incremental value of a harvested animal is the ratio of the increase in net WTP to the increment of animals harvested.

By observing visitation rates to different hunting areas that have different harvest success rates, a coefficient relating the combined effect of quality on participation and trip frequency can be estimated. Essentially, the origin and destination data from several sites with varying quality are pooled and one equation of the following form is estimated:

$$\ln(V_{ij}/POP_i) = B_0 - B_1(\ln DIST_{ij}) + B_2(\ln INC_i) + B_3(\ln THVST_j) \quad (10.2)$$

where

V_{ij} = visits from county i to hunt area j

POP_i = population of county i

$DIST_{ij}$ = round-trip distance from county i to hunt area j

INC_i = county i's per capita income

$THVST_j$ = hunt area harvest of either deer or elk

Estimation of equation (10.2) will allow calculation of the marginal value of harvesting an additional animal. Therefore it was desirable to estimate the hunting demand equations using the double-log model. This functional form produces a diminishing marginal value per animal when the coefficient on harvest is less than 1.

Study Area

To calculate the marginal-value product of the forage in producing elk and deer requires site-specific knowledge of the production relationships. Hunt areas 36 and 36B near Challis, Idaho, are ideal for this purpose. The Challis area was designated by the *Natural Resources Defense Council* v. *Morton* court decision as the area for BLM's first grazing Environmental Impact Statement (EIS). The Challis area has been the scene of substantial controversy over grazing versus wildlife prior to, during, and after the preparation of the EIS (Nelson 1980).

BLM's Final EIS (BLM 1977:Chap 3:21) states that during May and June there is spatial and dietary competition for grasses between cattle and antelope and between deer and elk in the area. Elk and cattle have strong dietary similarities (particularly in the spring) in terms of their preferences for consuming grasses. Therefore the potential dietary competition from increasing elk or cattle populations may be the greatest. Substantial evidence of social avoidance of cattle by elk also exists, with cattle (and associated humans tending the livestock) causing elk to leave an area of otherwise desirable habitat (Lyon 1985:17; Nelson 1984).

Data Sources

Those hunting in Idaho in 1982 were surveyed to collect the data for this model. To ensure that the assumptions of TCM were met, only hunters stating that hunting was the primary purpose of the trip and that this hunt unit was the primary destination were included in the analysis. The survey contacted, via telephone, a total of 1,629 elk hunters (Sorg and Nelson 1986) and a total of 1,445 deer hunters (Donnelly and Nelson 1985) during January and February of 1983 regarding their 1982 hunting season.

Calculating Marginal Productivity of Forage

BLM's Final EIS states that at least a 30% increase in deer is sustainable with additional forage. The potential for increased carrying capacity of elk habitat due to new grazing systems is about 20% (BLM 1977:Chap 3:29). Although there are many important components of habitat for elk and deer in the Challis area, forage on winter and spring ranges appears to be limiting populations in the area. The purpose of these estimates is to provide a benchmark of what the potential improved condition might be. The remaining analysis calculates marginal values of wildlife and forage using current harvest and a 25% increase in elk and deer populations that results from range improvements and reallocation of forage.

A 25% increase in elk harvest in unit 36 requires twenty-eight more bull elk. According to information provided by the Idaho Fish and Game, production of twenty-eight more bull elk for harvest (surplus production) annually would require that the elk herd in unit 36 increase by a total of 378 elk. The composition of the increase is 19% bulls, 54% cows, and 27% calves. The available literature (BLM 1977:1–2; Thomas 1984) suggests that each adult elk consumes between 0.4 and 0.67 AUMs of forage each month. This analysis uses the average of these two estimates, or .54 AUMs per adult elk and half this amount per calf. This information is combined with the herd structure to generate a simple production relationship relating the number of elk available for harvest to quantity of forage. Using unit 36 to illustrate the calculations, the relationship is

$$
\text{EH} = 1/[(9.85\text{AE} \times .54 \text{ AUM} \times 12 \text{ months}) \\
+ (3.65\text{CE} \times .27 \text{ AUM} \times 12 \text{ months})] \tag{10.3}
$$

where:

    EH = bull elk available for harvest
    AE = adult elk (bulls and cows)
    CE = calf elk

Carrying out the calculations in equation (10.3) yields the simple elk/forage relationship for unit 36 of

$$
\text{EH} = .0132 \text{ AUM} \tag{10.4}
$$

For unit 36B the elk/forage relationship is as follows:

$$
\text{EH} = 1/[(10.22\text{AE} \times .54 \text{ AUM} \times 12 \text{ months}) \\
+ (3.78\text{CE} \times .27 \text{ AUM} \times 12 \text{ months})] = .0127 \text{ AUM} \tag{10.5}
$$

The simple deer-forage relationship for unit 36 is as follows:

$$
\text{DH} = 1/[(6.935\text{AD} \times .25 \text{ AUM} \times 12 \text{ months}) + \\
(2.565\text{F} \times .12 \text{ AUM} \times 12 \text{ months})] = .0408 \text{ AUM} \tag{10.6}
$$

where:

DH = deer harvested
AD = adult deer
F = fawn

For unit 36B the simple deer/forage relationship is as follows:

$$DH = 1/[(5.548AD \times .25 \text{ AUM} \times 12 \text{ months}) + (2.052F \times .12 \text{ AUM} \times 12 \text{ months})] = .051 \text{ AUM} \qquad (10.7)$$

Statistical Results: Estimated Demand Equations

The elk TCM demand equation estimated using the two-stage least squares procedure described in Loomis and colleagues (1989) is as follows:

$$\ln(V_{ij}/\text{POP}_i) = 24.173 - 1.629(\ln \text{DIST}_{ij}) - 3.126(\ln \text{INC}_i) + 0.431(\ln \text{THVST}_j) \qquad (10.8)$$
$$t \text{ values } (20.85) \ (-30.28) \qquad (-24.09) \qquad (5.51)$$

The R^2 was .74. All the individual coefficients are significant at the 1% level. The size of the t statistics shows that the double-log functional form offers a good explanation of the relationships between the variables.

The deer demand equation was estimated using the same type of two-stage least squares procedure as that used for elk. The results are given in equation (10.9).

$$\ln(V_{ij}/\text{POP}_i) = 47.19 - .649(\ln \text{DIST}_{ij}) - 6.381(\ln \text{INC}_i) + .327(\ln \text{THVST}_j) \qquad (10.9)$$
$$t \text{ values } (11.33) \ (-11.88) \qquad (-13.14) \qquad (2.21)$$

The R^2 was .47. The distance and income coefficients are significant at the 1% level. The harvest variable is significant at the 5% level.

Calculation of Marginal Values

In unit 36 a 25% increase in bull elk harvest (twenty-eight more) generates a rightward shift in the elk-hunting demand curve. The area between the new and old curves for unit 36 is an increase in net economic benefits of $14,075 annually. The marginal value of a harvested bull elk is $502. The marginal value per elk and deer in unit 36B is $647 and $310, respectively. Table 10.2 displays marginal values per animal under current and improved conditions.

Combining the marginal product of forage calculated from equation 10.7 (.051) with the marginal value of a deer in unit 36B ($310) yields a value marginal product of $15.81 per AUM. The $15.81 represents the maximum amount hunters would bid per AUM for the increased forage to produce 25% more deer in hunt unit 36B. Calculation of the marginal-value product (MVP) for elk follows this same procedure used for deer. Table 10.2 presents marginal values per animal and per AUM

Table 10.2
Marginal Values (MV) of Wildlife in Challis, Idaho

| | Unit 36 | | Unit 36B | |
	Elk	Deer	Elk	Deer
Current Herd Size				
MV per animal				
harvested	$535	$167	$685	$333
MV per AUM	$ 5.70	$ 6.82	$ 7.04	$ 17.00
25% Increase in Herd Size				
MV per animal				
harvested	$502	$155	$647	$310
MV per AUM	$ 5.35	$ 6.32	$ 6.65	$ 15.81

for big-game units 36 and 36B. Although marginal values per animal are higher for elk than for deer, the preceding production functions reveal that a standardized AUM produces about four times as many harvestable deer as elk. This is reflected in the MVP figures. The large difference in forage value for deer in the two units relates to differences in marginal value per deer and the higher marginal productivity of unit 36B in producing deer. Specifically, it takes only an increase of 7.6 deer to produce one more available for harvest in unit 36B compared with 9.5 deer to produce one more for harvest in unit 36. The higher marginal value per deer in unit 36B appears to reflect the higher harvest rate in that unit.

Economic Value of Livestock Forage

As discussed in chapter 6, a variety of techniques can be used to estimate the value of public land forage to cattle ranchers. The joint Forest Service and Bureau of Land Management Appraisal Report (Tittman and Brownell 1984) states that fair market value of public land grazing in the region where Challis is located would be $7.60 per AUM. Wilson and colleagues (1985) use a linear programming approach with ranch budget data to estimate the value per AUM for the BLM land in the Challis area. The weighted average value of the forage across the four different size classes of ranches is $6.40 (where the weights are number of BLM AUMs used by each size class). However, the livestock value per AUM ranges from a low of $1.14 to a high of $10.10 in the two relevant areas studied by Wilson and colleagues.

Discussion and Conclusions

Comparison of the wildlife values in table 10.2 with these forage values shows that deer and elk are economically competitive with cattle in the Challis, Idaho, area. In particular, the marginal value of forage for wildlife in unit 36B is quite a bit larger than livestock forage values. A more economically efficient mix of uses would involve providing additional forage to wildlife until the marginal value to wildlife decreased to the marginal value of forage to livestock. Because the functional form of the demand equations estimated has the property of diminishing marginal value for each additional animal, in theory one can calculate the increase in wildlife herd size necessary to drive forage values down into equilibrium with livestock. The existing divergence in values of forage between wildlife and livestock in unit 36B shows the direction that resource management should be moving from an economic efficiency standpoint.

In addition, the wildlife values are useful for determining the economic feasibility of investments to increase forage production for wildlife. Incorporation of these more conceptually correct marginal values of wildlife and forage into BLM's SAGERAM would improve the accuracy of these analytical aids in suggesting economically efficient use of public rangelands.

Joint federal land management agency (Forest Service, BLM, and USFWS) and state fish and game agency surveys and analysis along the lines suggested here would go a long way to making forage allocations more sensitive to changing societal values.

Another View: Abuses of BLM Land Under the Anachronistic 1872 Mining Law

Although FLPMA cleared away many outdated public land laws, one of the oldest remained in force: the Mining Law of 1872. This law guides the prospecting and staking of claims for "locatable minerals" (i.e., hard-rock minerals, such as gold, silver, lead, and copper). The law allows anyone the right to "stake a claim' on BLM land by filing with the county and local BLM office notification that they have staked a claim for particular minerals. No formal proof that such minerals actually exist on the property or that it is economically feasible to develop minerals is required to stake a claim. In fact, the law gives BLM little control over what mineral exploration activities can be undertaken on BLM land. All that FLPMA provided was that the miner must notify BLM that mineral exploration would proceed and the degree of activity (in terms of surface disturbance expected).

The 1872 Mining Law requires that $100 a year of "annual development work," referred to as "diligence," be performed each year to keep the claim current. With thousands of mining claims in many BLM districts, it is not feasible for BLM to verify whether $100 of work is actually performed. In addition, $100 of work in 1872 was a substantial commitment of effort. In 1993 dollars, $100 worth of work may last only a few hours. More important, much of the annual development work that does occur results in needless damage to the land without promoting mineral development (U.S. General Accounting Office [USGAO] 1989:3). In addition, many miners make temporary homes on their mining claims. These "miners'' are living on BLM lands but performing little mining. In states like California with tracts of BLM land intermingled with expensive private land along rivers and in the foothills, such ''rent-free'' homesites are valuable to the miners.

A miner may receive ownership (i.e., title) to the BLM land by ''patenting'' the claim. Once patented, BLM will transfer ownership to the miner for as little as $2.50 an acre (USGAO 1989:4). The miner is then free to do with the land as he or she pleases. Since the 1970s, thousands of acres of BLM land have been transferred to private ownership, much of which ended up not as mining operations but being resold as resort properties near ski areas in Colorado or near growing western towns (USGAO 1989:4). The federal government received a few thousand dollars for lands that were resold for millions of dollars (USGAO 1989:4).

Even when the patented lands are used as the law intended—for mining—the federal government receives little payment for the mineral values associated with the property. The U.S. Government Accounting Office found ''the federal government in 1986 sold under patent 17,000 acres of land for $42,500. Weeks later, the patent holders sold these lands to major oil companies for $37 million'' (USGAO 1989:2). Unlike such energy minerals as coal and oil, to which the federal government both auctions off the mining rights (to ensure it receives fair market value) and assesses a royalty fee on production, the federal government receives no mineral royalties or fees associated with any production of hard-rock minerals from lands acquired under the 1872 mining law. Thus millions of dollars of gold or silver may be produced from what was formerly BLM land, yet the federal government (i.e., the taxpayers) receives nothing.

Finally, the few hard-rock mines that are actually developed often become the liability of the federal government after the mine is abandoned. That is, many abandoned hard-rock mines have become public safety hazards in at least two ways: (1) the abandoned mine shafts and subsiding land pose a threat to users of surrounding public and private land; (2) numerous mines result in toxic discharges and mine tailings that pollute streams and groundwater. Many hard-rock sites have resulted in serious fish kills or contamination of drinking water supplies. Several have already been listed as federal toxic ''Superfund'' sites, and many more are being considered for that status. The use of cyanide in gold-mining operations continues to be a major threat to fish and wildlife anywhere near such operations.

The USGAO recommends that the entire patenting system for hard-rock minerals on BLM land be eliminated, as such land disposal programs are inconsistent with national natural resource policies (USGAO 1989:5). USGAO further recommends that claim holders be required to pay an annual ''holding fee'' in place of the $100 of assessment work currently required. This would be far easier to monitor and enforce and would greatly reduce the environmental damage associated with some assessment work that actually does take place. Legislation introduced into Congress would also require payment of royalties from minerals produced from these BLM lands.

Although efforts at legislative reform come closer and closer each year (the latest effort was in 1991), BLM has been working to implement administrative regulations on restoration of hard-rock mining claim land and to curb flagrant ''squatting'' on mining claims on BLM land. For an interesting treatise on the 1872 Mining Law see John Leshy's (1987) book on the subject.

11

Tailoring the Cost of Planning to the Value of Resources at Stake

This chapter will serve as a critique of the planning efforts of the Forest Service and BLM. In addition, it will offer suggestions for improvement, as well as evaluating the proposed second-round Forest Service planning regulations.

Management and Planning Costs on National Forest Land

At the crudest level of analysis, the Forest Service received about a $2 billion budget in 1989 to manage 192 million acres. This translates into a cost per acre of $10.50, which includes the costs from the district ranger and forest supervisor levels and from the Regional Office and the Washington Office.

The Forest Service accounting system was not designed to keep track of NFMA planning costs as distinct from field-level activity planning and general management costs of the forest. To provide some preliminary information on the cost of NFMA Forest Plans in Idaho, Senator McClure of that state asked the U.S. General Accounting Office (GAO) to investigate. GAO studied the Boise and Clearwater National Forests in 1986. It tallied up the direct salary, supplies, computers, utilities, and so on, specifically associated with just NFMA Forest Plans. The costs include forest-level interdisciplinary teams and Regional Office and Washington Office assistance and review costs from 1981 to 1985. This covers step 2 (development of the planning criteria) to step 8 (Draft Plan) on the Clearwater to step 7 (valuation of alternatives and identification of preferred alternative) on the Boise National Forest. The costs for step 1 (identification of issues, concerns, and opportunities) were incurred prior to 1981 and are not included. The costs for the Final Plan for the Clearwater and for publishing the Draft and Final Plan, as well as those for responding to public comment on the draft, were not included for the Boise, as they had not been incurred at the time of the GAO report.

These subsets of costs were $2.8 million for the Boise and $2.3 million for the Clearwater National Forest in 1985 dollars (U.S. General Accounting Office 1986). Although this is at least five times what the BLM spends on its Resource Management Plans, the direct Forest Plan costs represent about 2% to 3% of the total operating costs of the respective forests (U.S. General Accounting Office 1986). On a per acre basis the costs are $1.06 for the Boise and $1.36 for the Clearwater. This difference makes some sense, as the Clearwater's costs include the Draft Plan, whereas the Boise National Forest has yet to reach that stage.

Management and Planning Costs of BLM Land

BLM receives about $720 million to manage 270 million acres, or $2.67 per acre. This is about one-fourth of what the Forest Service receives for its multiple-use management. This fraction of Forest Service funding is a little less per acre than Clawson's very rough estimate of the relative capital value of Forest Service land versus BLM land (Clawson 1983:111). A former BLM planning official (Crawford 1986:406) states that BLM was spending between $300,000 and $600,000 per Resource Management Plan covering an average of 1.1 million acres. The average cost per acre would vary between 30 cents and 60 cents. In some ways, it is good to see a range in the costs per RMP. One would hope the high-end costs would be associated with high-value Resource Areas and that Resource Areas with relatively low values would have low-end costs. For many Resource Areas where livestock grazing and wildlife require 20 acres per AUM, these planning costs are in line with the values at stake. In line with the differences in overall management intensity, the Forest Service's spending is also about four times that of the BLM.

Conformity Within Agency Planning

Although this book has presented detailed case studies of just one Forest Plan and one BLM Resource Management Plan, these plans are typical of each agency's uniform level of effort. After a personal review of data from nearly thirty Forest Plans and a dozen BLM Resource Management Plans, I note a generally consistent pattern within the agencies.

Forest Plans

The Forest Plans, although certainly not identical across every region, generally represent very detailed efforts centered around FORPLAN. This was true even on the Humboldt National Forest in Nevada, where the only forest product harvested was pine nuts, and was true in many of the recreational National Forests around the Los Angeles area (i.e., the Angeles, Cleveland, and so on). The Forest Service employees in California had to be fairly creative to adapt FORPLAN, which is essentially a timber-scheduling model, to many southern California forests. This intensive degree of planning is probably out of line with the returns from such Herculean effort for several of these National Forests. A review of forest planning suggests that FORPLAN should be relied on primarily in forests that have major timber programs (Shands et al. 1990:52) and that simpler analysis techniques should be adopted on other nontimber National Forests. Alternatively, some forests, where it was felt that an even more detailed and rigorous analysis was called for, were discouraged by the Regional and Washington offices, since it would "make other forests that did not do such a detailed analysis look bad." The implication here was that uniformity and consistency (even in the face of heterogeneity among forests) are more important than tailoring the level of analysis to the resources and issues at hand.

Bureau of Land Management's Resource Management Plans

A review of a dozen RMPs in eight different states indicates that BLM, much like the Forest Service, makes only a small effort to tailor the analysis within the planning process to the resource values at stake. For example, in the Grand Junction Resource Area in western Colorado, such minerals as coal, oil, and gas are significant resources, along with Wilderness. Not surprisingly, there is more discussion of these resources in the plan, but little in the way of additional analysis. As is common in RMPs, emphasis is placed on inputs such as the number of acres available for mineral leasing in each alternative are emphasized, but no information is provided on expected mineral production with the alternative. Many RMPs simply describe alternatives in terms of restrictions on acres of land available for mineral leasing, with no information on what those restrictions mean for production of minerals. Since in some cases BLM goes so far as to show that a bulk of its lands have low mineral potential, the actual opportunity costs of these mineral-leasing restrictions may be quite minor. But the public or the decision makers will never know unless BLM invests more in developing this type of information. At present, the manager and the public are given the impression of significant trade-offs, but these may be minor if the minerals in these low-potential areas are in uneconomic quantities.

Finally, it is worth noting that in the few cases where BLM has developed detailed information on a few selected outputs to perform regional economic analysis, this information on output levels has not been used fully in the rest of the RMP. That is, the information on resource outputs is not shown in summary matrix tables; hence it is not readily available to guide resource trade-off analyses. In addition, BLM does not take the resource output levels the next step in performing an efficiency analysis of the costs and benefits of a given alternative. In one or two Resource Areas, a half-hearted attempt was made only by confusing economic impacts in terms of jobs and local income as benefits to be compared to costs of the alternative. (See the San Miguel RMP in Colorado and Kemmereer RMP in Wyoming as examples of this error.) Nonhunting recreation activities and wilderness were the most commonly slighted resources in the regional economic analysis of income and employment. This was the case even in some Resource Areas that had very valuable recreation resources (e.g., several candidate wild and scenic rivers, dozens of candidate wilderness areas, and so on).

The other major inadequacy of BLM plans relates to the lack of a decision analysis framework to evaluate rigorously and summarize impacts so as to facilitate trade-off analyses and search for the "best" alternative. Given that no formal summary statistic, such as net present value, of each alternative is developed, an agency must be careful to allow for comparison of disaggregated impacts. However, in BLM's case this frequently was not done. Although a matrix approach was used, the elements in the matrix were a mixture of acres allocated to different uses, environmental indicators, and a few indicators of output levels (e.g., AUMs, MBF of timber, and so on). In some cases the summary matrix simply listed direction of the change (plus, minus, or zero) and its relative size (high, medium, and low). Although this provides an impressionistic overview, it is difficult to see how a decision maker or the public (as this information was the core of what was presented in the EIS on the RMP) could really make meaningful comparisons among alternatives or determine exactly what the trade-offs were among alternative management actions. For example, how does a low negative for wildlife trade off against a medium positive for oil and gas? Thus a key criticism is that BLM did not even portray the information it did develop in a framework that would help the manager or public to make meaningful trade-off analyses.

Does the Disparity in Multiple-Use Planning Processes Make Sense?

If one accepts Clawson's rough estimates of average capital values of Forest Service and BLM lands (Clawson 1983:111), then on *average* (only) this disparity in planning efforts between BLM and Forest Service might make some sense. However, when one recognizes the heterogeneous nature of many of the National Forests and BLM lands, this bipolar consistency makes less sense. Although political appointees and Washington bureaucrats love "top-down consistency," this makes little sense when the natural resources within the National Forest System are heterogeneous. As was noted in the mid-1985 proposed land interchange between BLM and the Forest Service, many areas of National Forests ecologically match BLM lands and warrant a non-FORPLAN planning approach. In the same way, portions of some BLM Resource Areas in Colorado, Oregon, Montana, Utah, and Wyoming are rich in timber or minerals or recreational opportunities that warrant far more than the superficial planning BLM has sometimes engaged in.

The key is that the intensity of planning must match the multiple- use resource values at stake. In some sense, this is just the logic of risk analysis the Forest Service has come to use in determining optimum fire suppression response: a comparison of the values at risk to the costs of response to determine what level of resource values are saved with different degrees of effort. Unlike planning, the Forest Service is more careful not to apply a uniformly excessive response to every fire that comes along. Rather, using information on the value of the timber and other resources at risk and the costs of response, the appropriate scale of fire suppression effort can be launched in a timely manner. This saves the Forest Service and ultimately the taxpayers both funds (from avoiding too much fire suppression effort) and valuable resources (from too little fire suppression effort). It is to the application of this framework to multiple-use planning that we now turn.

Tailoring the Intensity of Planning to the Values at Risk: Determining the Economic Value of Better Information

If we think back to chapter 8 and the production possibilities curve depiction of resource trade-off analysis, determining the optimal intensity of planning becomes clearer. Figure 11.1 presents a production possibilities curve to illustrate the trade-off between salmon production and timber. In the absence of collecting detailed inventory data and performing any analysis or planning, society may often find that the no-action public land output levels are substantially inside the boundary of the production possibilities frontier (PPF) at, say, point *A*. As will be recalled from chapter 8, more efficient use of this given amount of land, labor, and capital may allow us to produce out to the frontier itself, say, point *C*.

The essence of planning is to search out systematically the more efficient combinations of land, labor, capital, and natural resources to produce more of both outputs, or more of one, without giving up any of the other (as in point *D*). As we saw, the Forest Service tends to collect more data than the BLM and to use sophisticated analysis techniques, such as optimization models, in searching out this more efficient mix of inputs to produce alternative combinations of outputs. Thus the Forest Service offers the public a menu of both input combinations (e.g., acres allocated to different uses) and different output mixes (e.g., amounts of timber, visitor days, wildlife, fish, and so on). In many Forest Plans there are even alternatives that provide similar mixes of output but that are produced using different combinations of inputs (e.g., more emphasis on select-cutting, or a package with more total timber harvest and more wildlife mitigation projects versus one with less timber harvest and wildlife mitigation projects, both of which produce about the same amount of timber and wildlife). In the Siuslaw National Forest case study

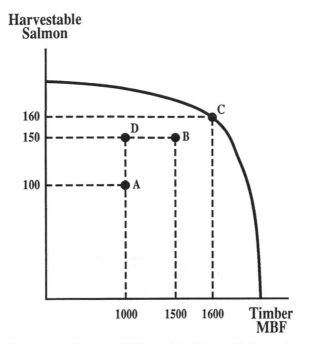

Figure 11.1. Diminishing Marginal Effect of Additional Information

in chapter 9, the Forest Service displayed in pie charts how close each alternative was to the maximum output (i.e., edge of the PPF) for each resource under each alternative.

Not only does BLM not perform any formal economic efficiency analysis, but what is worse, often it does not even quantify most of the multiple-use outputs produced from different combinations of inputs. Without knowing what outputs are produced it is not clear how the BLM resource planners even know if an RMP alternative is moving the agency toward the frontier or toward the origin of the PPF! In one sense, BLM's alternatives in the RMP can be viewed as simply different arrangements of inputs (i.e., different amounts of land allocated to different uses). But neither the manager nor the public is provided with a comprehensive picture of what different allocations of land mean to such multiple-use outputs as recreation visitor days, fish, mineral production, and so on.

Determining How Large an Investment to Make in Better Information

However, we can learn much more from this simple PPF about the benefits of searching out more productive combinations of inputs. It allows us to compare the benefits of finding a more productive combination of these inputs versus the costs of the additional inventory data and analysis (i.e., to determine how much investment in data and analysis is worthwhile). This may sound heretical to a range scientist, biologist, or silviculturalist, all of whom always prefer more data to less and more analysis to less. There is no doubt we know very little about how ecosystems function and respond to management. More time spent studying the forest, collecting larger numbers of samples, and performing more sophisticated statistical analysis would certainly allow the agency to formulate more productive combinations of inputs to produce more outputs. A very thorough inventory and analysis could move society to point *C* on the PPF, for example.

But beyond the sheer intrinsic joy that doing this type of applied science provides the scientists, it is quite costly to society in at least two ways: (1) the direct cost of the greater number of workers, more supplies, capital, and so on, needed to collect and analyze more data; and (2) the opportunity costs of delaying decisions while we wait for this perfect information and therefore remain at the inferior status quo (say point *A*). Often the net present value of the benefits is greatest using a cursory evaluation to move from *A* to *B,* rather than remaining at *A* for years and finally moving to *C*. In some ways, this trade-off is reflected in the pragmatist saying, "The best is the enemy of the better." Of course, some people would argue that one can always move from point *A* to *B* as soon as point *B* is determined to be superior, but one can keep inventorying and analyzing so as eventually to move to point *C*. This has its advantages (but also its disadvantages, which will be discussed later). In one sense, having several rounds or cycles of planning over several decades provides this opportunity to continue to build toward more efficient combinations of inputs. But in the process, society must allow the agency to accept points like *B* or *D* as improvements over the current state and not seek redress in the courts because point *C* has not been attained.

A Framework for Comparing Benefits and Costs of Additional Information

The purpose of this section is to present a simple method for determining whether it is even worth pursuing complete information (point *C*) or not. Using figure 11.1, we will calculate the net return associated with better inventory data and analysis. This information will be tabulated in table 11.1 by comparing the costs of additional inventory and analysis to the value of the output generated by such additional effort.

In figure 11.1, at point *A* the forest currently produces one hundred harvestable salmon and 1,000 mbf of timber with little inventory data or analysis. If the salmon are caught by sport anglers, then, using the type of travel cost method analysis discussed in the Montana fishing example in chapter 9, one might obtain a marginal value per fish harvested of $50. Suppose the stumpage value of timber is $100 per mbf. For the purpose of the example, these are net values of the outputs over and above the current management costs. As the example proceeds, we shall keep the level of management constant for two reasons. First, we want to evaluate how better inventory data and analysis allow the agency to combine the existing level of inputs

Table 11.1
Comparison of Benefits and Costs of Additional Inventory and Analysis

Pt.	Salmon	Timber	Total Benefits	Costs of Added Data	Net Benefits
A	$5,000	100,000	105,000	0	105,000
B	7,500	150,000	157,500	35,000	122,500
C	8,000	160,000	168,000	70,000	98,000

(including land and natural resources) more efficiently to produce more output. Second, increasing the level of labor or capital input would shift the PPF. This would only complicate the example, but the same comparison of incremental benefits and costs that we shall perform later still applies, with the addition of incremental investment costs added to inventory and analysis costs.

As shown in table 11.1, the current value of output is $105,000. Since we incur no new data collection or analysis costs under continuation of existing management, the cost of added data is zero. Thus the net return is $105,000.

Now of course all the staff scientists suggest the forest can do better. If we performed careful inventory of tree stands to pinpoint the ones not near anadromous fish streams, we could harvest more timber without damaging fisheries. If we better understood road design and layout, we could reduce sediment coming from the roads into the streams. The fisheries biologists could then use some current funds and labor to enhance the fishery rather than mitigate the effects of timber harvests and roads. Taken together this would allow the forest to produce output combination *B* (150 harvestable salmon and 1,500 mbf of timber) in figure 11.1.

If the estimated costs of acquiring this additional information via more detailed inventories, road design–sediment studies, and so on, is $35,000 a year (an annualized figure to keep the example simple), then as shown in table 11.1 this initial increase in intensity of planning is worthwhile. The net benefits rise from $105,000 to $122,500 with the better data and analysis, for a net gain of $17,500. In a benefit-cost ratio sense, every dollar invested in better inventory and analysis yielded $1.50 in incremental benefits ($52,500 gain in benefits divided by the $35,000 incremental costs).

Diminishing Marginal Value of Information

Of course, the staff comparing points *A* and *B* is really excited now. If a little better information yields $1.50 worth of benefits for every dollar of inventory costs, just think what increase in benefits nearly perfect information would provide. Suppose that such information would allow fine-tuning of timber and fisheries management (again assuming the same level of inputs). Unfortunately, at some point, better and better information is likely to produce less and less response. As any statistician knows, increasing one's sample size from fifteen to thirty significantly reduces the standard error, but increasing it from thirty to forty-five adds much less additional precision. In a Geographic Information System, reducing the size of the smallest grid cell from 200 acres to 100 acres might help a great deal, but reducing it further to 50 acres might not increase precision enough to warrant the large increase in computer memory costs and computational time. Thus continued intensification of sampling effort, adding years to study times and more detailed analysis, will increase output only to point *C*.

At point *C*, salmon harvest rises to 160 fish per year and timber increases to 1,600 mbf per year. However, if the costs of doubling the sample sizes, reducing the GIS cell size, and so on, also double, then, as shown in table 11.1, net benefits actually drop. In this example, the added information is worth about $9,500 of more output, but the increment in annualized cost is $35,000. Thus net benefits fall by $24,500. *Essentially, the additional precision of information and analysis is not worth the cost.* At times, some critics of Forest Planning are guilty of demanding that the Forest Service carry its inventory and analysis to this intense but uneconomic level.

Tailoring the Intensity of Planning Using the PPF

But the use of this type of production possibilities framework is also useful to guide agencies in determining for each forest or Resource Area what intensity of planning is sensible. For example, just because on the White Mountain National Forest in New Hampshire a moderately intensive inventory and analysis is warranted does not mean that a moderately intensive inventory analysis is needed in every National Forest in the eastern United States. Determining the appropriate intensity involves a comparison of just how much more productive the land is likely to be if additional information was ''bought'' through better inventory and analysis.

Of course, the same is true of many BLM Resource Areas. For example, in figure 11.2 we present a production possibilities curve for two typical nonmineral resources that receive much attention in many BLM districts. In this figure we have animal unit months (AUMs) of forage for deer on the vertical axis and AUMs for livestock on the horizontal axis. Suppose that at present the Resource Area finds itself at point *SQ* (for status quo). The botanists, wildlife biologists, and range conservationists argue that with information on dietary competition between deer and livestock as well as information about plant species response to intensity of grazing it is possible to design a grazing system to increase AUMs to both wildlife species without impairing the plant communities. The staff states that this increase can occur by better timing of livestock grazing and that no additional labor or materials are required (i.e., it is more efficient use of existing land, labor, and capital). Therefore the PPF remains in place.

This improvement in knowledge moves the Resource Area from point *SQ* with 1,000 AUMs of forage allocated each to deer and livestock to point *I,* where 1,100

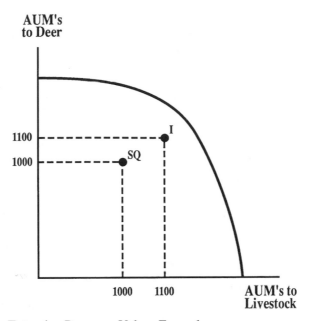

Figure 11.2.   Extensive Resource Values Example

Table 11.2
Comparison of Benefits and Costs of Additional Information on Grazing

Pt.	Deer	Livestock	Total Benefits	Costs of Added Data	Net Benefits
SQ	$10,000	10,000	20,000	0	$20,000
I	11,000	11,000	22,000	4,500	17,500

AUMs are allocated to each species. Although such an improvement would please both ranchers and wildlife enthusiasts, the key question is whether the additional inventory and analysis costs are worthwhile. That is, could a solid justification be included in the agency's budget and survive the scrutiny of Office of Management and Budget in Washington?

To determine this we compare the benefits of the additional forage to the costs of the additional inventory and analysis necessary to provide the information needed to adjust timing of livestock. Table 11.2 provides such a comparison of the benefits of additional AUMs and the costs of added information. To value AUMs we use techniques discussed in chapter 6 and illustrated in the case study in chapter 10 on livestock grazing in Challis, Idaho. Specifically, a value of $10 per AUM for both deer and livestock is close to the areawide average estimated net economic values generated in that case study.

As we can see in table 11.2 (above), the additional $2,000 in value of AUMs does not warrant the inventory and analysis costs of \$4,500. Although AUM benefits go up by \$2,000, costs go up by more, resulting in a reduction in net benefits of \$2,500 (\$20,000 − \$17,500). In fact, the data in table 11.2 imply that unless the dietary competition studies and analysis can be done for less than \$2,000, the additional data and analysis are not worth it. In this sense, this approach can be used in an anticipatory way by the agency to determine which resource studies (beyond what is legally required to implement their respective laws) are justifiable. Of course, in determining the costs of the studies an important consideration will be the added or incremental costs to the agency for the studies versus the fixed costs, such as many permanent full-time staff salaries that are unaffected by the study. For evaluating one study, just the incremental costs to the agency of performing the inventory and analysis would be relevant.

Lessons from the First Two Examples

The point of the last two examples has not been to create the impression that additional inventory and analysis are not worth the costs. Rather, it has been to illustrate two important points. First, figure 11.2 illustrated that when resource values per acre are low (or extensive), intensifying data collection and analysis to increase outputs may not be warranted, even if the costs are relatively modest. For example, in the BLM case study of the San Juan Resource Area, by far the most detailed analysis was provided for livestock grazing. Even at \$10 an AUM, with nearly 20 acres necessary to produce an AUM on most BLM land, this annual value of 50 cents an acre may not justify such detailed analysis when other resources, such as minerals or recreation, with higher values per acre were not analyzed in the RMP in equivalent detail. Much like the Forest Service, with its detailed analysis of timber regardless of the timber values at stake, agencies tend to analyze in detail resources with which they are familiar, rather than those that justify such analysis. We shall return to this point in figure 11.3.

Figure 11.1 illustrates the point that, with relatively high values per acre, *some* additional inventory and analysis provides a reasonable payoff over and above costs, but continued expansion of studies in a quest for perfect information is rarely warranted. Alternatively, as the next example shows, there are probably cases where the agency, in its quest for uniformity and consistency in planning, is doing too little analysis of some resources, especially those the agency is unaccustomed to evaluating.

Determining When to Collect More and Better Information

Figure 11.3 compares the production of wildlife habitat (in acres) and minerals (measured in barrels of oil, tons of coal, or ounces of ore) from a 250,000-acre resource area. At the beginning of the planning process the agency is at the status quo position (SQ), having 200,000 of the 250,000 acres of potential wildlife habitat available (i.e., not being mined) and therefore being used by wildlife. The agency is also currently producing 30,000 tons of coal annually from the 50,000 acres that are under coal lease in the resource area. Thus of the 250,000 acres, 50,000 are used for coal mining and hence are unavailable as wildlife habitat.

Suppose the agency geologist believes that, because of the similarity in geologic structure of 150,000 acres not in coal production and the 50,000 acres in coal production, coal might exist in the currently unleased 150,000 acres. In geologists' terms, there may be "inferred reserves" in that 150,000 acres. Often when they are informed that a "mineral might be present," agencies immediately take that as an indication that mining is the highest-valued use of the land. However, the mere presence of minerals says little about whether mineral production is the highest-valued use of the land. Nonetheless, suppose the geologist believes, based on his or her geologic inference, that coal production could be increased to 100,000 tons per year.

If the agency adopts the geologist's recommendation in the RMP and leases the additional 150,000 acres, undisturbed wildlife habitat falls to 50,000 acres. Therefore the agency moves to point RMP in figure 11.3.

Of course if more information was available on the distribution of the inferred coal reserves, less acreage might need to be leased. The key information relates to just where on that additional 150,000 acres there was actually recoverable coal deposits and of those recoverable coal deposits, which ones will in fact be economic (benefits greater than mining costs) to develop. There are millions of acres with

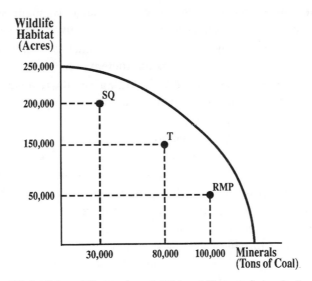

Figure 11.3.   High Values Warranting Additional Data and Analysis

Table 11.3

Comparison of Benefits and Costs of Additional Information on Mining

Pt.	Wildlife	Coal	Total Benefits	Costs of Added Data	Net Benefits
SQ	$400,000	105,000	505,000	0	505,000
RMP	100,000	350,000	450,000	5,000	445,000
T	300,000	280,000	580,000	25,000	555,000

minerals present but relatively few where the large costs of mine development and mining (let alone the social costs of the accompanying air and water pollution) justify mineral production. Much of the economic coal production could occur in one or two large deposits, and many small, scattered deposits might not be fully developed.

If the agency is artificially constrained by planning time limits and budget restrictions on how much it can spend to gather new inventory data, then much of this needed information will not be collected. Further, if the Washington office suggests (as has often been the case) that it cannot allow one Resource Area or forest to collect detailed information without allowing all such areas to collect it (because the plans must be consistent), then the agency is stuck at point RMP.

If the agency geologist was able to perform more site-specific investigations of the individual rock formations, he or she might be able to determine that nearly all the additional reserves would probably (in a statistical expected-value sense) come from just 50,000 of those additional 150,000 acres that he or she proposed for leasing. Building upon this additional information on location and size of more proven deposits, the economist might determine that only about 80,000 of the 100,000 tons of production would even be economical, given expected coal prices and costs of production. Taking these two pieces of data together, point *T* in figure 11.3 would probably be identified as a more efficient combination of coal leasing and wildlife habitat. As is illustrated in table 11.3, given the relatively high values at stake, acquiring this additional information and reallocating land to point *T* have a positive economic return. For the purposes of this table we have assumed that this is highly productive wildlife habitat providing $2 an acre in benefits for hunting and viewing and that coal has a net value of $3.50 a ton (Forest Service 1990a).

As table 11.3 illustrates, the emphasis on using existing data and consistency of data and analysis across plans results in the recommended RMP being inferior to the point *T*. Given the relatively high values at stake here, it makes sense to go beyond the minimal inventory and analysis in the RMP. As is illustrated in our hypothetical example in table 11.3, the minimal inventory and analysis of coal potential that leads the geologist to recommend leasing an additional 150,000 acres actually reduces net benefits of these values per acre of wildlife and coal. At least in this example, a little knowledge is a dangerous thing.

Does BLM Underplan and the Forest Service Overplan?

It is time to turn from these hypothetical examples that illustrate the principles of comparing the cost of added information to the benefits of the information to see whether BLM is underplanning and the Forest Service overplanning. The first section of this chapter indicated the Forest Service spends at least $1 to $1.30 per acre on planning, a figure that if even close to correct is a small fraction (2% to 3%) of their overall forest management budget. This figure seems to dispel the notion that forest planning is diverting substantial resources away from on-the-ground management. Although we do not have figures for a wide range of National Forests, such costs would seem high only for "high-desert" forests, such as the Humboldt, or remote (i.e., nonurban) chaparral forests. The BLM, on the other hand, spends 30–60 cents per acre. For some Resource Areas this is probably not enough, given the per acre value of some areas in Colorado, New Mexico, Oregon, Utah, and Wyoming. However, further empirical research on the possible returns to more detailed planning on both BLM lands and National Forests is warranted, using the framework provided earlier, before final conclusions can be reached.

Internal and External Critiques of Forest Planning

After completing the first round of forest planning under NFMA, the Forest Service worked with Purdue University and the Conservation Foundation to perform a formal critique. The goal was to evaluate the lessons learned and develop recommendations for the second round of forest planning, which would begin for some forests in the early 1990s. Having developed comprehensive "zero-based" Forest Plans that started from scratch to evaluate a wide range of alternatives, should the Forest Service build upon these plans or again start from scratch? Should the planning process continue its reliance on FORPLAN as the main analytical tool? Most important, given what was learned in the first round of forest planning, could pitfalls be avoided in this next round? As discussed in a series of forest planning critique documents (Larsen et al. 1990, Shands et al. 1990, Hoekstra et al. 1990, Teeguarden 1990), several lessons were learned from this first experiment in comprehensive integrated natural resource planning. Despite the frustrations encountered, there was general agreement that the Forest Plans were helping to lead the Forest Service into integrated multiresource management and away from single-resource management (Larsen et al. 1990).

Matching Integrated Natural Resource Planning and Management to Budgets

The theme that occurred most often in the critiques was the need to match budgets (both level and function) to the Forest Plans. The one element that has most hindered implementation of the integrated National Forest Plans has been the continued resource-by-resource budgeting in the appropriations process (Larsen et al. 1990, Shands et al. 1990, Teeguarden 1990). To quote Shands and colleagues (1990:58–59):

> The Forest Service annual budget request now is based largely on the forest plans. Because the plans represent an integrated approach to managing all the resources of the national forests, the budget information is also integrated. . . . Congress, however, requires that this information be translated into dozens of budget line items representing the individual components (resources). During budget review by the Secretary (of Agriculture), the Office of Management and Budget, and Congress, money is added to or subtracted from these line items with little or no recognition that they are interrelated. . . . This budget structure significantly impedes the Forest Service's efforts to accurately implement the forest plans and meet the requirement of NFMA.

Thus one long-term change required is to overhaul the way in which the administration and Congress deal with the Forest Service budget. If Congress truly wants integrated natural resource management of the National Forests, as its passage of NFMA would imply, then it must match the mechanics of funding to the functioning of the plans. The Forest Service cannot directly make this change, but it has pointed out the present inconsistency and will no doubt work to assist the administration and Congress in modifying their budgetary process.

Recommendations for the Second Round of Forest Planning

The Forest Service is empowered to change the planning regulations governing the second round of forest planning. The critique provided seven major recommendations (Forest Service 1991:6511). One theme is that where the newly developed Forest Plans are basically sound (i.e., no major changes have occurred in the resource base such as fires, massive wilderness designations, listing of endangered species, and so on), incremental revisions should be allowed rather than starting from scratch. This incremental revision would differ from BLM incremental planning, discussed in chapter 10, in that all forest resources and associated interactions would be evaluated in relation to changes proposed in the second round of forest planning. Of course, if the Forest Service or the public believe that the first plan must be heavily revised because of major resource changes or because it is inadequate, the regional forester can instruct the forest supervisor to develop a detailed second-round Forest Plan.

Another major recommendation of the critique was that the National Forests should be provided with a range of analysis tools that could be used and allow a more flexible FORPLAN structure tailored to the issues.

Proposed Second-Round Forest Planning Regulation

On February 15, 1991, the Forest Service published in the Federal Register its Advanced Notice of Proposed Rulemaking on Forest Planning Regulations for the second round of forest planning. These regulations may be considered a preliminary draft, because the public will have opportunities to comment and the Forest Service will revise the proposed regulations in response to the comments. These proposed regulations, when they become final (a process that may take several years), will apply only to second-round Forest Plans. The planning regulations described in chapters 2 and 9 still apply in National Forests still completing their first round and those beginning their second round before the new regulations become final. Of course, these proposed regulations must still comply with the National Forest Management Act of 1976, since no legislative changes have been proposed. The Forest Service proposes to revise the regulations within the legislative requirements of NFMA.

Major Changes in Direction for Revised Plans

Before discussing in detail the changes these regulations propose, their major thrust should be mentioned. The revisions aim at reducing the time required and the complexity associated with producing the second-round plans. The two most profound changes are as follows:

1. Tailoring the Forest Plan revision to just the issues and resources that demonstrate a "need for change." The emphasis here is on incremental changes to existing direction rather than on wholesale reevaluation of the forest's entire management direction. However, even the effects of incremental changes on all multiple-use resources and environmental values would be analyzed. If adopted, this emphasis on incremental planning would represent a significant departure from current zero-based planning, where significant departures from existing management are considered.
2. Giving the regional forester discretion on type and degree of analysis required in the revised Forest Plan. This would allow the flexibility to customize the analysis methods to the resource values at stake in the revision. This last provision moves away from national direction mandating that every forest perform the same type of detailed analysis using the same analytical models, regardless of the issues at stake (Forest Service 1991:6532). As was argued earlier in this chapter, this change probably makes sense for several reasons. First, not all forests have the high resource values per acre that warrant intensive planning. Second, the FORPLAN-type linear programming models are better suited to forests with significant timber output than to those with a primary recreation emphasis.

NFMA Concepts Given More Emphasis in the Second Round of Forest Planning

It is worthwhile to list the features of NFMA planning that are emphasized more in the second-round planning regulations.

1. *Greater emphasis on integrated natural resource management.* The proposed rules repeatedly emphasize that the second round of NFMA forest planning will integrate consideration of multiple-use resources and environmental and amenity values using interdisciplinary planning. Resource interactions and ecological interrelationships between resources appear to be stressed even more in the proposed planning regulations. Ecological indicators include not only individual wildlife species but also biological communities and special habitats (Forest Service 1991:6516).

2. *Greater emphasis on public participation and governmental coordination.* In part, this is in response to the criticism that the public did not understand the first round of forest planning. In addition, greater attention to conflict resolution and consensus building is called for. Coordination with local governments is also stressed.

3. *Refinement of NFMA planning principles.* Many of the planning principles are similar to what has been reviewed earlier. For example, revised forest plans must conform to multiple use and sustained yield. Timber suitability continues to combine biological and economic factors. However, the latter have been refined so that they now identify lands where "timber production is clearly not feasible now or in the future," as well as those where timber production would provide a positive return but for which there is not an adequate market for the timber (Forest Service 1991:6527–28). In addition, calculation of net present value of each alternative is still required under the proposed second-round forest planning regulations. Also continuing is the need to quantify local income and employment associated with implementing the revised forest plan.

Similarity and Dissimilarity of BLM and Forest Service Planning Regulations

The similarity between the Forest Service's proposed second-round planning regulations and those of the BLM is worth noting. Both emphasize incremental, flexible planning. Analysis is to be tailored to the issues at hand. These two multiple-use agencies may have appeared to move closer in their approach to multiple-use planning, and to an extent that is true. However, the Forest Service adopted its incremental approach to fine-tune its comprehensive plans, which were reached after a complete "zero-based" analysis of a wide range of alternatives. If this foundation is solid, making an incremental change in ten years is reasonable. In

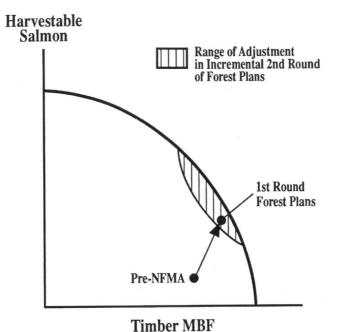

Figure 11.4.   First and Second Round of Forest Planning

the terminology of the production possibilities curve, the first round of forest planning moved the Forest Service close to the frontier of the production possibilities curve. The second round simply allows for updating where the optimal point is on the frontier, given the change in society's values. This is illustrated in figure 11.4 by comparing the major changes in the first round of forest planning to the range adjustments in the second round.

In BLM's case, the incremental planning was applied to a much weaker foundation. BLM has rarely performed a "zero-based" comprehensive analysis of a wide range of alternatives in either its pre-FLPMA planning process or its FLPMA Resource Management Plans. As such, there is little foundation in many Resource Management Plans to fine-tune. In some sense, the Forest Service is correctly allowing for less analysis on forests and situations where it is appropriate. In a few aspects, the second-round Forest Plans may resemble BLM's Resource Management Plans. However, the Forest Service continues to compare the expected outputs and net economic benefits of each alternative, whereas BLM does not.

The Forest Service has recognized that consistent intensive planning is not warranted on every forest. It is reducing analysis requirements and tailoring the type of analysis on many noncommodity-type forests. Its Resource Management Plans have the opposite problem: uniformly too little analysis, and key types of analysis are missing completely. Unfortunately, BLM does not appear to be proposing reevaluation of its planning process to increase its intensity and type of analysis for Resource Areas where high natural/environmental values warrant it.

Conclusion: Is Complexity of Analysis Always Bad?

Many of the features in the proposed second-round Forest Planning regulations represent desirable changes from those already in existence (reviewed in chapter 9). The potential to tailor the type and intensity of analysis to the natural resource issues on a particular forest is quite good. The benefits and costs of additional intensity of planning appear to have been explicitly considered. Moreover, the attempts to link budgets better to plans is also very desirable.

However, some features of the first-round Forest Plans seem to be underemphasized solely because of their complexity. The Forest Service acknowledges that developing a Forest Plan is complex (Forest Service 1991:6513). It also recognizes that it must balance this complexity with need for the public's desire for simplicity and comprehensibility of planning (Forest Service 1991:6513). However, sound planning requires solid analysis. Society does not require engineers to build bridges or airplanes using only principles easily understood by the public. Rather, safety is paramount, and engineers are allowed to use whatever principles or computer models are necessary to give the optimal bridge or airplane design. The same type of logic seems to apply to National Forest Planning. Society should not limit the Forest Service to easily understandable methods if this results in failure to identify superior management alternatives.

12

Wildlife Planning and Management in U.S. Fish and Wildlife Service

As discussed in chapter 2, the U.S. Fish and Wildlife Service (USFWS) has numerous responsibilities as the principal federal wildlife agency. One of its main missions is to manage refuges in the National Wildlife Refuge System. This agency's planning and management of these refuges is the primary focus of this chapter. However, as the lead federal wildlife agency, USFWS also is involved in planning and managing fish hatcheries, managing migratory waterfowl under the Migratory Bird Treaty Act (discussed in chapter 2), developing mitigation plans on numerous development projects the agency evaluates under the Fish and Wildlife Coordination Act, and establishing recovery plans for species listed under the Endangered Species Act. The diversity of tasks assigned to the USFWS suggests that the agency must tailor the principles of planning to the specific tasks at hand.

Much like the Forest Service, the USFWS has planning systems for the national, regional, and field levels. In some cases these are independent plans and are not linked with the various levels. For example, regional plans are formulated to cover the entire range of a species or habitat system, such as that of the Columbia River (Verburg and Coon 1987:22). National plans are sometimes strategic plans to direct the overall agency in attaining a particular mission. In other cases, such as the North American Waterfowl Management Plan, this national plan is linked to regional (i.e., multistate) flyway plans and then to specific field-level plans for acquisition of specific parcels to meet the overall national and regional objectives. We shall review this tiered planning approach later in this chapter.

The USFWS is also active in mitigation planning with other agencies. This type is also reviewed at the end of this chapter. We first start with refuge management and planning, which is the most well defined of any USFWS planning approaches.

Wildlife Refuge Management

The Size and Scope of the National Wildlife Refuge System

The Refuge System consists of 452 individual refuges spread among forty-nine of the fifty states (Robinson et al. 1989:10). The total acreage is 88.6 million acres, which is larger than the combined area of New York and Pennsylvania but slightly smaller than the state of California. The greatest concentrations of wildlife refuges are in Alaska and along the four major north–south migratory bird flyways (Pacific, Central, Mississippi, and Atlantic). Combined with the fact that nearly a third of the refuge acreage is wetlands, it is clear that one of the central focuses of the USFWS Refuge System is migratory birds, especially waterfowl. Nonetheless, more than six hundred species of birds and hundreds of species of mammals, reptiles, fish, and plants are supported by the diversity of habitats within the Refuge System (Robinson et al. 1989:10).

The administration and management of the Refuge System are largely decentralized, with managers having substantial responsibility for day-to-day management of the refuge (Robinson et al. 1989:10). The manager is guided by a common *Refuge Manual,* refuge-specific goals in the authorizing legislation, and often a Refuge Plan. Thus before discussing how planning allows management to reach its mission, it is worthwhile to review the mission of the Refuge System.

Common Objectives of National Wildlife Refuges

The USFWS *Refuge Manual* (Fish and Wildlife Service 1982) describes the objective of the Refuge System as preserving, restoring, and managing a national network of lands and waters to provide and enhance the widest possible spectrum of wildlife benefits. Within this overall mission statement are several goals that help meet this mission. These include (1) perpetuating migratory bird resources; (2) preserving natural diversity and abundance of fauna and flora on refuge lands; (3) preserving, restoring, and enhancing within their natural ecosystems all endangered species or those threatened with becoming so; (4) providing refuge visitors with opportunities to understand fish and wildlife ecology and people's role in that ecology; (5) providing refuge visitors with high-quality wildlife-oriented recreation experiences, to the extent that they are compatible with the purposes for which the refuge was established.

This mandate of wildlife conservation combined with human uses (some of which need not be wildlife oriented as long as they are compatible with wildlife management) demonstrates that refuges are not legally considered single-use lands; rather, they are lands where wildlife is the dominant use (Bean 1983:125). This anthropocentric view of National Wildlife Refuges is consistent with the mission of the administering agency, the USFWS. Specifically, this mission is to provide the federal leadership to conserve, protect, and enhance fish and wildlife and their habitats for the continuing benefit of people (Fish and Wildlife Service 1988).

Of course, many refuges were given individual mandates at the time of their designation that must be met in addition to these national goals. For example, lands acquired with Duck Stamp money may have specific directions regarding waterfowl management priorities. Other areas designated during the early 1900s were originally set up as game refuges and, in some cases, for specific species, such as elk or bison.

The largest expansions and additions to the National Wildlife Refuge System have been in Alaska. These came in 1980, when the Alaska National Interest Lands Conservation Act (ANILCA) created nine new Wildlife Refuges and expanded six of the seven existing ones. This act tripled the amount of acreage in the National Wildlife Refuge System and made the USFWS the largest federal landholder in Alaska. In addition, it specified that all Alaskan Refuges complete a ''comprehensive conservation plan'' (Bangs et al. 1986; Bean 1983:134). Finally, it refined the management objectives of refuges in Alaska, the details of which will be presented in the case study of the Arctic National Wildlife Refuge.

General Management Guidelines for Refuges

What Is the Range of Allowable Management Choices in Refuge Planning?

As emphasized earlier, the main use of refuges is to conserve native wildlife. About 75% of the recreation visitation to refuges is related directly to the opportunity to view wildlife, hunt, and fish (Reed and Drabelle 1984:60). Hunting has been controversial on many refuges, because many view it as harming, not conserving wildlife. Although the merits of this debate are covered elsewhere (Reed and Drabelle 1984:45–50) and the controversy is still very much alive (witness the lawsuits to stop hunting on refuges), we wish to focus on hunting as a recreation and management activity.

Hunting. Hunting occurs at about half the refuges in the National Wildlife Refuge System (Reed and Drabelle 1984:48), and most is for migratory birds. Many refuge lands were bought with funds from the waterfowl hunters' purchase of Duck Stamps. Of the refuges open to hunting, usually about half the acreage is available for hunting, with the other half being ''sanctuary.'' The refuges can also be open for hunting state-regulated resident species, such as deer and upland game. The determination to open a refuge to hunting requires that the refuge manager comply with the National Environmental Policy Act by preparing an environmental assessment and developing a hunting plan (Reed and Drabelle 1984:48). Such a determination must be consistent with the *Refuge Manual*'s direction that hunting contributes to or at least is compatible with management objectives of the refuge. For example, hunting and trapping are often used to regulate animal numbers, in line with the refuge's carrying capacity and national/state goals for species numbers.

Fishing.   Although hunting is a visible and controversial consumptive recreational use, five times more fishing takes place on most refuges than hunting (Reed and Drabelle 1984:50).

Nonconsumptive Uses.   Of course, the major nonconsumptive wildlife recreation activities are wildlife viewing, photography, and nature study. However, formal programs for wildlife and nature interpretation are active at only about 10% of the refuges (Reed and Drabelle 1984:61).

Incompatible Recreational Uses.   Some of the other recreational uses of refuges are often incompatible with wildlife. Historic use of speedboats and off-road vehicles has had to be curtailed at some refuges because these activities damaged habitats or resulted in birds abandoning it.

Commercial Activities on Refuges.   Although wildlife is a dominant use of refuges, it is not the exclusive one. Agriculture is often practiced to provide food or cover for wildlife. Livestock grazing often occurs. In some of these cases, it is used as a management tool to keep the vegetation at a level more beneficial to particular wildlife species. In others, USFWS has been trying to reduce historical but incompatible livestock grazing in the face of political opposition. One example of this will be presented in the case study on refuge planning, which deals with the Charles M. Russell Refuge.

Other commercial uses that sometimes occur on refuges include timber harvesting and mining. Unless mining claims or leases were filed prior to the establishment of the refuge, new ones would rarely be allowed. Debate over potential oil and gas leasing in the Arctic National Wildlife Refuge will be part of our case study of planning in Alaska's National Wildlife Refuges. However, when mining claims or leases exist at the time of the establishment of the refuge, the USFWS often attempts to place strict controls on the mining or extraction of those minerals to protect wildlife. Additional details of how refuges are to be managed can be found in the USFWS's detailed *Refuge Manual.*

USFWS Approaches to Refuge Planning

Whether developing a comprehensive conservation plan for an Alaskan refuge or performing a "master plan" for a particular refuge, USFWS personnel must blend the specific objectives of that refuge into the overall national objectives of the National Wildlife Refuge System. With refuges established in the early to mid-1900s, the plans must resolve conflicts associated with past incompatible uses (e.g., use of ORVs, livestock grazing, and so on).

When a new refuge is created, the legislation gives wildlife the dominant use of the lands. However, when an existing refuge updates its master plan, it rarely has the recent clear signals that new refuges have in their enabling legislation. Thus the planning tasks are somewhat different in each case. Refuge planning in Alaska (where most of the refuge acreage is located) is quite different from that for other refuges, because motorized access is common there, due to the vast distances (Reed and Drabelle 1984:29). In addition, Alaskan natives rely extensively on refuge lands for subsistence. For this reason there will be two case studies, one that deals with developing a comprehensive conservation plan in Alaska under ANILCA for new or expanded refuges and one dealing with a master plan for the long-established Charles M. Russell Refuge in Montana.

It is worth noting that the director of USFWS under former President Bush (John Turner) has reactivated national planning direction for the Refuge System. National direction applicable to all refuges is being developed through a program in the Washington, D.C., office called "Refuges 2003: A Plan for the Future of the National Wildlife Refuge System." The plan will present the general direction of management for the Refuge System from 1993 to 2003. Issues to be addressed include whether to prohibit hunting, fishing, and commercial uses on refuges; whether to increase environmental education on the refuges; and how to protect biological diversity. The Draft Plan and accompanying Draft Environmental Impact Statement came out in January of 1993; the final reports will be released in 1994.

Rationale for Planning on Refuges

The USFWS states that there are several reasons for comprehensive planning on National Wildlife Refuges. Most important is the proactive ability the plan provides to direct public use of the refuge to compatible areas rather than to react to public requests to access certain ones for specific uses (Bangs et al. 1986:421). The plan assists the manager in deciding which parts of the refuge should be open to vehicular traffic, which should be open for hunting, and which should be wildlife "sanctuary." It also helps him or her decide on the appropriateness of commercial uses (e.g., agriculture, grazing, or timber harvesting) and on the types, location, and timing of recreational access, so as to not be disruptive to wildlife on the refuge. Thus a wide variety of human uses can often be accommodated, but the manager determines where and when they will take place by considering the effects on wildlife populations and their habitats on the refuge. In terms of recreation, this often means providing a variety of settings (wilderness, roaded, and so on) in which to engage in wildlife-oriented recreation.

One of the other key features of comprehensive plans is that they can deal with cumulative impacts from a series of small actions. Treated individually, each public request for special uses (such as a filming permit) may not have a serious impact, but their cumulative impact on the refuge can be substantial. The refuge plan allows the manager to consider the potential cumulative impact and provides a basis for decisions about whether a particular use will be allowed at all, no matter how small the individual effect. This rationale for refuge planning is best summarized by Bangs and colleagues (1986:420), "By . . . designating each respective area on the Refuge, a manager can quickly determine what activities may be compatible with any specific area on the Refuge." Wildlife Refuge planning also provides an opportunity for the USFWS to blend its traditional focus on managing animals and habitat with an emerging concern of managing wildlife for people. A growing group of wildlife biologists within both USFWS and state Fish and Game agencies views planning as a method to determine how better to serve its customers: hunters, anglers, and wildlife viewers (Amend 1990). This is probably best embodied in the view of one of the leading wildlife planners that the planning process should focus on which outputs and benefits are produced from wildlife management, rather than solely on wildlife management as an end in itself (Crowe 1984:4).

National and Regional Direction in Refuge Planning

The USFWS developed a nationwide approach to refuge planning in the late 1970s and early 1980s in the form of a *Planning Workbook* and a section on master planning in the *Refuge Manual*. However, before a national planning system could be implemented fully (except in Alaska, where it was required by law), the Reagan administration's first director of USFWS decentralized planning out to the Regional Offices. During the 1980s, many regions unofficially used the *Planning Workbook* as their basic guide in refuge planning. Nonetheless, with a regional approach there was much opportunity to tailor planning to the issues and funding at hand.

An example of tailored regional planning approach is Region 5 in the northeastern United States. This region developed a streamlined consensus-building approach as an alternative to detailed master plans for refuges where the management issues were clear and noncontroversial (Laffin 1990). The goal is to provide timely short-term (covering the next three to five years) management guidance to refuge managers without incurring the costs associated with detailed master plans and Environmental Impact Statements. These short documents are called Station Management Plans for refuges. They contain descriptions of existing resources, management priorities, management objectives, and a single list of management strategies designed to meet the objectives in the best way possible. The key element missing from this process is consideration of alternative management strategies or actions to attain the objectives and any evaluation of the consequences of their implementation.

Unfortunately, the variety of USFWS planning approaches makes it more difficult to describe how refuges plan. We shall attempt to describe the common core in what follows.

Master Plan Approach to Refuge Planning

The Master Plan approach is the primary approach used by most refuges in the continental United States and Hawaii (i.e., outside of Alaska). It is also the most detailed approach to refuge planning outside of Alaska. In some regions of USFWS, such as Region 1 (which covers Washington, Oregon, California, and Hawaii), the Master Plan format appears to have been the sole approach used for all refuges, ranging from small island refuges in Washington to those for waterfowl in the San Joaquin Valley in California. In addition, as our case study of the Charles M. Russell Refuge will illustrate, many other planning efforts followed the spirit of this Master Plan approach, even though some of the steps have different names or labels. Therefore we shall describe the Master Plan approach, as it generally represents USFWS refuge planning outside of Alaska.

Specific Objectives of Refuge Master Planning

The USFWS *Refuge Manual* section on master planning states there are six major objectives or purposes: (1) to ensure that national policy direction is incorporated in the management of individual refuges; (2) to determine the capability of individual refuges to further USFWS goals; (3) to provide a systematic process for making and documenting refuge decisions; (4) to establish broad management strategies to guide refuge management programs and activities; (5) to provide continuity in the management of individual refuges and the Refuge System as a whole; (6) to provide a basis for budgeting requests to implement management programs leading to achievement of refuge objectives.

Thus refuge Master Plans are the building blocks on which the overall management of the National Wildlife Refuge System rests. How well the system meets the national goals spelled out earlier in this chapter depends on how well they are translated into on-the-ground management via the refuge Master Plans. The funding to attain these goals depends in part on how well the Master Plans justify the wildlife management investments required.

Steps in Refuge Master Planning Process

The USFWS groups its steps in the planning process into three main phases: inventory, capability analysis, and synthesis. The inventory phase involves determining refuge objectives in terms of resource uses and describing which existing resources are present. Capability analysis determines which uses different resource combinations can support. Synthesis selects the best combination of uses to meet the objects. Figure 12.1 provides a schematic overview of master planning and shows the interactions between steps. The details are discussed next under each step.

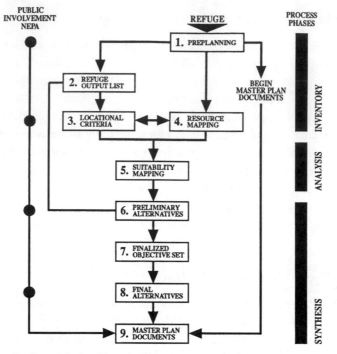

Figure 12.1.   Refuge Master Planning Process

As shown in figure 12.1, the USFWS Master Planning Process involves a series of sequential actions or steps. The first step in the Inventory Phase is Preplanning.

Step 1: Preplanning

Preplanning involves identifying the specific legal objectives of the particular refuge in its enabling legislation, reviewing the general goals of all refuges, and identifying resource management issues and concerns. The latter requires employee and public involvement. Field employees can provide input on what they see as the refuge management problems. For example, is there habitat deterioration? If so, is it due to historic incompatible uses that continue today or to overuse by the public, even though that use is, in principle, compatible with the purpose of the refuge? At refuges with a strong waterfowl production orientation, lack of a dependable source of clean water to maintain and expand wetland acreage is often a problem. Another category of management issues that employees may bring up relates to lack of budget or personnel to meet the goals and objectives of the refuge.

Of course, public input is needed to ensure that a wide variety of issues or concerns is addressed and that the public's views are heard. Often the public may not even perceive something as a problem or may see it quite differently from the biologists. A refuge manager should be aware of these differing perspectives early in the planning process to avoid being surprised at public meetings later on. In addition, the public might provide refuge personnel with insights on problems with current public use, such as that conflicting uses (e.g., hunting and wildlife viewing) are located too near each other or that access is difficult, either because surrounding private landowners control roads or because the refuge lacks trails.

This first planning step also involves identifying analysis methods to be used in the subsequent steps in the planning process. As discussed in chapter 5, this identification of the analysis methods will guide the data collection efforts in the next step, Inventory and Analysis.

Step 2: Refuge Output List

The Refuge Output List is a list of uses of refuge lands and types of outputs expected to be produced to address the issues raised in the first step, Preplanning. For example, on National Wildlife Refuges in the San Joaquin Valley of California, these Refuge Output Lists were priority rankings of current and potential wildlife species, habitats, human uses (e.g., types of recreation), facilities (e.g., camping, parking, roads, buildings, and so on), and wildlife programs that would be provided on the refuges. The listing by priority order helps the manager in later stages of planning, when trade-offs must be made. In some sense, later steps in the planning process determine the capability to support these uses and the best management strategies by which to do so. The USFWS stresses that output levels should represent quantified production goals by species or activity where possible. A review of several Refuge Master Plans shows that some refuges have made this extra effort. This quantification will be especially helpful to the refuge when monitoring the effectiveness of plan implementation. For with quantitative output targets the staff can determine if the plan is actually producing the desired results.

Step 3: Locational Criteria

"Determination of Resource Requirements" might be a better title for this step than "Locational Criteria." In essence, step 3 describes what inputs are needed to produce the outputs desired (step 2) to address the key refuge issues (step 1). As stated in the USFWS *Planning Workbook,* this step involves the following: "For wildlife-oriented outputs, locational factors are stated in terms of habitat or functional needs such as nesting, sanctuary, brood sites, or food and water sources" (Fish and Wildlife Service 1980:3.9). This process is often assisted by using a species habitat model such as that presented in chapter 5 for the mallard. Other inputs for public uses and programs include both direct facilities required (e.g., campgrounds and roads) and support facilities (road maintenance).

Step 4: Resource Mapping

Step 4 involves describing which of the habitat components, facilities, and so on, listed in the locational criteria currently exist on the refuge. That is, how much of the habitat and other inputs required to produce the desired outputs currently exist on the refuge. With mapping, the intent is to quantify not only the amount of these resources but their location relative to one another. USFWS offices are encouraged to use either an overlay system (similar to BLM) or a computerized Geographic Information System. Thus information on soils, hydrology, different types of vegetation, and so on, is compiled onto a series of transparent map overlays to visualize resource interrelationships on the ground. Of course, mapping also includes landownership patterns within and adjacent to a refuge. This is often important, because frequently portions of critical habitats within the refuge occur on private inholdings or adjacent to the refuge on state or private lands.

When subsistence is a legal requirement of the refuge, dependency of villages on particular areas of the refuge must be documented. The role the refuge plays in supplying the necessities of life, such as food and clothing to native populations, must be examined. This is important because in the planning process, the total resource capability must be evaluated in light of all legally permissible uses of refuge lands.

The Inventory Phase of the USFWS Master Planning Process is made up of the first four steps. The next phase is the Analysis Phase, which involves suitability mapping.

Step 5: Suitability Mapping

In the Analysis Phase of planning we move beyond a mere description of "what is" to a determination of what the refuge's current and potential resource capability to produce the desired outputs is. In step 5, suitability mapping, the resource requirements in step 3 are matched with the resource maps from step 4. One determination from this step is where on the refuge the various outputs can be produced. For example, which areas on the refuge are currently suitable for waterfowl production?

But this step also looks at potential suitability. Given the desired output levels and resource requirements, which additional management inputs are needed so that the refuge can meet its potential? An example might be determination of the habitat factors that are absent or below optimum for a particular species in the Refuge Output List compiled in step 2. Would the land respond well to vegetation treatments, such as prescribed burning or attempts to create more wetlands? The linkage of the habitat models to management variables as shown in figures 5.2 and 5.3 illustrates how this process works for mallard production.

The Synthesis Phase of the Master Planning Process contains the last four steps.

Step 6: Preliminary Alternatives

In step 6 the refuge personnel propose alternative ways for managing different types of lands within the refuge. These alternatives represent different ways of maintaining current resource capability (where it is at acceptable or optimum levels) and in some cases of expanding current capability to its production potential so as to meet the output goals established earlier in the planning process.

This step of developing preliminary alternatives involves two parts. First, possible management policies regarding intensity and type of wildlife management actions and allowable uses suitable for different land types identified in the resource mapping step are developed. For example, one management strategy might take an intense wildlife management approach and allow high levels of wildlife-oriented recreation on much of the refuge. It might involve vegetation manipulation (e.g., prescribed burns, selective timber harvests), creation of wetlands, opening up of fish passageways, and so on, along with improved public access for fishing, hunting, and wildlife viewing. The theme of this approach might be maximum wildlife production and human use; another might be more biocentric in terms of emphasizing natural processes in wildlife management and allowing only as much human use as would be consistent with unenhanced wildlife production. Another management strategy might be to allow many historic uses, such as power boating or ORVs, but only where resource damage would be acceptable.

Once these candidate management strategies are formulated, they are matched up with the potential amount of refuge land that is suitable or "available" to support that management strategy. This suitability is based on the land's resource capability mapping. For example, not every land area on a refuge might be able to support intense wildlife management. Some areas of the refuge might not be suitable for particular habitat management tools, such as burning or wetland creation. However, others may be suitable for many different types of strategies. A preservation strategy might work on several of the land types in the refuge.

The second part in developing refuge plan alternatives is to allocate different amounts of suitable lands to different management strategies. Each alternative represents a different refuge theme or emphasis by allocating more of the suitable land to a particular strategy. For example, a wildlife production alternative to maximize recreational, commercial, and subsistence use might allocate all the lands suitable for intense wildlife management to that strategy. The only areas receiving a preservation strategy would be those unsuitable for intense management. A different alternative theme could be one stressing preservation everywhere. A third might emphasize nongame instead of game wildlife, so that management strategies that emphasized nongame animals would be applied to more of the land in that alternative. Finally, as required by NEPA for the EIS, a "continuation of current situation" or "no-action alternative" is also required.

Thus each alternative reflects a different mosaic of land management strategies. Some might implement all the strategies, but the amount of land allocated would be quite different under each one. Others might not include any lands in one or more management strategies. The resulting differences between alternatives can be summarized in many ways.

The clearest of these is through a mapping approach, such as that shown in figure 12.2. This figure shows Alternative B (the most development-oriented alternative) for the Arctic National Wildlife Refuge in Alaska. Lands are zoned to four

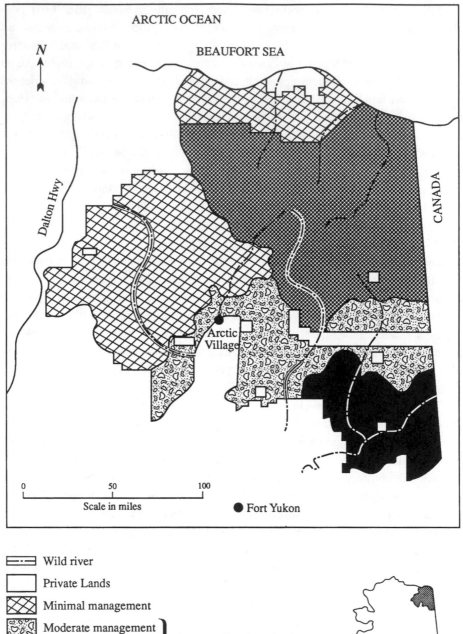

Figure 12.2. Example Land Management Map, Alternative B, Arctic National Wildlife Refuge

major levels of management (minimal, moderate, intense, and wilderness) and within these zones are designated as open or closed to oil and gas leasing. This alternative will be contrasted later with the preferred alternative when the Arctic National Wildlife Refuge case study is discussed.

The refuge *Planning Workbook* also suggests displaying information about the alternatives on a "Compatibility Chart." This is a matrix that classifies which uses are compatible, semicompatible, and noncompatible. When a particular refuge land area is used concurrently for two different purposes, so that one use would preclude the other (e.g., ORVs and ground-nesting birds), then only one of the noncompatible uses can be accommodated in a given alternative at that time of year. Thus one alternative might resolve the ORV–ground-nesting bird incompatibility in favor of recreational use and another might resolve it in favor of wildlife production.

The USFWS includes in the preliminary-alternatives step two additional components that most other agencies have as separate steps: (1) estimate the intended and unintended effects of implementing each alternative; (2) evaluate the alternatives using criteria such as environmental impacts and costs (Fish and Wildlife Service 1980:6.2). By *evaluation* we mean, given the estimated effects, how well each alternative produces the listed outputs, resolves the planning issues, and achieves the legal objectives of the refuge.

Refuge Impact Analysis. Because an Environmental Assessment or EIS is normally prepared on the Refuge Master Plan, the categories of estimated effects include the standard ones used in an Environmental Assessment or EIS. Therefore the general impact categories are physical (e.g., soils, water, and so on), biological (vegetation, wildlife, and so on), and socioeconomic. The effects are estimated by using a number of methods rather than by applying any single model or approach. For example, those on wildlife may be estimated by combining species habitat and population models with professional judgment. The USFWS Habitat Evaluation Procedures (HEP) may be used to quantify changes in habitat quantity and quality for key indicator species. Thus some of the effects are described in quantitative terms, such as wildlife population densities, bird days (number of birds times number of days on the refuge), and (sometimes) recreation visitor days. Many are described either in qualitative terms, such as whether a species would be rare or abundant, or in narrative terms, such as "improvement, slight increase, negligible effect, adverse effect," and so on. The drawback to reliance on such qualitative terms is that what is "slight" is not clearly defined. Is a 10% change slight? It may depend on the sensitivities of the person making the judgment as much as on the degree of effect.

Evaluation of Alternatives. Once the effects of the alternatives are estimated, the alternatives are formally evaluated. The key factors are how well the alternative resolves the planning issues and concerns listed in step 1, meets the legal objectives of the refuge (also detailed in step 1), and produces the priority resource outputs (listed in step 2). The USFWS uses a matrix approach similar to that discussed in chapter 4 to do this. The alternatives are listed across the top of the matrix, and the planning issues and legal objectives of the refuge are listed down the left side. If estimated effects are in quantitative terms, then the degree to which a given alternative meets the criteria and resource output objectives or goals can be clearly spelled out. Although this quantitative assessment was present in some Refuge Master Plans reviewed, in several others a simple narrative statement was used to state whether the alternative meets that objective. These statements range from "Yes, meets objective" to "high, medium, or low" potential to meet the objective. The lack of quantification makes precise comparison of the alternatives difficult. The basis for determining that one alternative provides a "high potential" and another provides "good opportunity" is often unclear. Impact analyses sometimes describe simply whether the impact is major or minor. Once again, the basis for deciding between the two is often not presented.

If conflicting output goals or issues are to be addressed by the planning process, the production trade-offs between habitats and species should be developed. For example, more intensive management for upland birds may reduce the amount of land available for expanded waterfowl management. This is where having the output list in priority order greatly assists the refuge manager in evaluating the alternative that best meets the key issues.

Step 7: Final Objective Set or Draft Plan Selection

After examining the results from step 6 on evaluation of preliminary alternatives, the alternative that best resolves the planning issues and meets the objectives of the refuge is selected as the draft preferred alternative. Such a determination is made by discussion between the refuge staff and managers and Regional Office staff and managers. The Master Planning Process calls this the final objective set, because in one sense, the draft preferred alternative will become management objectives and means to attain those objectives for the refuge. The preferred alternative is then described as the Draft Refuge Plan. The estimated effects of implementing this alternative, along with the other alternatives, would be described in the Draft Environmental Assessment for small or noncontroversial refuge's or a complete EIS for large or controversial refuge's. This would then be sent out for public comment. If appropriate, public meetings would be held.

Step 8: Final Alternatives

The planning team reviews the public comments from the Draft Environmental Assessment or EIS and responds to them. The public comments provide some indication of overall support for the preferred alternative versus the other alternatives. Public comments also allow USFWS to gauge public reaction to very site-specific policies about allowable public uses in particular areas. In response to public comments, the preferred alternative or other alternatives might be modified to address some of the public's concerns. The refuge staff and manager and the Regional Office staff and manager fine-tune the final preferred alternative for the Final Environmental Assessment or Final EIS. A great deal of effort goes into this final preferred alternative, as it will be blueprint for the Final Master Plan developed in detail in the next step in the planning process.

Step 9: Master Plan Report

If the final preferred alternative survives public scrutiny in the Final EIS, then it is developed into a detailed Master Plan. At this step the exact locations, layout, configuration, and timing of each management action necessary to meet final objectives and refuge outputs (from step 2) are developed. The necessary actions are phased in over time in an explicit time schedule. Specific project description worksheets are prepared on the materials, personnel, and costs of each project needed to implement the Master Plan. These become the basis for budget requests associated with implementing the Refuge Master Plan.

Finally, plan implementation involves taking the specific actions identified in the plan to attain the objectives of the refuge.

Updating of Master Plans.   It is worthwhile to note that Master Plans are periodically updated when new data or information becomes available that might suggest changes in the current plan. In addition, as unanticipated issues arise that require minor changes in the plan, such changes should be incorporated after public review. The public and affected agencies should be notified of such changes. A full review and major update or complete revision is made every ten to fifteen years.

Importance and Role of Public Involvement in Master Plans

Public involvement is significant in several steps in the planning process. It is encouraged in the preplanning step to obtain public input on the issues that should be addressed in the refuge management plan (i.e., the scope of the plan). It also occurs during step 7, when the public is asked to respond to evaluation of alternatives and the selection of the preferred alternative. Since most Refuge Plans are also EIS's, they must meet the NEPA requirements for public involvement. This is done at step 7 by having a public comment period on the draft plan and by holding public meetings. Of course, the public has one last opportunity to make itself heard in which it may protest the selection of the final management plan in step 8.

Analytical Planning Techniques Used to Implement Steps in the Planning Process

The USFWS planning technique combines a zoning/compatibility system with evaluation of alternative zonings by a matrix approach. Specifically, all lands in the refuge are administratively designated or zoned into different management intensities, ranging from wilderness to developed (i.e., a multiple-use mix, including firewood collection, fish hatcheries, and developed campgrounds). The zoning/compatibility approach is similar to grouping compatible multiple uses, but the multiple uses to be grouped are restricted to wildlife and compatible wildlife uses (recreation and, in specific cases, subsistence uses or livestock grazing).

The effects of different amounts of land being allocated to different uses are displayed in a matrix, which relates the effects of different land allocations to a variety of categories, such as wildlife populations, public uses, and economics. Such matrices also indicate how the alternatives meet the objectives of the refuge, although such determinations are often not made quantitatively.

Consideration of economic factors in refuge planning is integrated into the plan in several ways. The most common is translation of hunter, angler, and nonconsumptive user's expenditures into local income and employment using input-output models and associated multipliers (discussed in chapter 7). If other economic uses of the refuge, such as timber harvesting, agriculture, grazing, or mineral development, would be affected by different alternatives, these too would be translated into local income and employment effects using an input-output model. On occasion, the net economic value of wildlife recreation is also computed to compare to the cost of obtaining water for wetlands, acquisition of private inholdings, or major public investments in facilities.

Transition from Refuge Master Plans to Comprehensive Conservation Plans

As can be seen from the details of the Master Planning Process, this approach can be quite involved. Consequently, it is not surprising that some regions have developed the more abbreviated Station Management Plans mentioned earlier. In other cases, the Master Plan approach has been tailored to specific refuges, as in California and Washington. The current structure of this Master Planning Process has both strengths and weaknesses. The information developed at certain steps is quite good (like the refuge output lists in step 2), but at other steps the Master Planning requirements are weak. (For example, there is little structure for objective evaluation of alternatives.)

With the reactivation of a National Wildlife Refuge planning effort associated with Refuges 2003, both USFWS and Congress have considered comprehensive refuge planning (Olsen 1991). The outline of the planning process that would be used nationwide would likely follow more closely the one developed in response to the last mandate of Congress for refuge planning in Alaska (Olsen 1991). Thus we turn to a review and case study of refuge planning on the Arctic National Wildlife Refuge both to illustrate this approach to refuge planning and to provide a glimpse of what future refuge planning may resemble. In addition, this same planning process is being used to update the existing refuge plans in Alaska, starting with the Kenai National Wildlife Refuge near Anchorage, Alaska.

Planning on New Refuges in Alaska: An Example of a Comprehensive Conservation Plan for the Arctic National Wildlife Refuge

As required by ANILCA, each of the sixteen NWRs in Alaska must prepare Comprehensive Conservation Plans. These management plans, which also serve as an evaluation of refuge lands for suitability as USFWS-administered Wilderness Areas, must be in an EIS format to comply with NEPA requirements.

The Arctic NWR (ANWR) in the northern corner of Alaska provides a case study of refuge planning in Alaska. The process used is similar to that for other Alaskan refuges that developed Comprehensive Conservation Plans. Figure 12.3 shows the general location map of the Arctic NWR relative to other refuges in Alaska. As can be seen from this map, the Arctic NWR is the second largest refuge in Alaska and in the United States. It totals more than 19.5 million acres, about 19 million of which are federally owned; the rest of the land is state and privately owned. About 9 million acres of the refuge were designated by Congress as Wilderness, although part of the objectives of this plan is to determine the Wilderness suitability of other areas within the refuge. The ANWR is the source of some controversy, since Congress did allow limited oil and gas exploration on the coastal plain. However, development and production of oil and gas would require congressional authorization. The eight steps in the Comprehensive Conservation Plans of Refuges are the following:

1. Preplanning.
2. Inventory and Analysis.
3. Formulate Management Alternatives.
4. Evaluation of Alternatives.
5. Plan Selection (Initial or Draft).
6. Select Comprehensive Conservation Plan (Final Plan).
7. Plan Implementation.
8. Periodic Updating.

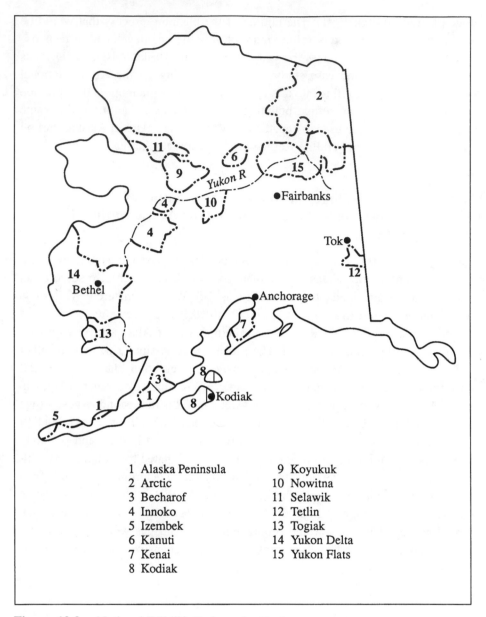

Figure 12.3.   National Wildlife Refuges in Alaska

As can be seen from this list, this planning process follows more closely the standard planning steps used by other public land management agencies, such as BLM and Forest Service, than it does the USFWS Master Planning Process. We shall now look at each step in more detail to see how it was applied to the Arctic National Wildlife Refuge (ANWR).

Step 1: Preplanning

Step 1 is similar to that in Refuge Master Planning: a blending of specific legal objectives of the refuge established by Congress with local issues and concerns over management. The specific legal objectives of the ANWR, in addition to the general goals of the National Wildlife Refuge System, are found in the Alaska National Interest Lands Conservation Act, Section 303(2B). This act provides four major purposes for which ANWR is to be managed: (a) to conserve fish and wildlife populations and habitats in their natural diversity, including, but not limited to, the porcupine caribou herd, polar bears, grizzly bears, musk-ox, Dall sheep, wolves, wolverines, snow geese, peregrine falcons and other migratory birds, and Arctic char and grayling (two fish species); (b) to fulfill the international treaty obligations of the United States with respect to fish and wildlife and their habitats; (c) to provide, in a manner consistent with the purposes set forth in (a) and (b), the opportunity for continued subsistence uses by local residents; and (d) to ensure, to the maximum extent practicable and in a manner consistent with the purposes set forth in subparagraph (a), water quality and necessary quantity within the refuge.

Through public meetings, solicitation of comments from the public (including native peoples, the oil industry, Alaska Department of Fish and Game), and discussion with USFWS management staff, eleven major planning issues and concerns were identified. These include

1. Protecting the porcupine caribou herd.
2. Protecting wilderness values in existing Wilderness.
3. Designating additional Wilderness.
4. Providing for aircraft and other motorized access into the refuge.
5. Providing for recreational uses, including commercial guiding.
6. Providing for oil and gas activities.
7. Providing for mining of active claims on refuge lands.
8. Providing for commercial timber harvesting.
9. Not restricting subsistence uses of fish and wildlife.

Step 2: Inventory and Analysis

Over a two-year period, USFWS staff collected data on vegetation, fish, wildlife, subsistence uses of the refuge, types of recreation uses (sport hunting, river rafting, backpacking, and wildlife viewing), and other activities, such as mining. The current land status and access systems (primarily airplanes, boats, and snowmobiles) used by people were also cataloged and mapped.

The USFWS also identified special values that were relatively unique to the ANWR, including the "arctic wilderness ecosystem," geological values, a continuum of arctic and subarctic ecosystems in the space of 150 miles, and the northernmost extension of the continental divide (i.e., the Brooks Range). The USFWS identified two areas of the refuge that meet the minimum criteria to be considered for Wilderness status.

Step 3: Formulate Management Alternatives

The enabling legislation set three standards that all alternatives had to meet:

1. They must assign all lands to a particular management strategy based on that land type's resources and values.
2. Alternatives had to provide a program to maintain and enhance the values of resources present on those lands.
3. Alternatives were required to determine which uses were compatible with each land type within the refuge.

To meet this challenge, the USFWS assembled five management strategies that it could apply to the lands within the ANWR. Each was a set of directions or programs to maintain and enhance the resource coupled with a list of compatible uses. The five management prescriptions ranged from "intensive management" to Wilderness.

The *intensive-management strategy* involved a program of "highly manipulative techniques, such as mechanical manipulation of vegetation, construction of artificial impoundments and dikes and the construction of permanent fish weirs and hatcheries. Public use facilities, administrative sites, transmission lines, pipelines, and transportation systems may be allowed. Increased public use would be encouraged . . ." (Fish and Wildlife Service 1988:37). In addition, timber harvesting and removal of sand and gravel would be allowed. This management strategy is much like multiple-use management on BLM lands. That is, the allowable uses span a large range of human uses, with wildlife being one. Obviously, this type of management could occur only on lands that could sustain such high levels of public use and respond to vegetation manipulation in the desired manner.

The next strategy along this continuum was the *moderate-management* prescription, which would allow for "managed" public use in a relatively natural setting. Public use facilities, such as campgrounds and boat launches, would be provided. Roads would be built, and motorized vehicles would be allowed. There would be some active management of fish and wildlife habitats, such as manipulation of vegetation or water impoundments. The emphasis would be on wildlife management and limiting other nonwildlife uses.

The third management strategy involved *minimal management*. It maintains areas with already high fish, wildlife, and other values and prevents incompatible activities, such as road building and motorized vehicles. The only motorized access would be motorboats, airplanes, and snowmobiles. All types of wildlife recreation would be accommodated. Subsistence uses, such as house log and firewood collection, would be allowed. However, vegetation management techniques would be more nonmechanical and nonstructural, such as prescribed burning.

The fourth general management strategy was *Wilderness management.* This was reserved for areas on the ANWR that were already designated as Wilderness, but it could also be applied to areas recommended to Congress by USFWS as Wilderness. That is, most federal agencies are required to manage a recommended area as Wilderness until its disposition by Congress, as any other management might destroy its wilderness values, thereby denying Congress the opportunity to make the final decision. Although this land management strategy would be the most restrictive, it is less restrictive than Wilderness in the lower forty-eight states. Because of subsistence access requirements, the USFWS permits snowmobiles, motorboats, helicopters, and airplanes in these Wilderness areas and allows for firewood collection (Fish and Wildlife Service 1988:39).

The fifth management strategy is a very specialized one dealing only with *Wild River management.* This too is a restrictive category, similar to minimal management. Only recreational uses that do not adversely affect the natural values of the rivers are permitted, although motorized access would be allowed for subsistence activities. Incompatible economic uses, such as timber harvesting or mineral core sampling, would not be allowed.

The USFWS developed a total of seven alternatives. Four of these were developed by USFWS directly (A, D, E, F), two were developed in response to resource development interests in Alaska (B, C), and one was developed in response to conservation groups (G). Alternative A is both the "current situation" and the USFWS "preferred alternative." Table 12.1 compares the acreage allocation under each alternative. These land use designations were also displayed on maps similar to that in figure 12.2.

Table 12.1
Percentage of Refuge Land by Management Strategy

Management Strategy	Refuge Plan Alternatives					
	A	B	C	D	E	F,G*
Intensive mgmt.	0%	11%	0%	0%	0%	0%
Moderate mgmt.	0	15	26	0	0	0
Minimal mgmt.	56	30	30	29	12	6
Wild river	2	2	2	2	2	2
Wilderness	42	42	42	42	42	42
Proposed wilderness	0	0	0	27	44	50
Total	100	100	100	100	100	100

*G is identical to F except that Wilderness would be managed more like Wilderness in the lower forty-eight states, in that there would be no helicopter access, limitation on aircraft landings, and no new administrative or recreational developments.

Step 4: Evaluation of Alternatives

The first part of step 4 was to estimate the effects of the land use allocations in table 12.1 on the physical, biological, and human environment. This was done largely in narrative form in terms of which uses would be permitted and, if allowed, whether they would be above or below current levels.

The formal evaluation of alternatives was done in two matrices. The first was set up to compare alternatives against the four legal objectives of the ANWR presented earlier. The second was an evaluation of how well each alternative did in resolving the eight management issues. This is an excellent example of clearly linking the evaluation criteria to the alternatives. Table 12.2 displays the matrix, illustrating the linkage of alternatives to refuge purposes. Table 12.3 displays the evaluation of Alternatives A–C, based on resolution of the planning issues.

Unfortunately, in both matrices effects are described in very general narrative terms, so it is difficult to ascertain to what degree each contributes to the objectives and resolves the issues. For example, in table 12.3, the porcupine caribou herd and wilderness values are described as "maintained." The only item bordering on economics relates to the relative cost of implementing the alternative plans. Little information is provided on which to judge the merit of providing commercial timber harvests. What is the direct cost of such harvests relative to the benefits? What are the effects on other resources, such as fisheries or wolverines, of timber harvesting? Certainly a rough estimate could be made of the effect of restricting helicopter and aircraft access on recreation visitation by subtracting visitors arriving using these modes from current visitation levels. If neither is known, then a simple visitor contact or guide survey would supply the information.

The associate director of Habitat Resources of USFWS stated the following in reference to the similar lack of quantification in the Kenai NWR plan four years prior to the ANWAR effort: "In particular, sound decision making is not well supported by the superficial treatments of recreational demand and values . . . " (Wallensnoin 1984).

Step 5: Draft Plan Selection

The preferred alternative (A) was identified and put forth in the Draft EIS to obtain public comment. During the ninety-day public comment period the USFWS received 961 written comments from local, state, and federal agencies; private industry; corporations owned by native peoples; conservation groups; and other interested parties. Five public meetings were held in Alaska in which 110 people attended and forty-two testified.

Step 6: Select Final Comprehensive Conservation Plan

The USFWS Final EIS notes that the public comments were "taken into consideration during the preparation of this final plan. It is important to note that the selection of the preferred alternative is not based solely on how many people support a particular alternative. Public comment is only one of several criteria used in the selection of the Service's preferred alternative" (Fish and Wildlife Service 1988b:45).

Nonetheless, the Fish and Wildlife Service (1988b:46) displayed the number of comments and organizations supporting each alternative in the Draft EIS. Alternative G, the most preservationist, which (if implemented by Congress) would put 92% of the ANWR into Wilderness, received by far the most support from individuals writing letters and giving testimony. Although Alternative A (the USFWS preferred alternative) received less than one-tenth the number of supportive comments and testimony, the few groups supporting it included the state of Alaska, Resource Development Council, Atlantic Richfield (an oil company, also known as ARCO), and one of the village public meetings. The key distinction in public perception of the two alternatives (A and G) was that in Alternative G, nearly all lands are permanently protected (from oil and gas development and intensive management) as Wilderness. In Alternative A, on the other hand, the USFWS would provide only "minimal management" to retain the lands' natural character (hence similar to Wilderness), but these lands could be reallocated administratively to intensive management (including oil and gas development) in future revisions. Given that the U.S. Department of Interior oversees the USFWS, there is some concern that the more development-oriented Department of Interior might pressure USFWS to accept more development in the future to the detriment of wildlife resources.

Thus although Alternative A received far less broad support from individuals, conservation groups, and several villages, the preference of the USFWS, the state of Alaska, and development interests resulted in its being retained as the Final Comprehensive Conservation Plan in the Final EIS and Record of Decision.

However, because of public comments, the Final EIS was revised. Proposed Wilderness boundaries in alternatives recommending Wilderness were modified. Several sections dealing with the porcupine caribou herd and endangered species were rewritten. Management practices proposed for ORVs and mechanical manipulation of vegetation in ANWR have been revised to be consistent with USFWS policy on these resource uses.

Table 12.2

Evaluation of the Alternatives Based on Refuge Purposes

Refuge Purposes			
(1) To conserve fish and wildlife populations and habitats in their natural diversity	(2) To fulfill the international treaty obligations of the United States with respect to fish and wildlife and their habitats	(3) To provide, in a manner consistent with purposes (1) and (2), the opportunity for continued subsistence use by local residents	(4) To ensure, to the maximum extent practicable, water quality and necessary water quantity within the refuge
Alternative A (Current Situation and Preferred Alternative)			
High potential for maintaining natural diversity and abundance of wildlife while continuing to provide for current levels of traditional uses and access	High potential to protect sensitive fish and wildlife habitats in compliance with international treaties	Good opportunity to provide for continued subsistence use of refuge resources; no significant restrictions of subsistence use by local residents	High potential to maintain water quality and quantity
Alternative B			
Natural diversity and abundance of fish and wildlife maintained overall; potential for localized reductions in populations and habitats if developments are permitted (e.g., oil development, timber harvesting)	Protects most sensitive fish and wildlife habitats in compliance with international treaties	Maintains opportunities generally for continued subsistence use of refuge resources; potential for significant restrictions of the subsistence uses of a few residents in a localized area	Maintains overall water quality and quantity; permitted developments could adversely affect water quality and quantity on a localized basis
Alternative C			
Natural diversity and abundance of fish and wildlife maintained overall; potential for localized reductions in populations and habitats if developments are permitted (e.g., timber harvesting)	Protects most sensitive fish and wildlife habitats in compliance with international treaties	Maintains opportunities generally for continued subsistence use of refuge resources; no significant restrictions of subsistence use by local residents	Maintains overall water quality and quantity; permitted developments could adversely affect water quality on a localized basis

Refuge Purposes			
(1) To conserve fish and wildlife populations and habitats in their natural diversity	(2) To fulfill the international treaty obligations of the United States with respect to fish and wildlife and their habitats	(3) To provide, in a manner consistent with purposes (1) and (2), the opportunity for continued subsistence use by local residents	(4) To ensure, to the maximum extent practicable, water quality and necessary water quantity within the refuge
Alternative D			
High potential to maintain natural diversity and abundance of fish and wildlife while continuing to provide for current levels of traditional use and access	High potential to protect sensitive fish and wildlife habitats in compliance with international treaties	Good opportunity to provide for continued subsistence use of refuge resources; no significant restriction of subsistence use by local residents	High potential to maintain water quality and quantity
Alternative E			
High potential to maintain natural diversity and abundance of fish and wildlife while continuing to provide for current levels of traditional use and access	High potential to protect sensitive fish and wildlife habitats in compliance with international treaties	Good opportunity to provide for continued subsistence use of refuge resources; no significant restriction of subsistence use by local residents	High potential to maintain water quality and quantity
Alternative F			
High potential to maintain natural diversity and abundance of fish and wildlife while continuing to provide for current levels of traditional use and access	High potential to protect sensitive fish and wildlife habitats in compliance with international treaties	Good opportunity to provide for continued subsistence use of refuge resources; no significant restriction of subsistence use by local residents	High potential to maintain water quality and quantity

Table 12.3

Evaluation of Alternatives Based on Issues

Issue/concern	Alternative A (Current Situation and Preferred Alternative)	Alternative B	Alternative C
Protecting the porcupine caribou herd	Maintains the caribou population; negligible impacts from public and economic uses	Potential for minor impacts from oil development; cumulative impacts from this use and other human activities within and outside of the refuge could lower herd's productivity	Maintains the caribou population; negligible impacts from public and economic uses
Protecting wilderness values	Maintains wilderness values overall; in localized areas increasing public use could diminish wilderness values	Wilderness values generally maintained; in localized areas wilderness values would be diminished or lost due to increasing public use, oil development, timber harvesting, and other economic uses	Wilderness values generally maintained; in localized areas wilderness values would be diminished or lost due to increasing public use, timber harvesting, and other economic uses
Designating additional wilderness	No additional areas proposed for wilderness designation	No additional areas proposed for wilderness designation	No additional areas proposed for wilderness designation
Providing for aircraft and other motorized access into the refuge	Maintains existing opportunities for traditional access (aircraft, snowmachines, motorboats)	Maintains existing opportunities for traditional access (aircraft, snowmachines, motorboats)	Maintains existing opportunities for traditional access (aircraft, snowmachines, motorboats)

Providing for recreational use, including commercial guiding and outfitting	Opportunities maintained for hunting, fishing, and nonconsumptive uses; no restrictions placed on increased public use, provided it is compatible with refuge purposes	Opportunities maintained for hunting, fishing, and nonconsumptive uses; no restrictions placed on increased public use, provided it is compatible with refuge purposes	Opportunities maintained for hunting, fishing, and nonconsumptive uses; no restrictions placed on increased public use, provided it is compatible with refuge purposes
Providing for oil and gas activities south of the "1002" area	Oil and gas studies permitted with restrictions in the Arctic Wilderness and Wild River corridors; no oil and gas development permitted	Oil and gas studies permitted with restrictions in the Arctic Wilderness and Wild River corridors; oil and gas development may be permitted (with congressional approval) on the south side of the Brooks Range	Oil and gas studies permitted with restrictions in the Arctic Wilderness and Wild River corridors; no oil and gas development permitted
Providing for commercial timber harvesting	No opportunities provided	Commercial timber harvesting may be permitted in 26% of the refuge	Commercial timber harvesting may be permitted in 26% of the refuge

Step 7: Implementation and Revision of the Comprehensive Conservation Plan

The Record of Decision in November 1988 by the regional director of USFWS concurred with selecting Alternative A as the Final Plan. The USFWS would begin to implement the proposed action through development of site-specific and resource-specific action plans. These plans include a fishery management plan, a fire management plan, and a public use plan. The tasks to be performed on the ground will be identified in these site-specific plans. Those identified will be included in the USFWS annual work plans and budgets submitted to the Department of Interior. However, the USFWS recognizes that, "Implementation of the proposed actions in this plan will depend upon the availability of funds and personnel and upon the coordination of many governmental activities" (Fish and Wildlife Service 1988b). Thus the Final Plan will be implemented as funds are made available to USFWS.

Lastly, the USFWS views the Final Plan as a "dynamic document" that will be reviewed every three to five years in light of informal comments from the public, other government agencies, and research studies. A full review and updating of the plan will occur every ten to fifteen years.

Planning on Existing Refuges in the Lower Forty-eight States: Case Study of Charles M. Russell Refuge Plan and EIS

To provide a case study of major refuge planning in the continental United States and illustrate how refuge planning can be improved, we choose the Charles M. Russell Refuge (CMR) in Montana as an example. First, we shall look in detail at what the USFWS staff did as it adapted the refuge Master Planning Process described earlier to the CMR. The USFWS effort on the CMR also has some strong features that were lacking in the ANWR case study just reviewed. In addition, we shall discuss how the USFWS process could be further improved using a simplified approach to benefit-cost analysis.

The 1-million-acre Charles M. Russell Refuge in eastern Montana is a good example of updated planning on a refuge established in 1936 as the Fort Peck Game Range. This refuge surrounds the large Fort Peck Reservoir operated by the U.S. Army Corps of Engineers. From 1936 to 1976 the area was under joint management by BLM (responsible for livestock grazing) and the USFWS. In 1976, Congress directed the USFWS to manage CMR as a unit of the National Wildlife Refuge System. However, the ranchers who had grazing permits were not pleased with this move from joint management with BLM, where livestock had coequal priority with wildlife conservation, to one where wildlife was the priority. A series of court decisions later reaffirmed that wildlife had priority to the forage resources and that CMR was to be managed as a Wildlife Refuge. An EIS was prepared to evaluate management of the refuge under this new direction. The format follows that of the NEPA EIS and the master planning approach of the USFWS.

Step 1: Preplanning

Members of the staff of the CMR compiled the legal objectives of the refuge from its enabling legislation as well as from its general mission under the National Wildlife Refuge System. More important, they detailed the CMR goals in priority order, the wildlife objectives (also in priority order), the range objectives, and the recreation objectives. Putting the CMR goals and wildlife objectives in priority order will help considerably in evaluating alternatives when different ones may help attain one objective but make the attainment of another less likely.

The overall CMR goals, in priority order, are as follows:

1. Attain a balanced, natural diversity of plant and animal communities favoring threatened and endangered (T&E), then all other native, and finally, desirable exotics.
2. Protect wilderness, historic, cultural, and natural areas unique to the Missouri River Breaks.
3. Restore and maintain habitats to sustain optimum populations of mammals and nonmigratory birds.
4. Manage migratory bird habitats first for production and then for use during migration.
5. Provide for grazing of domestic livestock when compatible with wildlife.
6. Provide for public understanding and appreciation of fish, wildlife, recreational, cultural, and scenic resources on CMR through high-quality programs in environmental education, interpretation, wildlife observation, hunting, fishing, and other forms of wildlife-oriented recreation compatible with wildlife goals.
7. Coordinate and integrate, where feasible, management of the CMR with objectives of other federal and state agencies and private landowners in and around CMR.

Clearly, CMR places wildlife first, then livestock grazing, recreational use, and finally consistency with adjacent landowners objectives. "The wildlife objectives drive the planning process, dominate the other objectives and may modify or preclude the others if conflicts arise that cannot be resolved any other way" (Fish and Wildlife Service 1984:135). This statement reflects the legislative intent of all wildlife refuges, that although wildlife is not the sole use of a refuge, it is the dominant one. This clearly differentiates wildlife refuges from true multiple-use lands such as National Forests.

Within the wildlife objective, the USFWS has set the following objectives (in priority order):

1. Reintroduce peregrine falcons and black-footed ferrets and maintain existing migration habitat for bald eagles.
2. Improve and maintain habitat for sharp-tailed grouse in ponderosa pine/juniper and grass/deciduous shrub land types to support thirty spring-breeding birds per square mile by the year 2005.
3. Improve and maintain pronghorn winter habitat in the juniper and sage-grassland land types to support fifteen hundred wintering animals by the year 2005.
4. Improve and maintain riparian habitats in good to excellent condition by the year 2005 to benefit wildlife species, such as deer, raccoons, beaver, waterfowl, kingbirds, mourning doves, elk, American kestrels, pheasants, and turkeys.
5. Improve and maintain deer habitat to support overwintering populations of ten deer per square mile by the year 2005.
6. Improve and maintain elk habitat to support overwintering populations of 2.5 elk per square mile (in coniferous and associated grassland communities) by the year 2005.
7. Improve waterfowl habitats to good or excellent condition on all suitable ponds.
8. Reintroduce Rocky Mountain bighorn sheep and swift fox in suitable habitats.

There are two points to note about these wildlife objectives. First, threatened and endangered species receive the number 1 objective. The second, and most important point, is the nature of these objectives: the quantity, land type, and timing of the objective are spelled out. This makes monitoring easier, since an unambiguous quantitative target is spelled out and so is the time frame for achieving the target. This priority listing and the quantitative goals listed in this first step are similar to the Refuge Output List in Refuge Master Planning.

The initial targets in the objectives were revised after step 2, inventory and analysis. In particular, the capacity of the refuge to meet these targets (even under intense management) was checked and the targets were revised if necessary (Fish and Wildlife Service 1984:135). This illustrates a common concept in planning: planning is an iterative or cyclical process. That is, some initial objectives are quantified, and the effects of meeting those objectives on other resources are determined. This learning allows the planner to go back and reevaluate the original objectives in light of their feasibility or effects on other resources. Thus several passes through the first five steps in a planning process might be made before finally selecting the preferred alternative.

The USFWS identified three range objectives: (1) improve range condition and stability of soil resources; (2) provide forage beyond the needs of wildlife to livestock; (3) provide stability to livestock users and their operations consistent with wildlife objectives.

The USFWS also identified three recreation objectives: (1) to inventory and protect all cultural resources; (2) to continue to provide wildlife-oriented recreation (e.g., hunting, fishing, viewing, interpretation); and (3) to support other agencies (e.g., the U.S. Army Corps of Engineers that run the dam and the reservoir) to provide developed recreation and nonwildlife recreation.

Step 2: Inventory and Analysis

The USFWS inventoried landownership, soils (including soils capability classes and soil mapping), water resources, vegetation type (six general types were used in mapping, but eleven were used in habitat models), grazing, and recreation use (both wildlife related and nonwildlife related).

Habitat Modeling

The habitats and population distribution of twenty-four wildlife indicator species were analyzed. Wildlife indicator species were chosen on the basis of representing a particular habitat type (e.g., waterfowl representing wetlands), of being a good indicator for many other species, and of being important economically to the public (Fish and Wildlife Service 1984:137). The USFWS utilized information on vegetation type with the 1976 version of the Habitat Evaluation Procedures (HEP) modeling approach described in chapter 5 for each of the twenty-four indicator species. These models are word models that rate the overall suitability of habitat for a species based on a series of habitat criteria that the scientific literature and local experts indicate are important habitat requirements for that species. For example, prime red-tailed hawk habitat is described as being ponderosa pine forest with 30% or less canopy closure, near areas of sage and grasslands. The optimum size of trees for nesting is 20–30 feet tall and 8–10 inches in diameter. Habitats meeting all these conditions would receive the maximum score.

Components that are critical to habitat suitability for a species but that are deficient would then be prime candidates for management prescriptions to augment them. For example, in the elk model, both western wheatgrass and water sources every 640 acres are required for maximum elk carrying capacity. If these two were the only habitat factors that were very low in a particular area, the model would indicate a low current carrying capacity in that area. But this would also suggest that an appropriate management prescription would be to seed western wheatgrass and develop water sources by building small catchments or well-gullzers.

Rangeland Vegetation Modeling

Determination of forage availability for livestock grazing followed a screening or residual approach. Soils, vegetation, and precipitation were mapped into range condition classes. Then the Soil Conservation Service *Montana Grazing Guide* was used to determine the amount of forage that would be available in a typical year, if livestock use was distributed evenly in the area and livestock was the only objective. This set the initial amount of AUMs of forage for livestock. This number was adjusted downward to account for the fact that livestock would not be distributed evenly but would avoid steep slopes and would congregate near water. The erosion potential of certain soils required that more residual vegetation be retained than only what was calculated in the first step for plant vigor. Finally, the need for additional vegetation and forage for particular wildlife species, as indicated in the species HEP habitat suitability models, was examined. The final amount of AUMs available to livestock was that remaining after these additional screens or criteria were met.

Economics

The current socioeconomic situation is described as an agrarian economy, with ranching on the CMR being one major use. Recreation visitor spending is one of the other major "industries" associated with the CMR.

Step 3: Formulate Alternatives

The USFWS developed several management prescriptions that reduced the limiting factors facing the wildlife indicator species and brought up their habitat values as close to their potential as practical. These prescriptions were also used in different combinations to form the various alternatives. They included prescribed burning, fencing (for livestock control), small pond water developments (for wildlife and livestock), grazing (as a vegetation management tool), and planting of trees and shrubs. The amount of AUMs available for livestock was largely a management variable also.

Various combinations of these actions were combined to produce five different alternatives or management themes. The range of the alternatives included "no action," proposed action, intensive wildlife management, multiple use, and no livestock grazing. The no action involved continuation of the current amount of livestock grazing and management of the area as it had been. The proposed action stressed meeting the wildlife objectives stated in the preplanning step. Native species would be reintroduced, livestock grazing reduced, and so on. Intensive wildlife management would involve maximum wildlife production using a wide range of structural and nonstructural measures. Maximum wildlife production would be stressed instead of naturalness. Lastly, the multiple-use alternative gives equal consideration to wildlife, livestock, and recreation.

The pattern of management prescriptions (i.e., the inputs) associated with each alternative is presented in table 12.4. The other key characteristic of each alternative besides management prescriptions relates to what public uses would be allowed on the refuge. In the multiple-use alternative, for example, more developed and roaded recreation would be permitted. In the preferred alternative, more wildlife-oriented recreation would be favored.

Table 12.4
Management Prescriptions by Alternative in Year 2005

Prescriptions	No Action	Proposed Action	Intensive W/L Mgmt.	Multiple Use	No Grazing
Vegetation Mgmt. (Acres)					
Burn	0	7,700	15,000	1,500	11,300
Plant trees	0	0	25,000	0	4,700
Plant shrubs	0	500	3,000	300	500
Soil ripping	0	0	900	38,000	10,000
Fencing (miles)	56	124	502	122	456
Water develop. (#)	150	176	223	255	163
Livestock AUMs	56,524	40,628	22,823	61,260	0

Step 4: Evaluation of Alternatives

Step 4 involved first estimating the effects of implementing the alternatives and then, given this information, evaluating how the alternatives met the objectives of CMR and resolved the planning issues.

Table 12.5 presents the effects of implementing the management prescriptions in table 12.4. Effects on wildlife habitat quality are quantified using HEP. Specifically, the index number represents how close to maximum carrying capacity the habitat would be for that species. A 10 is the maximum wildlife density under the best habitat conditions for all factors.

Recreation Use

Recreation use estimates for each alternative were made by recreation activity using data from a refuge visitor survey, traffic counts, and the state of Montana's State Comprehensive Outdoor Recreation Plan. Recreation visitation was projected into the future, based on growth in human population in the counties and cities within 150 miles of the refuge. Refuge personnel then adjusted these use estimates for each alternative based on the amount of recreational access, facilities, and emphasis of the particular alternative. This process appears to be somewhat subjective because no formal recreation demand models were used that linked visitation to wildlife quality (similar to models described in chapters 9 and 10). This is not too surprising, as the original recreation work on this EIS was done in the late 1970s and early 1980s, before these types of demand models were widespread.

Recreation Economics

Socioeconomic effects were estimated for recreation and range primarily in terms of sales in the region and employment. Recreation visitation associated with each

Table 12.5
Effects and Outputs by Alternative in Year 2005

Effects/Outputs	No Action	Proposed Action	Intensive W/L Mgmt.	Multiple Use	No Grazing
Habitat Quality					
Sharp-tailed grouse	4.7	7.2	8.0	5.3	8.6
Mule deer	5.8	7.2	8.0	6.5	7.5
Pronghorn	5.3	7.4	7.1	6.6	6.3
Elk	6.2	7.6	8.0	6.4	7.8
Waterfowl	3.4	7.0	9.0	7.5	8.3
Visitor Days					
Viewing scenery	107	107	106	110	107
Viewing wildlife	5	7	7	7	9
Hunting	11	14	16	11	15
Fishing	77	83	89	95	85
Camping	35	34	25	35	34
Picknicking	80	79	68	81	79
Ft. Peck Lake Rec.	101	95	76	103	95
Economics (Change from No Action)					
Rancher income ($1,000's)		− $205	− $382	$66	− $722
Total range jobs		− 13	− 24	− 3	− 43
Total recreation jobs		11	− 24	40	13

alternative was converted to total sales in the six-county study region using an expenditure of $30 per person distributed between six different industries providing recreation- and tourism-related services.

Using an input-output model similar to that discussed in chapter 7, total sales (direct plus indirect linkages) and total employment were calculated for recreation. This figure is shown in table 12.5.

Livestock Grazing Economics

Since earlier efforts to account for economic effects of ranchers had proved inadequate under public scrutiny, a ranch budget approach was used to generate the changes in rancher income associated with forage reductions under each alternative. A range economist was contacted to provide the specialized data, modeling, and expertise unavailable within USFWS. A linear programming model (LP) was used with ranch budget data to simulate how ranchers would reallocate their available forage supplies and change their livestock operation in response to reductions in forage under the five alternatives. Average reductions in rancher gross income were calculated by herd size using the LP model for each alternative. This reduction in income is the initial direct effect used in the input-output model to calculate the total direct and indirect economic effects on all ranch-related sectors in the six-county area. The economic efficiency cost of reallocating forage away from livestock is only the direct change in rancher income, as the indirect effects merely reflect transfers of economic activity. Finally, to address distributional equity among ranchers with above-average dependencies on the CMR for their livestock forage requirements, an LP analysis showing the impact by rancher, according to grazing allotment, was run.

The results of these analyses to estimate the effects of implementing each alternative are shown in table 12.5, which is adapted from Fish and Wildlife Service (1984).

Although table 12.5 presents the estimated effects, the next step is to use this information to evaluate how the alternatives do in meeting the objectives of CMR and resolve the planning issues. Table 12.6 presents this information from the USFWS table 1 (Fish and Wildlife Service 1984:27). Three points should be mentioned about the "evaluation approach" reflected in table 12.6. First, the basic format is sound: explicitly rating the alternatives against the objectives to determine the dominant alternative. The second point relates to the less than rigorous use of this format: it is not clear how the USFWS made such determinations of whether an alternative met the objective. No numerical thresholds are presented that would allow the reader to understand on what basis one alternative met the objective and another did not. Third, meeting an objective with as many dimensions as wildlife, range, or recreation (each of which had several goals) is rarely a yes-or-no decision. As the data in table 12.5 illustrate, each alternative contributes to a different degree or magnitude to each objective. For example, the multiple-use alternative with over 60,000 AUMs allocated to livestock better meets two of the three range goals than the no-grazing alternative with zero AUMs.

Step 5: Plan Selection

Much like both BLM and the Forest Service, the proposed action was developed in step 3, formulation of the alternatives. As is evident from table 12.6, the USFWS believes only the preferred alternative meets all the objectives. The basis on which this determination is made will be examined later.

As part of step 5, the Draft EIS was sent out for a sixty-day public comment period. In addition, three days of public testimony were held in Montana.

Table 12.6
Evaluation of Alternatives in Terms of CMR Objectives

Alternative	Objectives to be Met		
	Wildlife	Range	Recreation
No action	No	No	No
Proposed action	Yes	Yes	Yes
Intensive wildlife mgmt.	Yes	No	Yes
Multiple use	No	No	Yes
No grazing	Yes	No	Yes

Step 6: Select Final Refuge Management Plan

The USFWS received slightly more than fourteen hundred responses, although of these only four hundred were letters, with about one thousand identical postcards making up the other responses (Fish and Wildlife Service 1985:138). Eleven comments were received from federal and state agencies as well as legislators.

The USFWS made only minor changes to the Draft Refuge Management Plan as a result of public comments. The distinction between habitat condition and livestock range conditions was emphasized more, as was monitoring of management actions.

The Record of Decision (ROD) continued to choose the original proposed action (Alternative B) as the Final Refuge Management Plan. This alternative was selected as it met the key laws and court rulings regarding management of the refuge. This alternative is consistent with wildlife being a priority over livestock on refuge lands.

Unfortunately, the brief rationale for selecting Alternative B is so general that the statements made could apply almost as well to the intensive wildlife management alternative or the no-grazing alternative. Both of these give priority to wildlife. The intensive management meets all applicable laws and affords a substantial amount of livestock grazing as well.

Step 7: Final Plan Implementation

Both the Final EIS and the ROD sketch out the implementation of the Final Plan. Both stress that site-specific Habitat Management Plans will be completed for each grazing allotment by 1989. On average, the Final Refuge Plan results in a 33% reduction in livestock AUMs by 1990. However, an allotment-by-allotment reduction in livestock grazing AUMs is presented. A monitoring program will be implemented to measure effectiveness of attaining refuge goals in the Final Plan.

Conversations with the USFWS during summer of 1990 (Shrank 1990) indicated that the grazing reductions proposed in the plan have in fact been implemented on the ground. About two-thirds of all the Habitat Management Plans have been written, with about half of the remaining ones in draft form. The more-capital-intensive management actions, such as plantings and fencing, have not taken place at the rate anticipated in the plan. In general, this is due to lack of funding from USFWS to the Charles M. Russell Refuge. Some of this lack of funding is due to severe cutbacks in the funding USFWS received during the Reagan administration and due in part to USFWS prioritization of the limited funding it was receiving to only the most critical and highest-priority wildlife projects. One of those, which will be discussed later, is habitat management and acquisition for implementing the North American Waterfowl Management Plan. Efforts for the 1991 budget cycle to fund capital improvement projects on the CMR involved conservation organizations offering to provide matching funding if Congress would provide equivalent funds.

Criticism of CMR Refuge Planning and Resource Allocation Process

As mentioned earlier, the major criticism of the CMR's process relates to how the conclusions were reached in table 12.6. It is difficult to tie the conclusions in this table to data in the EIS. On what basis does USFWS conclude that multiple use does not meet the range objective when there is actually a gain in livestock forage in the year 2005 with this alternative? In table 12.6, the USFWS concludes that the proposed action meets the range objective; yet there is nearly a 20% reduction in livestock forage.

The most undocumented statement in the EIS may be the most critical statement and appears to provide the rationale for the optimality of the preferred alternative. "On a dollar for dollar basis, the Proposed Action provides a greater benefit to wildlife than Intensive Management" (USFWS 1984:25—evaluation of alternatives). If benefits are interpreted as local jobs (which, as discussed in chapter 6, was shown to be incorrect), then this statement is inconsistent with the estimated effects in table 12.5, as the multiple-use alternative has the greatest number of jobs. If USFWS means that the proposed action is more cost effective (where the costs of intensive management may increase more than proportional to the response of wildlife, indicating strong diminishing returns), then this statement is plausible (but again, not obvious from the EIS data in table 12.5). If cost-benefit comparisons are to be the primary criteria for evaluating alternatives, then they should be developed and presented in the EIS. However, benefits of wildlife recreation were not quantified and included in economic effects to show what the balancing economic efficiency gains would be relative to losses in rancher income. (The latter is an economic efficiency loss.) With near doubling of wildlife habitat quality for several game species, economic value of hunting would very likely increase substantially. Unfortunately, the economic value of this improved hunting is not developed in the EIS.

As the two case studies on elk hunting and deer hunting at the end of chapters 9 and 10 illustrate, changes in habitat quality can be translated into changes in economic values. The same is true of the other recreational outputs in table 12.5. Next, the costing of the management prescriptions in table 12.4 could be performed. Combining this type of information would allow USFWS to carry out, in its own words, "a dollar-for-dollar" comparison of the alternatives.

Such a comparison need not always involve detailed original studies. For the outputs listed in table 12.5, the analysts could rely on existing estimates of the economic value of recreation found in the literature (Sorg and Loomis 1984) or use the Forest Service standardized values per recreation visitor day found in the Resources Planning Act document. Cost information can be developed from past projects.

Development of this benefit and cost information aids the managers in determining which alternative was most beneficial. More important, comparing the costs of the management prescriptions with the benefits generated would give valuable feedback on what combinations of such prescriptions would maximize net benefits of refuge management. It would also lend economic justification to requests for the millions of dollars that some of the management prescriptions require.

North American Waterfowl Management Plan

The Migratory Bird Treaty Act of 1913 sent the predecessor to the USFWS and its Canadian counterpart, the Canadian Wildlife Service, into coordinated efforts to reverse the decline of waterfowl. In 1936, Mexico became a party to this treaty as well. An international effort is needed because many of these bird species breed in Canada but winter in the United States, particularly in the Central Valley of California, on the Mississippi Delta, and along the Eastern Seaboard.

In 1986, a significant step forward toward active international cooperative management occurred with the signing of the North American Waterfowl Management Plan. This plan set population goals for thirty-seven species of ducks, geese, and swans. Priority habitats and strategies to protect them were identified. If these population goals are not met in the future, reduction in sport harvest is also proposed. These population goals will be reviewed every five years, and the entire North American Waterfowl Management Plan (NAWMP) will be revised every fifteen years.

A review of the NAWMP provides an opportunity to see how national- and international-level planning in USFWS is translated into management actions at each organization level within USFWS.

The overall goals, guidance, and direction in the NAWMP are translated into operational programs at the flyway level. There are four main flyways: Pacific, Central, Mississippi, and Atlantic. The flyway councils set hunting regulations, including bag limits, season lengths, and so on. They work with their respective national and state governments to protect, through acquisition or easements, the habitats needed to meet the population goals.

The actual land acquisition projects or easement proposals, including funding, often come from ''joint ventures'' of local waterfowl associations, conservation groups, fish and game agencies, and the USFWS. Joint-venture projects are cooperative efforts to pool funds from the member agencies or groups, to identify particular parcels available and needed as habitat, and generally to do much of the legal and financial work associated with buying these parcels. In other cases, the joint venture works to provide financial incentives to farmers to modify their farming practices to benefit waterfowl. Finally, these agencies within the joint venture develop the management plans for these acquired areas in cooperation with the agency that will have the lead responsibility. If the land will ultimately be added to a state's wildlife areas, then the state Fish and Game Department would take the lead rather than USFWS. The opposite would be the case if the area was to be added to the National Wildlife Refuge System.

At present there are operating joint ventures for the Central Valley of California (Pacific Flyway), the Prairie Habitat Joint Venture (Central Flyway), the Gulf Coast Joint Venture (Mississippi Flyway), the Lower Mississippi Valley Joint Venture, the Atlantic Coast Joint Venture, and the Eastern Habitat Joint Venture in Canada. The Gulf Coast Joint Venture has acquired 9,600 acres to become the 453rd National Wildlife Refuge (located in Louisiana) and has obtained a rent-free lease from Amoco Oil Company to provide 7,400 acres of company land as a sanctuary for wintering waterfowl (Fish and Wildlife Service 1989:8).

As is evident from the preceding discussion, implementation of the North American Waterfowl Management Plan involves a three-tier planning approach. The top tier provides the international legal authority and broad goals. The flyway councils translate these into flyway-specific regulations and habitat management objectives. Finally, the joint ventures use the legal authority established at the top level and the specific habitat acquisition objectives from the flyway councils to acquire and manage specific parcels of habitat in their respective flyways. This three- tier planning approach seems to be an effective tool for managing species with ranges of 1,000–3,000 miles, crossing international and state boundaries. The quantitative goals, backed by breeding surveys and monitoring, provide a link between the planning system and what is occurring on the ground. The interagency and private and public partnerships ensure that goals are attained in a cooperative manner. The strengths of each member of the joint venture allow these groups to do far more than what each one could do independently. In addition, resources are pooled for priority acquisitions that make sense as a system, rather than for the protection of isolated but unintegrated habitats.

Both Canada and the United States have committed multiyear funding to implement the plan, with much of the money being used on a matching-grant basis with local funding sources. The cycle of implementation, monitoring, and revision of the NAWMP will likely continue over the next fifteen to thirty years.

Mitigation Planning Techniques Used by USFWS and Other Federal Agencies

Under the Fish and Wildlife Coordination Act, the USFWS recommends how projects might minimize their effect on wildlife and then how to compensate for the residual losses. NEPA continues this emphasis of requiring that residual losses be mitigated in some way. More recently, the USFWS has adopted a mitigation policy that outlines what types of compensation are acceptable to the agency. For critical species and habitats, only ''in kind'' is acceptable. That is, only the same type of habitat can serve as compensation for scarce habitats or habitats of critical species. Thus for permitting the filling in of wetlands in a development project, the agency or company must provide equivalent wetlands habitat in the same geographic area by creating them or by protecting existing ones threatened by other development. Other types of more abundant habitat can be compensated using ''out-of-kind'' habitats. Thus loss of 10 acres of prairie grasslands habitat could be compensated by equivalent habitat values for riparian, upland, or other habitats. In many cases, no new land may need to be acquired, if the habitat value on existing land can be increased by management for the species that lost habitat because of the project. Thus limiting grazing or planting vegetation might double the remaining habitat's carrying capacity for ground-nesting birds.

For each type of habitat, a wide range of ways exists to achieve an acceptable mitigation goal. Land or wildlife habitat easements may be purchased, a variety of intensive wildlife practices can be implemented, or entire habitats can be ''created.'' Each strategy has a different cost and return of habitat carrying capacity. Where land is quite expensive, intense management on a small parcel might be a more cost-effective way to compensate for habitat losses than outright purchase and protection of existing habitat with low existing carrying capacity. In large projects, such as major manufacturing plants, subdivisions, dams, military bases, and so on, many different habitats supporting dozens of species may be affected. Determining the most cost-effective mix of compensation practices is important, since millions of dollars could easily be spent on compensation in these large projects.

Optimization Approach

To provide a tool for identifying the cost-effective compensation strategies the USFWS has developed two techniques. The most elaborate relies on the same basic linear programming, or LP, approach discussed in chapter 9 for the Forest Service. In this case, the LP model is programmed to minimize the compensation costs subject to meeting the compensation targets for each habitat type or species. This type of model has been used to develop the mitigation plan for the Garrison Diversion Unit Dam in North Dakota. (For more details see Matulich et al. 1982.)

Cost-Effectiveness Approach

A less complicated tool developed by the USFWS for field biologists to determine cost-effective compensation plans is called the Habitat Management Evaluation Model, or HMEM. This microcomputer model links habitat models for user-selected wildlife species together with habitat management practices, including the costs of those practices. The model essentially compares the increase in habitat carrying capacity from each practice to the cost of the practice. This information is displayed so that the biologist can "optimize" the package of habitat practices by an informed "trial-and-error" process—specifically, by comparing the gain in carrying capacity per dollar from one management practice to the next.

HMEM relies on the USFWS Habitat Evaluation Procedures to quantify the change in carrying capacity or habitat value to a species. The Habitat Evaluation Procedures (HEP) is a habitat accounting system that is combined with a series of Habitat Suitability Index (HSI) models like those discussed in chapter 5.

These approaches provide a useful technique for performing wildlife habitat impact analyses and planning improvements. Although this tool was initially set up for compensation and mitigation, it is also quite valuable for planning wildlife habitat improvements made for enhancement purposes. The linkage of the habitat model to the practices allows the biologists to determine which management practice is most beneficial to wildlife in that particular area. (For more information on this HMEM model see Andrews, Sousa, and Farmer 1984 or Sousa 1987.)

Determining Minimum In-stream Flow

Finally, the USFWS has developed a set of software packages to guide the determination of minimum and optimum in-stream flows on rivers throughout the United States. This widely used software package provides a systematic methodology for evaluating the consequences of different flow regimes on fish populations. Once again, a measure of fish habitat carrying capacity is developed from these models. These models are widely used by all federal agencies and a variety of other state agencies involved in water resource issues. (See Bovee 1982, and Milhous et al. 1984 for documentation of these models.)

Comments on USFWS Planning

Historically, land and resource management planning has not been a major concern of the USFWS. Most of its employees are, not surprisingly, biologists who have little training as planners. Thus planning methods and resource allocation techniques are not as well developed as in such agencies as the Forest Service. In some respects this is appropriate. The range of resource trade-offs in a wildlife refuge is much less broad than that on pure multiple-use lands. Nonetheless, wildlife planning is growing as a legitimate activity within federal and state wildlife agencies. An Organization of Wildlife Planners has become very active in recent years. Training workshops in planning are being held. Crowe's book *Comprehensive Planning for Wildlife Resources* is into its third printing. With Crowe's appointment by the director of USFWS, one would expect to see improvements in the quality of planning in USFWS in the 1990s.

External Threats to National Wildlife Refuges

As illustrated by the North American Waterfowl Management Plan, many fish and wildlife species require large habitats that encompass many landowners and that may even cross international boundaries. Successful wildlife management requires that all habitat lands be managed in a coordinated fashion; otherwise the weakest link can undo the best management practices on the other habitats.

A common external threat to NWR located in the western United States stems from irrigated agriculture. Water diversion and groundwater pumping for irrigated agriculture have often resulted in reducing the natural flows of water to many refuges. Irrigated agriculture often poses an additional threat from the discharge of agricultural drainage water containing toxic trace elements from pesticides and fertilizers. Such contamination made national news in 1984 with the discovery at Kesterson National Wildlife Refuge that agricultural drainage water flowing into the refuge contained many toxic trace elements, such as selenium. This one element alone was responsible for hundreds of bird deformities. This discovery initiated an evaluation of other National Wildlife Refuges that received contaminated drainage water; similar, although less severe, effects were found.

The problem of too little water and of receiving primarily contaminated drainage water as the main water supply at these refuges has brought about more integrated water resources planning in these areas. The U.S. Bureau of Reclamation and USFWS have completed two interagency studies to deal with these problems (Bureau of Reclamation 1987, Department of Interior 1990). Water resources planning for traditional agriculture and municipal uses now explicitly recognizes the needs of fish and wildlife. These considerations attempt to go beyond minimum in-stream flow determinations to provide something closer to optimum flows for wildlife. This requires balancing the benefits of competing uses of additional water to agriculture, cities, and wildlife. The allocation of water was given a big boost with the passage of the Central Valley Project Improvement Act of 1992. This act requires that nearly 1 million acre feet of water be dedicated to fish and wildlife purposes in California. Of this amount, about 400,000 acre feet are allocated to National Wildlife Refuges in the Central Valley.

13

National Park Service

As discussed in detail in chapter 2, the National Park Service (NPS) manages a dozen or so different types of lands in the National Park System. These range from the "crown jewel" National Parks (e.g., Everglades, Glacier, Grand Canyon, Great Smoky Mountains, Mt. Rainier, Shenandoah, Yellowstone, Yosemite, and so on) to National Historic Sites, Battlefields, National Seashores, National Parkways, and several National Recreation Areas. This chapter emphasizes the management of the National Parks and Monuments units and mentions differences in management of the other units or categories of lands. As will be recalled from chapter 2, each of these many land designations has a somewhat different emphasis.

Mission of the National Park Service and National Parks

In its founding legislation in 1916, Congress spelled out the mission of National Parks. Their "purpose is to conserve the scenery and the natural and historic objects, the wildlife therein and to provide for the enjoyment of the same in such a manner and by such means as will leave them unimpaired for the enjoyment of future generations" (Everhart 1983:185).

This original mandate was refined in 1918, when the secretary of interior gave the first director of the NPS three general guidelines for park management: "First, that the national parks should be maintained in absolutely unimpaired form for the use of future generations as well as those of our own time; second, that they are set apart for the use, observation, health and pleasure of the people; and third, that the national interest must dictate all decisions affecting public or private enterprise in the parks" (Dana and Fairfax 1980:109). The recently revised NPS *Management Policies* manual repeats this quote and states that these directives remain as valid today as the foundation for the management policies and planning of the NPS as when they were first written (NPS 1988:5).

The differences between National Parks, Monuments, Recreation Areas, Seashores, Historic Sites, and so on, and the multiple-use federal lands are quite clear. First and foremost, non-multiple-use lands are to be preserved in a natural and unmodified state. Thus Congress generally precluded many traditional multiple uses, such as commercial timber harvesting, mining, and livestock grazing, on park lands. In contrast to BLM's emphasis on consistency with local plans, NPS management concentrates on national interest. Nonetheless, the management of National Parks is not without its own brand of trade-off analyses. Inherent in both the 1916 act and the directions to the first director of the NPS is the potential conflict between preserving the natural environment of the park and providing for visitor use. Some authors have referred to these twin missions of the NPS as a dilemma (Foresta 1984:100) or a fatal flaw (Everhart 1983:46), although the latter term probably goes too far.

One other key difference between the National Park units and National Forests and BLM lands relates to specificity in mandate for each individual unit. The enabling legislation of each National Park unit often spells out the specific allowable uses. This contrasts with the two multiple-use agencies, which receive general guidance in their organic acts (NFMA and FLPMA) for managing all the lands.

Before discussing how the NPS deals with this twin mandate, a brief note about its structure is in order. The NPS is headed by a director and has ten Regional Offices (each with a regional director) throughout the United States and its territories. Nonetheless, a sizable amount of control over the management of each unit is left to each park's superintendent. Although his or her long-term strategy is guided by the NPS national policy and the park's specific enabling legislation, Statement for Management, and any Master or General Management Plan, many everyday decisions about implementing and enforcing the provisions of the plan are ultimately made by the park superintendent. The NPS is in some ways more effectively decentralized than BLM or USFS.

Preserve Versus Use Dilemma: The NPS Balancing Act

The NPS recognizes that automobiles, campgrounds, and even hikers alter the naturalness of a park. But it strives to accommodate compatible public uses without permanently impairing the park's environment (NPS 1988:5). Some impacts (e.g., trails, campgrounds, and in some cases even buildings or roads) are acceptable because they can (in principle, although not frequently in practice) be reversed and the environment can be restored. Hence they are often classified as temporary if a long enough time horizon is taken. Therefore permanent impairment has not occurred and the NPS is consistent with its charge to leave the resource "unimpaired for future generations." But if the alterations (e.g., the roads, buildings, or campgrounds) would result in damage to unique features of the park that cannot be restored, this is not allowable.

Within this broad range of acceptable, temporary impacts and providing for visitor use and enjoyment, much room is left for interpretation and discretion. Much like the multiple-use mandate, the "preserve but provide for human uses" is a management philosophy that is often interpreted in light of both the scientific knowledge and public opinion of the times.

The early days of the NPS saw the preserve-but-use mandate as providing "modern" access to some of the most scenic features of the parks and providing "comfortable accommodations" to park visitors. This early view resulted in some of the grand hotels and lodges in Yellowstone, Glacier, Yosemite, and Grand Canyon National Parks. To a limited degree, early directors of the National Parks (interested in public support for funding existing parks and expanding the Park System) sought to encourage visitors to the parks and accommodate them on their terms (Forestra 1980). Concessionaires were allowed to provide a wide range of services, such as horseback riding, restaurants, gift shops, ski areas, swimming pools, and even golf courses. Thus the park would be modified to the visitors' terms. Of more importance for current management of National Parks is that once many of these uses became established in a park, it was difficult to remove them when scientific knowledge and public opinion suggested they were not appropriate with maintaining the park's natural environment.

This shift toward a preservation philosophy occurred in the early 1960s with the Leopold Report, which emphasized ecological management of National Parks (Sellars 1989:6), and selection of Stewart Udall as secretary of the interior. This trend continued (with a brief interruption in the early 1980s) into the 1990s. The balance shifted toward requiring the public to visit the park on the park's terms. The visitor must accommodate the park's environment and not vice versa. Thus camping was to be emphasized over hotels, and concessionaire services were to be reviewed in terms of their role in fulfilling the mission of the park, not in simply meeting the public's desire for a wide range of consumer services in a park setting. Although the NPS has had limited success in removing existing development, there have been notable gains in returning the parks and monuments to more natural management. It has eliminated ecologically incompatible practices such as "fire-falls" in Yosemite and allowing bears to feed at uncovered campground dumps so that visitors could view them easily. It has also been able to direct new visitor developments to areas outside the parks, improve blending of park facilities into the park environment, and ration visitor use levels to those more compatible with the park environment's carrying capacity.

As will be evident from our review of current NPS management practices, the trend of the last thirty years is toward more preservation, in the "preserve-but-use" balancing act. But our increasing knowledge of parks as ecosystems has recently resulted in reexamining what to preserve the scenery or the ecology of the parks. For example, the NPS has designated several areas of particular National Parks as "prescribed natural fire" areas. The fire policy allows a natural force, fire, that operated for centuries before to create and maintain the park environment, to continue. Many types of native vegetation thrive after a fire. However, as evidenced by the public response to large-scale fires in Yellowstone during the summer of 1988, many view the fires as "destroying" the park. Of course, what the fires affected was the scenery during that year and for a few years later.

What to Preserve: Scenery or Ecology?

The view that the purpose of the National Parks and Monuments is to preserve the scenery dates back to the 1916 mandate of the NPS, where scenery is explicitly mentioned as a component to be preserved. Coupled with this is the view that parks are protected for the pleasure of people. Attractive scenes are pleasing to people; burned areas or insect-infested trees are not. It was the photographs and paintings of the magnificent scenes that provided support for establishment of some parks and gave most people their first glimpses of them. The pictures of Ansel Adams froze in many people's minds the idea of what a particular park should look like when they arrived. And why not? It was an area preserved and protected from mining and timber harvesting, so many thought it should always look the same for them and their grandchildren.

This view resulted in what Sellars (1989) calls "façade management." Façade management stresses maintaining the most visible features of the park: grand vistas and large mammals. Management would concentrate on resources that were obvious to the public; other resources would matter only insofar as they affect scenery. Thus the concern over air pollution drifting into the park would focus on scenic visibility rather than on ecological effects. Preservation of the surface view would be emphasized even if natural forces had to be restrained. This implied that fires had to be suppressed, outbreaks of insects that kill trees had to be controlled, and visitor facilities had to be located to provide the best view. In this way, the park would be preserved and public use and enjoyment accommodated.

The problem is that the park environment is a dynamic ecosystem that is constantly changing in the face of such events as fires, outbreaks of insects, wildlife population cycles, and so forth. Trying to maintain the park as a static snapshot is very difficult in the face of such dynamic forces. But the early emphasis on scenery management was in part due to the infancy of ecological sciences when the National Parks were established. Therefore the "unnaturalness" of façade management was not apparent.

A different kind of ecological bias is introduced in the park environment if managers are successful in controlling some natural forces, such as fires, but not others, such as flash floods. In some respects, the park environment was created by these natural forces. In Grand Canyon National Park, the second-largest rapid on the Colorado River through the park was created in 1966, when a flash flood (caused by an intense thunderstorm) washed numerous large boulders into the river. Should the Park Service go in and remove the large boulders as they might to suppress a lightning-caused fire from the same thunderstorm? Managers' ability to counteract different natural forces varies, thus introducing an almost evolutionary selection process into the park's ecosystem. Those effects that managers can offset will be less frequent forces in the park's environments and vegetation, and wildlife associated with them will be reduced relative to the species favored by the manager's intervention. The sometimes disastrous consequences of the suppression efforts themselves compound this. Fire fighting, with its heavy machinery and chemicals, may do more lasting damage to the natural environment than the fire. Pesticides to control tree-killing insects may harm far more than the targeted pests. In the end, façade management, if carried too far, can border on impairing the park environment, thereby depriving the visitor of the opportunity to witness many of the ecological forces that helped create and sustain the park.

The ecological view of the NPS mandate to preserve the park aims at preserving ecological processes. These forces include the microorganisms in the soil, the insects, the naturally occurring fires, and all the native plant and animal species.

With the ecological view, public use would focus on the park as an educational tool to understand the forces of nature and their self-regulating mechanisms. Park exhibits would stress the interrelatedness of resources and how these continually shape the park. Visitor facilities and accommodations would be tailored so as not to interfere with critical ecological processes, rather than emphasizing scenic views.

After a fifty-year tradition of emphasizing visitor convenience, the Park Service has moved ahead in implementing a more ecological view of managing all units of the National Park System. Some of the changes have been dramatic, such as the elimination of garbage feeding of bears and management for a natural bear population in backcountry areas of parks. Other changes, such as the prescribed natural fire policy, have been less obvious until an event like the Yellowstone fires causes the public to become aware of them. Some policy changes, because of the legacies of the past, are resisted. As our case study of Yosemite National Park will make clear, unseating already established uses may meet with considerable resistance by those who feel that a certain (even though nonnative) wildlife species or facility is a "natural" part of the park. The heated debates over attempts to remove visitor facilities at such parks as Zion and nonnative wildlife (e.g., burros at Grand Canyon National Park, mountain goats at Olympic National Park, and exotic deer at Point Reyes National Seashore) point to continued resistance by some members of the public to this ecological view of parks. Nonetheless, as the next discussion of current policies of park management will attest, ecological management is being increasingly practiced in all units of the National Park System in the United States and, to the extent that the United States is a model for national parks in other nations, around the world.

Current Visitor Use, Facilities, and Access Policies in Park Management

The policies regarding allowable uses and acceptable modifications of park or monument environments are spelled out in the NPS *Management Policies* manual. It is important to establish these National Park System–wide policies for a number of reasons. First, they provide the menu of available management practices that might be chosen in park planning (discussed later). Alternative management practices that are inconsistent with general NPS policies cannot be included (unless they are specifically authorized in the park's or monument's founding legislation). In addition, these general policies provide an interim operating guide for a park, while the General Management Plan is being prepared. This section discusses accepted management practices for a variety of issues related to preservation and visitor use.

Types of Visitor Uses to Be Accommodated

NPS management policy encourages "nonconsumptive" recreation, which involves viewing, learning, hiking, camping, picnicking, cross-country skiing, mountain climbing, and swimming. The common thread in these activities is that they do not take anything away from the park environment; instead they involve a "direct association" with park resources. They involve appreciation of or direct contact with the natural resources.

A use with a special place within the National Parks is that of interpreting the park environment to the visitor. The guidelines for developing interpretive programs emphasize that interpretation should provide

1. Information and orientation about the specific opportunities available in that park.
2. Understanding of the special features and values of the park.
3. Opportunities for visitors to interact safely with park resources without damaging them (e.g., providing special areas to touch, feel, and so on, where that use will not damage the resource) and guidance on how to visit the park without damaging it (e.g., low-impact camping techniques, why it is important to stay on the trail, and so on).
4. Opportunities for visitors to develop nature-related skills, such as plant and wildlife identification.
6. Avenues for communication between the park managers and the general public.
7. Education on the National Park System (NPS 1988).

Consumptive uses, such as fishing and berry picking, are allowed only when they have been traditional, authorized by prior treaties, or authorized by federal law for that park. Many parks have attempted to convert potentially consumptive activities into relatively nonconsumptive ones by using regulations such as catch-and-release fishing (set in cooperation with local fish and game laws). Hunting or trapping are allowed only where authorized by federal law and when consistent with "sound resource management principles." In addition, hunting and trapping are allowed only when they do not compromise public safety. As might be expected, hunting and trapping are rare in most National Parks outside of Alaska.

Other recreation activities, such as use of off-road vehicles (ORVs), snowmobiles, bicycles, and aircraft and hang gliding, "require that special regulations be developed before these uses may be authorized in parks" (NPS 1988:chapter 8:6). If such uses are permitted by the unit's enabling legislation, determinations of exactly when and where they might be allowed are usually made through the park's planning process.

Recreation management is addressed as part of the overall Master Plan or General Management Plan for the park. The primary principles of recreation management in the parks and monuments involve four elements: (1) provide for public enjoyment; (2) ensure public safety; (3) protect park resources; (4) minimize conflicts with other visitor activities and park uses (NPS 1988:chapter 8:6). Once a General Management Plan has determined whether an area is suitable for certain types of recreation, an activity-specific recreation plan may also be developed, such as a river use plan. For example, although there is a Master Plan for the entire Grand Canyon National Park, a separate Colorado River Management Plan for regulating river rafting also exists.

Regardless of the type of recreational activity, the NPS will "regulate the amount and kind, and the time and place of visitor activities" so as "to prevent derogation of the values and purposes for which that park was established. Where practicable, such restrictions will be based on the results of study or research. Any restrictions imposed will be fully explained to the public" (NPS 1988:chapter 8:2). The management tools for controlling and directing visitor activities include public use limits and associated permit or reservation systems, closures of particular areas, and regulation of the conditions of recreational use (e.g., group size, location of backcountry camps relative to streams and lakes, and so on). In backcountry areas, public use levels will be limited in accordance with the natural system's ability to absorb human waste (NPS 1988:chapter 8:8).

This latter rationale for limiting use is based on what might be called an area's ecological recreational carrying capacity. The ecological recreation carrying capacity of an area may be determined by soils, vegetation's tolerance to trampling, availability of suitable camping areas, concerns about water quality, or sensitivity of wildlife to human presence. For example, parks with extremely short growing seasons (e.g., high-elevation or northern parks) or little precipitation often may restrict the location of visitor activity or limit visitation to protect plants.

However, the National Park Service may also employ a social carrying capacity approach to limit the number of visitors to an area, so that visitors may enjoy the solitude or wilderness experience associated with the sights and sounds of nature. In Grand Canyon National Park, social carrying capacity to provide for minimal contacts between rafting parties is a major component of NPS's quota system. The principles of determining social carrying capacity and their application to Yosemite National Park will be provided later in this chapter.

A recreation activity will not be allowed if it would result in any of the following: (a) inconsistency with park enabling legislation or deterioration of park values; (b) undue interference or conflict with other visitor uses; (c) consumptive uses of park resources beyond those authorized by legislation; (d) unacceptable impacts on park resources; (e) danger to the safety of participants or the public (NPS 1988:chapter 8:7).

Implementation of many of the criteria for compatible recreation activities in the park will require that these concerns be translated into observable and preferably quantitative indicators. Otherwise, determining what is "undue interference or conflict with other visitor uses" or "unacceptable impacts on park resources" may depend on the value judgments of managers.

Visitor Facilities

Visitor facilities include visitor centers and other interpretive facilities; recreation use areas, such as picnic areas; and, where appropriate, boating facilities, campgrounds, hotels, and restrooms.

Visitor centers may provide a variety of services, including a central focus for visitor contact, park information, museums, and other interpretive efforts that cannot be performed outdoors. Other interpretive facilities include roadside and trailside exhibits. In large campgrounds, amphitheaters may be constructed to facilitate evening interpretive programs. However, the building of permanent facilities for such events as concerts, plays, and art exhibits (even if related to the park's purpose) is restricted and discouraged. In general, these events may be held in the park only when they are directly related to the park's purpose, as in the case of Wolf Trap Farm.

Formal recreation facilities will be provided where this type of visitor use has been deemed appropriate (often through the planning process). Such facilities include boat ramps, marinas, breakwaters, and so on, to allow for safe visitor use of recreational waters in the park. Existing downhill ski areas in several parks will be allowed to operate unless closure or expansion is approved through the planning process. No new downhill ski areas will be allowed in any unit of the National Park System (NPS 1988:chapter 4:15). This would rule out any plan alternative containing the addition of a downhill ski area.

Overnight accommodations in a park range from "unimproved" backcountry campsites to developed campgrounds, hostels, and restaurant-hotels. The NPS management policy stresses that overnight accommodations and food services should be located outside a park in most cases. When they must be located within a park for people to have a meaningful visit, such visitor facilities "will be restricted to the kinds and levels necessary to achieve each park's purposes" (NPS 1988:chapter 4:13). The NPS states that commercial park concessions will be authorized only when such facilities or services are necessary to public use, and they will enhance the enjoyment of the park without substantial impairment of park resources or values. Finally, such facilities must be located where the least impact on park resources and values will occur (NPS 1988:chapter 9:1).

The NPS manual provides guidance on design and location of developed campgrounds as well. Both tent camping and recreational vehicle camping may be provided. However, campgrounds for recreational vehicles may be located only where existing road access can safely accommodate such vehicles. Campgrounds designed or designated for recreational vehicles need not provide all possible utility hookups. (For example, they need not provide electricity.)

Campgrounds should generally not exceed 250 individual sites (unless approved by the director of NPS). The camping areas should separate recreational vehicles, tent camping, and group camps, when desirable.

Visitor Access

NPS management policy is to provide "reasonable access to parks and ensure that the means of circulation within them are adequate to permit public enjoyment of the park resources" (NPS 1988:chapter 4:8). Visitor access may include a mixture of modes, including automobiles, buses, bicycles, foot trails, horses, and boats, depending on the size and purpose of the park. The park's planning process determines which modes are acceptable and which of the acceptable modes are appropriate for different areas of the park. Among the criteria to be used in making these planning decisions is the selection of modes that protect park resources and are cost effective. Nonmotorized travel will be emphasized (NPS 1988:chapter 4:8).

Roads in and through the park receive particular management attention. The intent is to provide access and maintain the quality of the visit, not to provide the fastest, most direct route. The NPS feels that for most established parks and monuments, existing roads are adequate. As part of the park planning process, the existing road structure is evaluated to see which roads might be closed, reoriented, or supplemented by other means of transportation (i.e., parallel hiking or bicycling trails) or, when a clear need exists, expanded. The fact that a road is consistently used to its capacity does not itself justify its expansion. In such situations, public transit or visitor use limits will be considered as alternatives to expanding road capacity (NPS 1988:chapter 4:8). Only when a new road or road expansion is the best means to provide the necessary access and expansion can be accomplished without significant effects on natural resources of the park will such a road be considered (NPS 1988:chapter 4:8).

The NPS provides a variety of trails to meet different park purposes. Interpretive trails are designed to inform and educate. Walks are short, high-standard trails (sometimes graveled or even paved) that provide access to attraction sites such as waterfalls, overlooks, or other points of interest. Hiking trails are means for foot travel through the park and provide backcountry access. Some hiking trails may be multipurpose and allow travel by horseback, or horses can be restricted to separate trails to minimize user conflict and resource damage. That is, horse paths may have to be routed around more sensitive areas that can be traveled by hikers. Bicycles will not be permitted on hiking trails, but bicycle lanes and bike paths may be constructed in the park. These forms of nonmotorized visitor travel are encouraged to, in one park superintendent's words, "get the visitors out of their cars."

Support Facilities

The NPS *Management Policies* manual stresses integrating facilities into the park landscape and environment. This is important so that the facility will have the minimum ecological impact and not compete with or dominate park features. The integration is to complement cultural, historical, or ethnic values that may be present in the park or that were given particular emphasis in the legislation establishing it. Thus design, location, and arrangement of facilities could reflect these other values.

The park is to rely on existing water and power systems, preferably connecting to systems outside it, rather than constructing their own. This is sometimes difficult, as many parks are in remote areas and thus must develop their own water systems.

So far we have focused on the broad guidelines regarding the "use and visitor enjoyment" part of the NPS mandate. Next we turn to a discussion of the natural resource management aspects of National Parks.

Natural Resource Management Guidelines

The NPS defines natural resources very broadly when determining which resources will be managed. Included as natural resources are "plants, animals, water, air, soils, topographic features, geologic features, paleontologic resources, and aesthetic values such as scenic vistas, natural quiet and clear night skies" (NPS 1988:chapter 4:18). These resources encompass nearly the total environment and in many ways define the total visitor experience. To provide clear night skies for star viewing, outdoor night lighting must be as little as is required for visitor safety and it must be designed to reduce skyward glare.

Plants and Animals

Consistent with its mandate, the NPS goal is "to perpetuate native plant life as part of natural ecosystems" (NPS 1988:chapter 5:8). Manipulation of vegetation is to be kept to a minimum in the park. To the maximum extent possible, only native plants are to be used for revegetation. The primary exceptions to this general rule are for historic sites, battlefields, and so on, where the vegetation is often kept consistent with the purpose of the unit.

With respect to native animal life (which includes mammals, birds, amphibians, reptiles, fish, insects, worms, crustaceans, and so on), the NPS goal is to provide for self-regulating populations. Human impacts on natural animal population dynamics are to be avoided. Removal of individual animals will be allowed only when population growth of one species threatens other park resources, when necessary for research, or when public safety requires it. The preferred means of removing animals will be live trapping and transplanting elsewhere (either in other areas of the park or outside of it).

Unlike hunting and trapping, fishing has been a traditional activity at many parks and monuments. The NPS still emphasizes native fish species and restricts fishing where necessary to provide for self-sustaining native fish populations. Fishing regulations may require "flies and lures only" as well as "catch and release" to ensure that enough fish remain in the stream to reproduce and to serve as food for bears, eagles, and other wildlife. In some cases, lakes and reservoirs within the park are stocked to enhance recreational sport fishing. However, this practice is becoming less acceptable as ecological processes are increasingly emphasized.

Management of migratory wildlife species presents the NPS with a special challenge, since only a portion of the animals' habitat requirements are met in the park. Since the animals depend partly on habitat and regulation outside park boundaries, the NPS may participate in regional land use planning and regulations on adjacent lands. Thus NPS personnel can provide input on hunting regulations and land use plans of other federal agencies or of state and local governments. Since the mid-1980s it has become increasingly recognized that parks are small islands that fulfill only part of many animals' habitat requirements. This has led to a call for integrated planning that encompasses all land ownerships in a particular ecosystem. (This topic will be discussed in more detail in chapter 14.)

Water Resource Management

The NPS stresses having water of adequate quantity and quality to fulfill its mission of protecting park resources for present and future generations. The NPS itself or any concessionaire will minimize withdrawal of surface water and groundwater for consumptive uses (visitor water systems, restaurants, lawn watering, and so on), and the return flow of all withdrawn water will be treated so that it can be returned to the same watershed. Such water withdrawals will only be permitted when absolutely necessary and when studies show they will not significantly alter ecosystems in the park.

The NPS will maintain the quality of water originating within the park by controlling as many human impacts as possible. These include ensuring that sewage treatment and disposal do not lead to water pollution and controlling erosion associated with roads and trails. Where pack animals are permitted in a park, corrals and pastures will be located away from streams and lakes. Motorized boating and marinas, when permitted, will be strictly regulated to minimize discharge of petroleum products into the water.

Often the quality of water flowing into the park or adjacent to it is influenced by water that originates outside the park or by activities in nearby watersheds. In these cases, the NPS will participate in planning outside the parks and enter into agreements with other agencies or governments to avoid water pollution in the park (NPS 1988:chapter 5:18). (This issue of interrelated resource management to protect parks will be discussed more in chapter 14.)

Air Quality

Maintaining excellent air quality is important for providing scenic visibility and protecting the ecological processes in National Parks and Monuments. Parks with more than 6,000 acres are given special protection under the Federal Clean Air Act. They have a Class I designation for prevention of significant deterioration of air quality. The associated strict air quality standards have helped direct large coal-fired power plants proposed near National Parks in Utah to other areas.

These standards also provide some authority to attempt to reduce air pollution from sources that are damaging air quality at National Parks. In the late 1980s there was confirmation that the coal-fired Navajo power plant in Page, Arizona (partly owned by the federal government), was a significant cause of the air pollution over the Grand Canyon. This led to active NPS involvement in efforts to update pollution control equipment on this plant. The NPS currently monitors air quality at many of its parks and participates in air quality permitting decisions.

Cultural, Historic, and Paleontological Resource Management

Many units of the National Park System, particularly National Monuments and Historic Sites, were specifically included in the NPS to protect certain cultural, historic, or paleontological resources. The NPS has several procedures for managing these resources. First, they are inventoried to determine their location, type, and physical condition. Next, each is evaluated in terms of its scientific, cultural, or historic significance. Cultural and historic resources judged to be of national significance can be nominated for the National Register of Historic Places as a National Landmark or even as a World Heritage Site (if the resource is internationally significant).

If the resources are deteriorating so that there is a risk of losing the site's scientific and educational values, then physical stabilization or, in rare cases, excavation may be appropriate. Strong preference is given to maintaining the site in its original location. Of course, cultural, historic, and paleontological resources can be visited as long as this does not compromise the integrity of the site or any scientific values.

Many of these same procedures for managing cultural, historic, and paleontological resources are used by other federal agencies, such as the BLM or the Forest Service. This is partly due to very specific federal laws to manage these resources, such as the Antiquities Act of 1906 and the National Historic Preservation Act of 1966, which apply to all federal lands, regardless of the agency.

Wilderness Management

Many unroaded natural areas in National Parks and Monuments have been designated by Congress as Wilderness areas and thus have become part of the National Wilderness Preservation System. As part of its planning process, the NPS evaluates the suitability of existing roadless areas within a park, monument, or National Recreation Area as candidates for Wilderness. It makes these recommendations to Congress, which has the final authority for designating areas as Wilderness.

Designated Wilderness areas are still managed by NPS, but following the legal guidelines prescribed in the Wilderness Act of 1964 (which applies equally to all Wilderness areas on any federal lands). In general, these preclude motorized access, timber harvesting, and most mining. The objectives of the Wilderness Act include maintaining land in its natural ''primeval character'' and providing ''outstanding opportunities for solitude or a primitive and unconfined type of recreation.'' In some ways, these principles mesh well with the Park Service's own preserve-and-use mandate, but they differ in that public use must be nonmotorized, there must be no permanent improvements, and opportunities for solitude must be provided.

Primary visitor access in Wilderness Areas is by hiking, horseback (where allowed in the park), and nonmotorized boating. Mechanical transport, such as mountain bikes, is not allowed. Trail bridges will be provided only where significant safety hazards exist during the normal use period or for resource protection (NPS 1988:chapter 7:4). No permanent shelters will be built, as Wilderness users are to be self-sufficient in providing their shelter (NPS 1988:chapter 7:4). Motorized access by helicopters or planes is authorized only for emergency situations. Monitoring of visitor use levels and associated impacts will be implemented to ensure that impacts do not exceed the thresholds established in park planning for Wilderness areas.

Fire Management

The NPS has an active fire-prevention and fire-containment program for human-caused fires and fires that would destroy cultural resources or developed facilities or that would cause human injury or death. A park's overall approach to fire is often presented in its Master Plan or General Management Plan. Determination of specific areas where active fire suppression will be performed and the type of suppression efforts to use is made in a park's Fire Management Plan, which is developed after the General Management Plan.

The original NPS policy toward naturally caused fires from 1916 through 1968 was largely one of total fire suppression (in many cases to preserve the scenery). However, with growing emphasis on perpetuation of ecological processes rather than scenery, the NPS in 1968 modified its policy regarding such fires. Specifically, given certain meteorological conditions, the condition and type of vegetation, and location of the fire, natural fires would be allowed to burn. But this tool had to be used with consideration of potential effects on neighboring landowners, as the spread of fires is not perfectly predictable and fires have often simply gone out of control. Determination of prescribed fire areas is part of the fire management plan of a park.

The main argument for this modern fire policy is that lightning- or volcano-caused fires are a natural part of many parks' ecological processes, and certain features of a park depend on such fires. In some areas where past fire suppression has left large amounts of unburned material and the ecosystem would benefit from burning, purposefully set "prescribed burns" can be used to create the effects of natural fires. Therefore, either through allowing certain natural wildfires to burn or through prescribed burns, fire has become an ecological management tool of the NPS to enhance wildlife habitats, maintain historic scenes, and so on, starting in 1968 and continuing today.

The prescribed natural fire feature of NPS fire policy received great public scrutiny after the massive fires in Yellowstone National Park in the summer of 1988. Several fires throughout the park (including both human-caused fires and wildfires) burned continuously from late June through late October, when fall rain and snow extinguished them. At their peak, they had spread to nearby National Forests and endangered nearby towns. Large areas of the park were closed to visitors and smoke spread for hundreds of miles. To a public raised on Smokey the Bear's mottos of the destructive power of fire, it seemed that Yellowstone was being destroyed right before their eyes.

These fires and the associated national media attention brought about a review of the National Park Service (and in Wilderness Areas, of the Forest Service) prescribed fire policy. Ecologists argued that fires were beneficial in the long run to Yellowstone and many other parks and that many of the wildlife habitats would be substantially improved after them. Other ecologists indicated that such large-scale fires may have been common in Yellowstone's past.

Nonetheless, the Department of Interior asked for a reevaluation of the NPS fire policy. With the completion of the reevaluation of fire policy and revision of fire management plans, fire as a management tool has been retained. Prescribed natural fire options and burning are still allowed. However, much tighter constraints and control have been imposed on fire as a management tool to reduce the possibility of such large-scale fires as occurred in Yellowstone in 1988.

Managing Other Public Uses Such as Grazing and Mining

Mining

All National Parks and Monuments (and all but five National Recreation Areas) are closed to establishment of new mining claims. However, mineral development may be allowed in National Parks and Monuments when there is an existing valid mining claim, mineral lease, or nonfederally owned mineral right. In general, mining will be allowed in this case only when it will not permanently destroy resources, unless Congress specifically mandates that such mining operations be allowed. When permitted by Congress, the NPS will set requirements over the manner in which the exploration, mining, extraction, storage, and transportation of minerals will be performed in the park or monument.

Agriculture

Limited agricultural activities may be permitted if they are allowed in a unit's enabling legislation. These activities are usually associated with a demonstration farm used for interpretive purposes to illustrate activities associated with the historic period of the park or historic site. Employees living in a park may also cultivate small gardens for personal use. In neither case are the agricultural activities to deplete or pollute water supplies or adversely affect visitor enjoyment or result in deterioration of park resources.

Livestock Grazing

In general, livestock grazing is not permitted in National Parks or Monuments. Commercial grazing or stock drives through the park or monument will be allowed only in those parks where they were specifically authorized by federal law at the time the park was created or where they were a reserved right. Limited livestock grazing may be allowed if it contributes to the historic scene of the area. Where livestock grazing is authorized it will be kept to levels that will not disrupt "composition or condition of native plant and animal communities" (NPS 1988:chapter 8:18). Livestock grazing must be done in a manner that minimizes its effects on public use. Even when concessionaires are permitted to provide horseback riding or pack stock, these animals may only graze as they pass through an area, and the operators are encouraged to pack in weed-free feed for the animals (NPS 1988:chapter 8:19). Finally, if authorized livestock grazing or trailing of animals through the park "conflict with public enjoyment of the park or interfere with the functions of the natural ecosystem, the National Park Service will eliminate grazing, whenever possible, through orderly and cooperative procedures with the individuals and organizations concerned" (NPS 1988:chapter 8:20).

Internal Threats to National Park Facilities

Several reviews of the "state of the parks" have found many to be short of meeting NPS management policies in one or more areas. The common culprit in most of these reviews is lack of funding to repair, maintain, and upgrade facilities to meet the current and expanding levels of visitor use. According to a 1988 study by the General Accounting Office, many facilities have simply worn out or have been used beyond their expected engineering life. Others are being used at a far higher visitor use level than they were originally designed for. In the early 1980s, efforts began to remedy glaring deficiencies in such facilities as water systems, and they are continuing today. But a long-term capital maintenance and upgrading funding mechanism appears needed to prevent deterioration of roads, sewers, water systems, campgrounds, and other facilities. In 1988, the General Accounting Office reported that the parks needed an immediate $1.9 billion to avoid complete loss of some roads, trails, and buildings.

General Management Planning for National Parks

As can be seen from the preceding discussion, the mandate of the NPS and the management policies of the NPS set some specific limits on uses of National Parks and Monuments. At the same time, the management policies give a reasonable amount of latitude to managers to decide which uses will take place in a specific park and where they will be allowed. As highlighted several times earlier, these determinations are part of the park's planning process.

Planning for the management of NPS units has a long history in the NPS. The first plans in the 1920s and 1930s were oriented to incorporating visitor access and facilities into the park environment (deFranceaux 1987:13). The Park Service began what it called Master Planning in the 1950s as it embarked on its ambitious development of park units under the Mission 66 program (deFranceaux 1987:13). In 1978, planning for NPS units became a legal requirement, with the passage of the National Parks and Recreation Act of 1978. That act "directed the NPS to prepare and revise in a timely manner general management plans (GMP) for the preservation and use of each unit of the National Park System" (deFranceaux 1987:14). When a new unit is added by the NPS, the requirement to prepare a comprehensive plan for its management is normally part of the enabling legislation.

The NPS states that the "planning process is an important problem-solving tool that will often be used in day-to-day operations and management" (NPS 1988:chapter 2:1). In particular, park planning is used to evaluate major decisions, such as construction or rehabilitation of facilities and suitability of different park uses, and to determine rules for day-to-day management decisions that account for their cumulative impact.

In some ways, park planning is a continuum. At one end, nearly all NPS units (e.g., parks, monuments, seashores, battlefields, and so on) have developed a Statement for Management that is based on each unit's enabling legislation. As we shall see later, such a statement is the first step in the more comprehensive end of the spectrum, the General Management Plan. Given the limited availability of planning resources within NPS, this Statement of Management may serve as an interim plan. Some park units rely on older Master Plans, which are in the middle of the continuum. These are not as comprehensive or as detailed as the General Management Plans that complete the other end of the NPS planning spectrum. Nonetheless, the Master Plan often suffices until major pressures relating to visitor uses, concessionaires, or natural resources require a comprehensive reassessment of park management. It is to the development of these General Management Plans (GMPs) that we now turn.

The GMP provides the basic philosophy and strategies for resolving issues and attaining park objectives (deFranceaux 1987). Once the long-term direction for park management is determined in the GMP, the GMP is often supplemented by more detailed site-specific plans for wilderness management, river recreation programs, facilities development, or grazing management. Taken together, these documents spell out how the NPS will accomplish its mission of preservation with public use.

Steps in Preparing a General Management Plan

There are seven steps in the National Park Service's planning process:

1. Statement for management.
2. Outline of planning requirements.
3. Development of information base.
4. Formulation of alternatives.
5. Evaluation of alternatives.
6. Selection of final GMP.
7. Plan implementation.

These steps are similar in spirit to what is used by Forest Service and BLM. However, their application to National Park units results in a somewhat different planning process for several reasons. First, the mandate in the Park Service's Organic Act of 1916 is far more limited than the broad mandates given the Forest Service and BLM. Second, unlike the general mandates given to the Forest Service or BLM, Congress in many cases has given each National Park unit a very specific purpose within the NPS. Finally, most National Park units are much smaller than National Forests or BLM Resource Areas, which makes more detailed planning feasible. We now turn to the details of the Park Service's General Management Plan process.

Step 1: Statement for Management

The first step involves three phases. The first phase is to determine the objectives of park management. This is done by considering the legislation or congressional purposes for establishing the park or monument. These are supplemented by general

NPS mandates for units in the NPS and the nature and significance of the park or monument's resources. In essence, the park superintendent determines what the park is supposed to be and, given the natural and cultural resources, what it could offer the public.

The second phase draws closer to reality and describes the existing conditions in the park. In particular, what is the current condition of the natural resources in terms of degradation, pollution, or ability to function as an ecosystem? What is the condition of facilities in the park in terms of their ability to support current use levels in a safe manner?

The final phase identifies the gaps between what is (identified in the second phase) and what should or could be (identified in the first phase). The problems and issues resulting from the inability of the park to meet its legislative purpose and attain the general goals of the NPS are identified. A determination is made as to what issues must be resolved and what problems must be solved if the park is to reach its potential and meet its general mandate. Thus, much like planning processes of other federal agencies, the NPS identifies the problems and issues to be addressed in the General Management Plan.

The completion of this first step results in a document called the Statement for Management. Besides providing the focus for the GMP, it provides interim guidelines for managing the park unit while the GMP is being completed.

Step 2: Outline of Planning Requirements

Step 2 is basically a "plan for doing the plan." In essence, the remaining steps in the NPS planning process are laid out and the requirements for accomplishing each step are detailed. What information must the NPS know to solve as well as possible the problems raised in the Statement for Management completed in Step 1? Does this information exist? Must existing data be compiled or are new studies needed? What types and amounts of personnel are needed for the planning team? Since the planning team is to be interdisciplinary, where will the park or monument obtain such personnel? (That is, are they already employed by the park and can they be made available, or must they be "borrowed" from the Regional Office? Must new hires or contractors be used?) The study must be funded as part of the unit's budget. What performance milestones must be accomplished each time period to ensure that the GMP is completed in a timely fashion? How will public involvement be carried out during the relevant steps in the planning process? All these issues must be given practical answers if the GMP is to be successfully completed.

Step 3: Development of Information Base

Step 3 involves collecting or assembling data on natural and cultural resources as well as regional demographic and socioeconomic data according to the Planning Requirements identified in Step 2. This database will serve as the foundation for formulating alternatives, evaluating them, and finally selecting one. This data collection step is expected to take about two to five years. No funds for subsequent planning steps will be provided to the park or monument until an adequate database is in place.

Step 4: Formulation of Alternatives

Step 4 formulates a range of alternative management strategies to solve the problems and issues raised in Step 1. The general rule is that any alternative must resolve the issues and attain the objectives of the park unit within fifteen years of the plan's approval. The alternatives must meet administrative feasibility; be practical, given the resource base; and represent "cost-effective solutions to the issues" (NPS 1988:chapter 2:3).

Each alternative should reflect a different strategy to attain the goals and objectives of the park. However, "Every general management plan (alternative) will include interrelated proposals for resource protection and management, land protection, interpretation, visitor use, accessibility for disabled visitors, carrying capacities, park operations and a general indication of location, size, capacity, and function of physical developments. A plan implementation schedule and cost estimates will be included" (NPS 1988:chapter 2:5). Where required to resolve particular issues, plan alternatives should also address suitability of lands for Wilderness, boundary adjustments to the park, or monument and visitor access.

One way each alternative management strategy can be developed and represented is through the amount of the unit's land and water that is allocated to one of four management zones. The management zones are literally just that: a land use zoning system of appropriate uses. Each zone represents a management theme that would make certain visitor uses appropriate and others inappropriate and implies different visitor carrying capacities. The NPS management policy manual on planning states that, as part of developing each alternative, the four primary management zones will be identified within the various alternatives (NPS 1988:chapter 2:4).

The four management zones include natural, cultural, park development, and special use. Within these four major categories more site-specific subzones can be defined if local conditions require slight modifications.

Lands in the *natural zone* are managed to give priority to conservation of natural resources and to allow ecological processes to operate. Visitor use is limited to types and amounts consistent with achieving this objective. Therefore visitor uses are limited to nonmotorized dispersed recreation, such as hiking, backpacking, and nonmotorized boating. Only minimum facilities are provided (e.g., trails, trailside information displays, and small, walk-in administrative facilities) when they are needed.

Lands in the *cultural zone* are managed to give priority to "preservation, protection and interpretation of cultural resources and their settings and to provide for their use and enjoyment by the public. In most cases, a property listed or eligible for listing in the National Register of Historic Places will be placed in this zone. . . . Development in the cultural zone will be limited to that needed for preservation and interpretation of cultural values'' (NPS 1988:chapter 2:4).

The park *development zone,* as its name implies, reflects lands that are managed for intensive visitor use and that provide the support facilities for accommodating that use (e.g., parking lots, developed campgrounds, hotels, and so on) and managing it (e.g., visitor centers, amphitheaters, and so on). Park administration buildings and facilities, such as a motor pool or stables for horses, will be located in the development zone. The NPS stresses that before new land is added to the development zone, location of that developed use outside the park must be seriously considered (NPS 1988:chapter 2:5).

Finally, the *special-use zone* involves lands within the park that are controlled by other agencies or private interests. Examples include private inholdings, rights of way, and reservoirs where the NPS has only limited control over uses.

Different alternatives are constructed by allocating different amounts and locations of land to each of these four management zones. However, any General Management Plan alternative must also address the mandatory issues quoted earlier in the second paragraph under Step 4. The process of constructing a particular alternative involves assigning all lands in the park to one of the four zones. This zoning will provide the general framework for appropriate uses. Given this, detailed specification is provided about where and at what level (i.e., carrying capacity) activities consistent with the zone will take place and what facilities will be provided. A range of alternatives, some of which provide more emphasis on preservation of a majority of lands in the natural zone and some of which allocate more land to the development zone, are formulated so that as many planning issues as possible can be resolved. In practice, an alternative that allocates more land to development zones may better satisfy issues related to public demand for access, campground availability, and so on, but not resolve others related to crowding, soil erosion on trails, and so on.

Step 5: Evaluation of Alternatives

At Step 5, planners and managers evaluate the environmental and other effects of particular alternatives. Specific effects that must be considered include (1) resource protection; (2) visitor safety; (3) visitor use and enjoyment of park resources; (4) interests of park-associated communities and groups; (5) short- and long-term cost effectiveness (NPS 1988:chapter 2:3). Since an EIS is often performed on selection of a General Management Plan, the other requirements of the National Environmental Policy Act, such as assessing whether any alternative results in irretrievable or irreversible environmental changes, must also be met.

Of course, as part of the NEPA process, the Draft EIS on the proposed GMP is circulated for public review and comment. Comments of other agencies, governments, and the general public are used to evaluate the desirability of each alternative.

Step 6: Selection of Final GMP

In principle, the alternative that best resolves the planning issues and problems becomes the one selected as the Final GMP. The reasons given for selecting a particular alternative as the Final GMP are given in the Record of Decision that accompanies the Final EIS.

Because the planning issues are rarely ranked and no overall summary performance indicator for each alternative is developed, it is difficult to determine objectively which alternative is the best. Often the selection of the Final GMP is justified in words that appeal to the NPS's mandate to preserve the resources while allowing public enjoyment. Much like the multiple-use motto, the "preserve-but-use" motto is consistent with any number of alternative management strategies and is by itself inadequate to discriminate between alternatives that are acceptable and those that may be "better" or "best."

Step 7: Plan Implementation

Plan implementation involves several phases. Existing, but under the new GMP, nonconforming uses are brought into compliance based on the schedule in the GMP. Management actions requiring special funding are requested through the annual budget cycle. A series of implementation plans is developed that provides the site-specific detail for construction or rehabilitation of facilities provided for under the GMP. If changes in wilderness or backcountry management are called for in the approved GMP, the details of phasing in such actions will be developed. This involves working out the logistics of permits, staffing to issue and check permits, and so on.

We shall provide a case study of a GMP for Yosemite National Park. However, even the preceding sketch of GMP planning should make it clear that NPS planning is a far less structured process than that used by the Forest Service. In part, this may be justified because the issues to be decided in a GMP are less wide-ranging than those under multiple use. Nonetheless, defensible actions require a close link between operational performance criteria and the decision about what is best. The operational criteria for the degree of desirability among alternatives appear to be lacking.

A Suggested Extension of Operational Criteria for Balancing Preservation and Visitor Use in GMP

The NPS does have a somewhat conflicting mandate of preserving parklands but providing for visitor enjoyment. How much land to allocate to the natural zone and how much visitor use to accommodate are largely questions of comparing the values to society that derive from different management emphases. To do this an agency can rely on several operational criteria. The first two relate to appropriateness of types of visitor uses and the amount of use. The third (to be discussed later) relates to benefits from visitor use versus more emphasis on preservation.

A 1988 CBS ''48 Hours'' broadcast shows the full range of visitor uses in Yosemite National Park on the 4th of July weekend in 1988. These include rock climbing, hang gliding, backpacking, developed camping, hiking, rafting, sunbathing, swimming and diving in the river, swimming in public or hotel swimming pools, golfing, bicycling, staying at resorts, going out to eat (shown in the program are gourmet meals at the Ahwahnee Hotel and pizza at concessionaire facilities), consuming alcoholic beverages, getting married, sitting around listening to radios, and general campground partying. All these take place in the park, with many in Yosemite Valley itself. Given that there is substantial excess demand to visit the park and that many people are turned away because of lack of reservations or lack of camping space, which uses should be allowed? If one is not careful to specify criteria to judge the merit of these uses in Yosemite National Park (not their merit in general!), such a determination could degenerate into personal biases and value judgments.

Developing Operational Criteria for Determining Appropriate Uses

The very first step in developing criteria on appropriate uses is already given by the NPS in its Statement of Management: combining legislative intent for the park and the goals of the NPS with the significant features of the park. This provides information on legislative intent and attributes of the park. Next, more emphasis should be given in the NPS database on the resource capabilities of the park. What uses can it sustain without permanent damage? The next step is the critical one: determining which sustainable uses can be met only by resources provided in this park or monument and which can be met in outside areas. This determination involves sorting all potentially suitable uses into three categories: (1) visitor uses that require the special features of this National Park or Monument; (2) visitor uses that could be engaged in elsewhere but that are enhanced substantially by being done in this National Park or Monument; (3) visitor uses that can be performed elsewhere and that are not enhanced significantly by being done in this park or monument. These state in rank order the priority visitor uses of the park or monument.

The first visitor activity demands to be met are those that require special resources that are unavailable in other parks or monuments. For example, Yosemite is known worldwide for its solid granite cliffs for rock climbing. There are few places in the United States or indeed the rest of the world where such cliffs exist. They are a natural endowment of the park and cannot be re-created by people elsewhere. Moreover, they are not a common feature of most other National Parks, Monuments, most state parks, or national forests.

Activities such as backpacking, rafting, hiking, and camping, although they do not require special attributes of a National Park such as Yosemite, are substantially enhanced by the resources in the park. The fact that cattle grazing is not allowed in most National Parks or Monuments increases the enjoyment of these recreation activities over engaging in the same activities on National Forests or BLM land where cattle grazing may affect backcountry campsites, hiking trails, and water supplies. In addition, all these activities are nature based, and the outstanding scenery or other natural features of the park significantly influence the quality of the recreation experience. Activities like week-long backpack trips require large areas that are not crisscrossed with roads or developments (i.e., hotels, timber harvesting, mining activities, and so on).

At the other end of the spectrum are low-priority uses that do not require resources unique to the park. These can commonly be performed elsewhere, and the special attributes of the park add little to enhance the enjoyment of the activity. For example, swimming in swimming pools, sunbathing, bicycling, staying at resorts, going out to eat, listening to radios, getting married, and partying neither require special attributes of the park nor are enhanced significantly by it. All these events focus primarily on activities that have little to do with the natural resources of the park or monument but that are centered instead on such human-related facilities as swimming pools or human institutions or social activities. Sitting in a lounge chair sunbathing or listening to the radio is a fine activity, but is this the highest-valued use of the scarce space in Yosemite Valley? Not when the sunbather displaces someone who wants to rock-climb in the park. The sunbather can find hundreds of equivalent environments, but the rock climber cannot. Although the sunbather might prefer (at the current entrance fee and considering their proximity to the park) to use the park to sunbathe, that mild preference is rarely matched by the intensity of preferences for the special attributes of a park that cannot be found elsewhere. Of course, the mix of appropriate uses will vary from park to park. At parks or monuments with strong historic and cultural themes, the priority uses would be different from those at parks with a strong natural orientation.

Thus in planning which ecologically sustainable activities receive priority there must be some method to determine which recreation activities require park resources or will be enhanced substantially by them. In long-term planning, the most obvious strategy is not to provide facilities for activities that do not require park resources and not to allow such activities.

Therefore golf courses, swimming pools, and wedding facilities would not be provided or, if they are present, would be removed. (The latter is quite difficult, as our case study of Yosemite Valley will make clear.) Directly selecting people to admit by their stated recreation activity is easier for backpacking, rock climbing, and hang gliding, where permits and safety checks are sometimes required. As for other activities, it will be difficult to observe what intended activity such visitors might be coming to the park for (e.g., if they state that they are there to hike, when it is their intent to sunbathe).

The most direct way to ration according to the priority order described earlier may be to let the visitors sort themselves out through a rationing system. This system must be capable of assessing just how badly they require the park resources for their activity or just how much their recreational activity is enhanced by performing it in this National Park or Monument as compared to areas outside the park. Some people have called for rationing by merit, where a person would be required to perform some service for the park to gain entry, such as clean up litter, maintain trails, replace signs, and so on.

Pricing is one way to ration when more people want access to a resource than there is space available. Those who really require park resources for their recreation activity or who would have their activity substantially enhanced by visiting the park would probably pay two to three times the current nominal entrance fee.

Those that plan to sunbathe, swim in the river, or sit around listening to radios would probably find that they would be better off engaging in those activities at public reservoirs, city parks, National Forest campgrounds, and so on. The greater the number of cheaper substitute areas available for that recreation activity, the less likely a person would be to pay substantially more to engage in essentially the same activity in a National Park or Monument. As the discussion on visitor carrying capacity will indicate, such price rationing may be needed only on weekends, with lower or current nominal fees charged during nonpeak weekdays to ensure that higher fees would not preclude lower-income users from visiting the park or monument. More details on price rationing will be discussed after the issue of how much of the appropriate type of visitor use to allow is presented. At present we simply want to mention that the equity effects of price rationing for National Parks are not as bad as they might seem. Most National Parks are located substantial distances from urban areas. The high travel cost and the need for a vehicle capable of driving up mountain roads and for camping equipment already eliminate many of the low-income households that might be affected by higher entrance fees. Most who visit these distant (but congested) National Parks have the incomes to pay higher fees.

How Much Use Is Optimal: Social Carrying Capacity

Even when we have prioritized the uses that are most compatible with a particular park, there is still the question of how much of that type of use to allow. In evaluating how much visitor use to accommodate, one can be guided by the theory of recreation carrying capacity. Specifically, a given area has three types of carrying capacities: ecological, facility, and social. Any one of these three can be the most binding constraint on total visitor use. The NPS cannot allow so many people that vegetation along trails or in backcountry campgrounds is permanently destroyed. The NPS cannot allow fifty people to camp at a site designed to accommodate ten. In the following discussion we assume that use can be ecologically handled, even if a series of campsites must be used on a rest/rotation basis, where a portion of the units are periodically closed to be rested (and soils aerated and revegetated, as is sometimes the case).

The following model of social carrying capacity is partly meant to allay fears that if the NPS were to tip the scales away from preservation and toward visitor use, this would imply unlimited public access and would double or triple current use levels. This need not occur and, as the example will show, should not be the case even if visitor satisfaction takes precedence over preservation.

The concept of social carrying capacity (Manning 1986) is at the heart of this restraint in visitor use. Even if the NPS was to emphasize visitor use, this does not translate into maximizing the number of bodies that can be squeezed into a particular park or monument. As more and more people enter a park, they begin to interfere with each other's ability to enjoy the its resources. Wildlife viewing, photography, and rock climbing all become less enjoyable if people cannot pursue their objectives because of interference by others. But how much use is too much? Once again, without some formal criteria or principle, this too becomes a value judgment. But there is a simple logical principle: Another person should be admitted if the recreational enjoyment gained by that person exceeds the loss in benefits to all others already in the area to which the new person would be admitted. Once we have illustrated this principle we shall turn to the challenge of implementing it in a park.

Figure 13.1 illustrates this principle of social carrying capacity. Total visitor enjoyment rises as more people are allowed in the park or monument, but only up to some optimal quantity of visitors per day represented in figure 13.1 as Q^* (150 people). To see why Q^* is the optimal number of visitors per day consider the following. With one hundred people in the park, allowing another adds to total visitor enjoyment: That is, allowing that one additional person in the park might provide him or her with twenty units of "fun" but reduce the fun of the one hundred people by only .05 each. Society gains fifteen units of fun since the gross gain of twenty for the additional person exceeds the sum of the losses to existing visitors ($100 \times -.05 = -5$). This comparison of gains to an additional person entering a more and more crowded park with the loss to existing visitors continues as more people are admitted. At Q^* (150) visitors, the gain received by the additional visitor is exactly equal to the loss to all other visitors. Thus the 150th person might get ten units of fun from entering the crowded park, but everyone else's fun would be reduced by .067 per person. Since now 149 persons are experiencing this loss in visitor enjoyment, the total enjoyment lost by these 149 existing visitors (-10 units of fun) just equals the benefits gained by allowing that 150th person in. Beyond Q^* as more people are let into the park or monument, total visitor enjoyment actually decreases as the loss to existing visitors exceeds the enjoyment gained by the new person entering the already crowded park. This is illustrated by the marginal "fun" curve in the lower panel of figure 13.1. The marginal or additional benefits of another visitor become zero at Q^* indicating that there is no net gain to letting that person in. Two forces are at work as crowding increases. First, the benefit to an additional person entering a more crowded park is falling. Second,

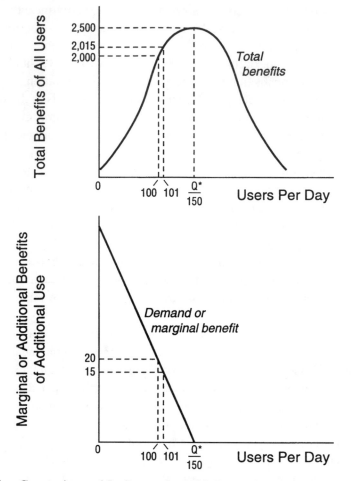

Figure 13.1.   Congestion and Its Interaction with Recreation Benefits

the external congestion costs of an additional person are imposed on a larger and larger group. Both cause total recreation enjoyment to fall.

This conceptual model has been operationalized and implemented by recreation researchers in several different ways, the most successful of which is to use visitor willingness to pay (WTP) as a measure of recreation satisfaction, or "fun." That is, people will not agree to pay more than the value of the enjoyment they are receiving. As crowding increases, their enjoyment goes down. Therefore a manager can trace out curves similar to those in figure 13.1 to indicate what the optimal visitation level is from the standpoint of social carrying capacity. This type of study has been performed using the contingent valuation method (CVM) described in chapter 6. CVM has been applied to determine optimal capacities for wilderness areas (Walsh and Gilliam 1982), ski areas (Walsh et al. 1983), lake recreation (Walsh et al. 1980b), stream recreation (Walsh et al. 1980a), and beach recreation

(McConnell 1977). An example of this type of CVM study is provided in our case study of Yosemite National Park.

If visitor demand at current entrance fees exceeds the social carrying capacity, then use must be rationed to that capacity. Visitor use can be rationed by advance reservation; on a first-come, first-served basis; or by price. Rationing by first come, first served is frustrating to visitors who have traveled long distances to reach a park and are then turned away. Both advance reservation and first-come, first served may be unable to ration the available social carrying capacity (Q^*) to those that would receive the most benefit from visiting a particular National Park or Monument. As has been demonstrated by several researchers (Mumy and Hanke 1975; Loomis 1982; Hof and Loomis 1986), there is a loss of public benefits if the rationing system is incapable of distinguishing high-valuing users from low-valuing ones. The pricing system will ensure that during peak periods high-valuing users of a park or monument enter before low-valuing ones.

During nonpeak periods such as off-season months or during midweek, visitor demand may be less than Q^*; hence no special rationing fees are needed. Thus differential pricing with peak and nonpeak fees helps to ensure that lower-income users still can visit the park during nonpeak periods.

How Much Preservation Versus How Much Visitor Use and Enjoyment?

So far we have discussed operational criteria for determining which type of visitor uses may be appropriate for particular parks or monuments and how to determine the optimal amount of use. Now we turn to the broader questions of how to deal with those situations where expanding facilities or intensifying management to accommodate more visitors necessarily reduces the naturalness of some areas of the park that is, the trade-off between having more or fewer areas developed. Once again, some formal operational criteria are needed to make such a decision without being overly judgmental about how much natural area is enough.

One operational approach that combines public involvement with measurement of the intensity of preferences to evaluate the desirability of more preservation versus development is to use a combination of contingent valuation method (CVM) surveys as follows. One would ask current and potential visitors about their willingness to pay (WTP) for visitor access, facilities, and crowding associated with each park alternative. This survey would quantify visitor benefits for each alternative GMP. The second would survey the general public. It would describe the natural resource impacts associated with each GMP alternative. Members of the public would be asked their WTP to have the park as described in each alternative. Since a large majority of the general public may not be current visitors, their WTP would largely reflect possible option values for their own future use, existence values for just knowing the park or monument environment would be preserved in its natural form, and a bequest value to future generations from preserving the park today. These three motivations for WTP for natural resources one does not currently use but would like to see protected are referred to as *preservation values*. They would be elicited for each alternative.

In essence, the preservation WTP represents the weight *the public would give* NPS's preservation mandate and the visitor's WTP for recreation represents the *weight the public would assign* the "visitor use and enjoyment" part of the mandate. A comparison of the net benefits (WTP benefits minus costs of each alternative) would be used to select among nonimpairing management alternatives, some of which emphasize visitor use more and preservation less, and vice versa. This could be done by comparing WTP of recreationists for expanded use (but within the optimal social carrying capacity described earlier) versus the general public's existence-bequest values for preservation of natural ecological processes. This would provide a more focused public involvement process and one yielding comparable information for both groups. Thus summing up WTP for recreation benefits and preservation benefits of each alternative and subtracting the costs would result in a performance indicator for each alternative. In this way, a more objective assessment could be made of which alternative represented the best achievement of both preservation and visitor use.

The general suggestion to use benefit-cost analysis in evaluating alternative GMPs is mentioned in the NPS's earlier *Planning Process Guide* (NPS-2 1978, cited in Walsh 1980), and there has been growing acceptance of benefit-cost analysis within NPS. Recent applications of benefit-cost analysis to NPS issues include benefits of protecting scenic visibility at parks and benefits of rafting and environmental protection along the Colorado River through Grand Canyon National Park. The case study of Yosemite National Park's GMP will emphasize the application of benefit-cost analysis sponsored by the NPS.

Yosemite National Park's General Management Plan as a Case Study

Yosemite National Park is located in east central California, about four hours by car from seven million people in the greater San Francisco Bay Area and within a day's drive of nearly all twenty-eight million people in California. The park encompasses 1,200 square miles, although the most frequently visited Yosemite Valley is a small fraction of the total area. Of the 1,200 square miles, 94% is undeveloped. About two to three million visitors per year enter Yosemite. There are 266 miles of roads and 750 miles of trails. The annual operating budget is $13 million, and concessionaire sales are $78 million. In Yosemite Valley, the concessionaire maintains three lodging complexes capable of accommodating nearly five thousand visitors at any one time. The campgrounds can accommodate four thousand visitors in Yosemite Valley. Yosemite Valley also contains four restaurants, two cafeterias, five bars, six gift shops, two fast-food stands, two grocery stores, two sporting goods stores, and a clothing store, car rental agency, beauty shop, bank, school, hospital, and dental office (Walsh 1980:3), as well as tennis courts.

Park Planning Issues

With this intensity of development in the most scenic part of the park it is not surprising that one of the key issues to be resolved in 1975, when the GMP process began, was the level of development, visitor use, and naturalness in the park. The Draft EIS may well summarize what NPS stated would be a ''new era at Yosemite . . . by rectifying an overzealous attempt to civilize the park'' (NPS 1978:5).

   Figure 13.2 illustrates how the Yosemite planning team developed the planning objectives in the 1976 GMP effort. Legislation was reviewed, including laws specific to the creation of Yosemite Park as it exists today, Acts affecting National Parks in general, legislation regulating concessionaires in the park and a variety of applicable federal environmental laws such as NEPA, the Endangered Species Act and the Wilderness Act. Next the significance of Yosemite National Park (YNP) was described in terms of inspiring waterfalls, giant sequoia trees and grand vistas. NPS lists two purposes of the park as ''preservation of the resources that contribute to Yosemite's uniqueness and attractiveness'' (NPS 1975:23) and making Yosemite ''available to people for their enjoyment, education and recreation, now and in the future'' (NPS 1975:23). The park experience was described as allowing visitors to ''structure their own personal experiences by choosing from a variety of high quality and complementary activities directly related to those environmental, spiritual and esthetic distinctions that were responsible for the establishment of the park'' (NPS 1975:25). Two themes are stressed here: a variety of activities, all of which are to relate directly to park resources.

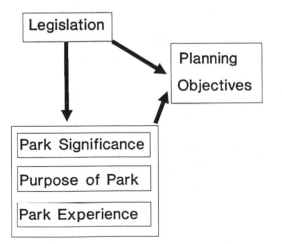

Figure 13.2.   How Planning Objectives Were Determined

As illustrated in figure 13.2, these three factors, along with legislation, create the planning objectives. The YNP planning team identified five major objectives under which it grouped numerous specific objectives and subobjectives:

1. Restore and maintain natural ecosystems so that they may operate essentially unimpaired. Subobjectives include: restore altered ecosystems, perpetuate natural processes, permit only those types and levels of use or development that do not significantly impair park natural resources and limit unnatural sources of pollution to the greatest degree possible.

2. Preserve, restore, and protect significant cultural, historic, and prehistoric resources.

3. Assist people in understanding, enjoying, and contributing to the preservation of the natural, cultural, and scenic resources. Subobjectives include providing visitor information and interpretation, ''Provide only for those types and levels of programs and activities that enhance visitor understanding and enjoyment of park resources. Permit only those levels and types of accommodations and services necessary to visitor use and enjoyment of Yosemite. Provide the opportunity for a quality wilderness experience. Provide transportation services that facilitate visitor circulation and enhance preservation and enjoyment of park resources'' (NPS 1975:28).

4. Maintain a safe and functional environment that provides compatible opportunities for resource preservation and enjoyment by visitors and employees. This involved several subobjectives, including several emphasizing public safety and boundary adjustments, to make the park a more complete ecological unit and provide more effective management.

5. Support integrated regional land use planning for recreation, community development, preservation, and economic utilization of resources.

Development of Alternatives

Alternatives were developed twice during the planning process. First, four alternatives (A–D) were developed to be evaluated by the public. Alternative A was a pristine YNP alternative calling for removal of roads, no automobile traffic in the valley, and removal of all hotels, lodges, and restaurants. Alternative B was aimed at providing just the necessary visitor services, emphasizing shuttle bus use in the valley and reducing the number of developed campgrounds and hotel accommodations. Alternative C described management close to what was the existing level and extent of transportation, visitor facilities, and uses. For example, conventions could occur only during the off season, current hang gliding would continue, but no snowmobiling would be allowed. Finally, Alternative D was the most pro-visitor development; it allowed snowmobiling, added more hiking and horse trails, expanded hotel and lodging accommodations, and imposed no visitor use limits. These alternatives were used as reference points in a checklist presented to the public to see what combination of features from each the public desired.

The YNP effort used maps of these alternatives in an aggressive public involvement campaign that provided much of the information discussed earlier in a workbook for the public. This workbook urged people to read the background information provided and then "design a management plan" for the ten major areas encompassing Yosemite National Park using the large, poster-sized maps included in the packet. The public was asked to address level of visitor accommodations, transportation, resource management, and park operations. Details of what is meant by each of these were printed on the map. The workbook was mailed to nearly sixty thousand people, of whom about twenty-one thousand returned it.

The alternatives evaluated in the Draft Environmental Impact Statement (EIS) were constructed from a "cluster" analysis of the most frequently favored items from the map checklists. Only a random sample of the twenty-one thousand returned questionnaires was used to keep down data entry and computational costs, but the map checklist filled out by the concessionaire was included.

The YNP planning team developed a preferred alternative and four other alternatives in the Draft EIS from the public comments. Both the preferred alternative and Alternative 1 reflected the combination of most frequently chosen but compatible features from the map checklists. These slightly different alternatives fit between the strong preservation alternative (Alternative 2) and the resort development alternative (Alternative 3). Alternative 4 was the required "no action" or continuation of existing management alternative.

The preferred alternative and Alternative 1 generally removed such services and facilities as bank, barber, beauty shop, auto rental, and clothing stores from YNP and emphasized shuttle buses over private automobiles. Grocery stores, gas stations, and equipment rental facilities would continue to be provided. Major YNP administrative functions and much employee housing would be relocated outside of Yosemite Valley. Wilderness designation was recommended for 666,600 of the park's 761,000 acres.

The preservation alternative (Alternative 2) would remove cars from YNP and would substantially reduce the amount and type of visitor accommodations, deemphasizing hotels and lodges and emphasizing walk-in campsites. The golf course and tennis courts would be removed.

The resort development alternative (Alternative 3) emphasized what the concessionaire at that time (the Yosemite Park and Curry Company) desired. It would increase overnight accommodations and retain all existing visitor recreation facilities, such as the golf course, tennis courts, and swimming pools. An additional two-hundred-car parking area would be constructed in Yosemite Valley.

Evaluation of Alternative GMPs

The NPS compared the preceding four alternatives in the Draft EIS on seven key factors shown in table 13.1. Most of the categories are self-explanatory, but a few merit brief explanation. Day use and overnight capacity are measured in number

Table 13.1
Comparison of Yosemite National Park GMP Alternatives in Draft EIS, 1978

Effects	Existing (Alt #4)	Preferred	Alt. 1	Alt. 2	Alt. 3
1. Day-use levels (# people)	40,584	39,644	39,879	42,945	43,284
a. Yosemite Valley	18,240	11,655	12,780	11,205	19,185
b. Rest of Park	22,344	27,989	27,099	31,740	24,099
2. Overnight capacity	15,420	15,615	16,713	9,127	18,576
a. Valley Hotel/Cabin	4,934	3,994	4,934	1,030	5,018
b. Valley Campgrounds	4,132	3,668	4,132	1,280	4,132
c. Other Hotel/Cabin	594	833	917	2,057	2,016
d. Other Campgrounds	5,760	7,120	6,730	4,760	7,410
3. Treated water use (1,000's of gallons)	2,359	2,079	2,098	1,372	2,418
4. Acres disturbed (sum of all vegetation types)	1,024	877	847	506	887
5. Impacts to archaeological sites (number of sites)	193	38	38	108	33
6. % changes in concessionaire profits					
a. Sales		4.2%	−7.0%	−13.0%	13.0%
b. Operating expenses		3.7%	−4.0%	−18.0%	11.0%
c. Total capital expenses		178.0%	−194.0%	188.0%	95.0%
d. Net income		−107.0%	−156.0%	−139.0%	−39.0%
7. Scenic quality Acres impacted views in Yosemite Valley	297	242.8	249.1	45	300.3

of persons that would be permitted or accommodated. Direct and indirect impacts to archaeological sites result from disturbance to the site from facility or roadway construction or removal of facilities in the area of or adjacent to the archaeological site. Scenic quality is measured as number of acres of land that have visual intrusions (e.g., buildings, parking, and shops) in the line of sight of such major park features as waterfalls or other scenic values (e.g., El Capitan). As table 13.1 shows, the preferred alternative reduces several impacts to the park and the valley portion of the park. Fewer acres are disturbed, slightly less water is used (and hence less wastewater must be treated), and direct and indirect effects on archaeological sites are substantially reduced.

Figure 13.3 compares the changes in visitor use to Yosemite Valley and the rest of the park for each alternative. Overall, the proposed GMP reduces total park visitation by only a small amount by shifting a good percentage of the use in the valley to other areas in the park.

The NPS also evaluated air quality effects and costs of each alternative, although these factors were never directly compared across alternatives. The initial cost estimate of the preferred alternative was $78 million for relocation and rehabilitation of NPS and concessionaire facilities.

Selection of Final General Management Plan

The Draft Proposed General Management Plan and Draft EIS went out to the public for comment and review in August of 1978. During that year, nine public meetings were held in California and one was held in Washington, D.C. Analysis of the large volume (about thirty-five hundred) of public comments resulted in a substantial enough revision in the proposed GMP that the NPS issued a new supplemental Draft EIS in January of 1980. The major change that prompted the need for a second public review was the greater emphasis in the revised GMP on relocating NPS administration and operations facilities and concessionaire employee housing from the valley to other areas of the park (such as Wawona and El Portal), which would create additional environmental impacts in those areas. The new proposed GMP resulted in continued cutbacks in day-use parking in the valley (to about two-thirds of the existing spaces) and elimination of the originally proposed valley parking area addition. The new proposed GMP speeded up the effort to remove automobiles from Yosemite Valley in favor of shuttle buses.

After public comment on the supplemental Draft EIS, a Final EIS and Final GMP were issued in October of 1980 (NPS 1980). This Final GMP continued to emphasize relocation of major NPS administration and operations facilities outside of Yosemite Valley, with El Portal (just outside the western park boundary) becoming the YNP headquarters. Concessionaire employee housing in the valley was cut

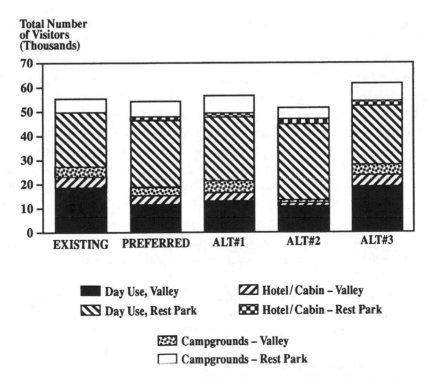

Figure 13.3. Yosemite National Park GMP Alternatives-Visitor Use Levels

to one-third of existing levels. In addition, commercial overnight visitor accommodations in the Valley were to be reduced by 17%. Several non-park-related shops and facilities would be removed from the park. Lastly, the amount of acreage recommended for designation as wilderness increased by 10,000 acres to 676,600 of the 761,000 acres in the park. Day-use parking was to be reduced and shuttle buses were to be emphasized. Revised and more complete cost estimates of the GMP put the investment costs at $162 million (with $25 million of that being concessionaire costs), more than double the first estimate in the first draft.

The Final EIS compares six alternatives. These are the original Alternatives 1, 2, and 3; the original preferred alternative that was in the 1978 Draft EIS; the January 1980 revised preferred alternative that was in the 1980 Draft EIS; and the Final GMP. Once again, the alternatives were compared on the same seven major factors listed in table 13.1. It is not clear how the selection of the final GMP relates to resolving the five major planning objectives initially identified by NPS.

Implementation of the Final GMP: 1980–1990

Many features of the Final GMP have been implemented, but several key features have not. In 1984, nearly all the acreage recommended for Wilderness in the park was formally designated as such. Recent minor modifications to Wilderness Area boundaries have resulted in 94% of Yosemite National Park being designated as Wilderness. Much (six hundred spaces) of the day-use parking to be removed under the Final GMP has been removed. A great deal of natural resource rehabilitation has taken place in the valley and in backcountry areas throughout the park. This includes discontinuance of golf at the Ahwahnee Hotel and returning that area to natural conditions (NPS 1989:26), restoration of numerous acres of land around valley facilities, and cleanup and revegetation of over fifty backcountry sites.

Unfortunately, the funding to implement several related items has not been forthcoming from the Department of Interior or Congress. One of the key shortfalls has been with the expansion of public transit in the valley to transport those people who, because of the reduction of six hundred parking spaces in the valley, can no longer park in the valley. Outlying satellite parking areas have not been built. The shuttle bus system has been expanded to include Mariposa Grove and Badger Pass, but it still does not serve much of the park and outlying areas. There appear to be several reasons for this, nearly all related to the large funding requirements to implement the outlying parking, purchase additional buses, and improve roads to accommodate them.

The first obstacle occurred with the change in presidential administrations in 1980, when the Reagan administration shifted federal funds away from programs aimed at environmental preservation in general and appointed a secretary of interior who wanted to ''open up'' the parks, not restrict automobile access to them. Second, during the 1980s several National Parks had crossed environmental thresholds that required immediate action. These included funding for endangered-species work and private land acquisitions to protect park resources from escalating outside threats (to be discussed in more detail later). In addition, much of the available funding to the NPS had to be used to rehabilitate deteriorating water, sewer, and electrical systems that put public health and safety at risk in many parks. Funding to relocate existing employee housing in Yosemite was also a lower priority than providing decent employee housing, which was completely lacking at many other National Parks. Visitor resistance to use of buses has proven more difficult than anticipated in the GMP.

Another stumbling block to implementation was the fact that certain features of the Final GMP were the responsibility of the concessionaire. For example, relocation of concessionaire employee housing and removal of certain overnight accommodations and visitor services were the domain of the concessionaire. However, under the existing contract (dating back to 1963), the NPS could not require the concessionaire to make those changes. Not until a new contract is negotiated in 1993 can the concessionaire's legal requirements be brought into conformity with the Final GMP. The prospects for obtaining a concessionaire more sympathetic to Yosemite's preservation-oriented GMP took a major step forward when the Yosemite Park and Curry Company's parent company (MCA) was sold to a Japanese company. As a result, the more conservation-oriented National Park Foundation and others now have a better chance of becoming the new concessionaire in 1993.

However, it should be noted that both NPS and the concessionaire have relocated some administrative functions out of the valley. Some relocation of employee housing out of the valley to El Portal has taken place, but the traffic created by these employees commuting back to the valley has added to road and parking congestion. At present, there has been only minimal progress toward the GMP's objective of reducing overnight accommodations in the valley and expanding overnight accommodations in Wawona (in part, because of lack of adequate water supply expansion potential at Wawona).

Another important element in the inability to implement many of the major features of the Final GMP may relate to poor cost-estimating techniques and lack of data on what the benefits of such capital investments are. For example, the total government investment cost of the Final GMP was estimated to be $137 million in 1980. More complete cost estimates for construction, bus acquisition, and operation of just the outlying-parking feature of the plan came to $74 million in 1986. In real terms (adjusting for inflation between 1980 and 1986), the cost of just this one feature of the plan was 40% of the original total government cost estimate in the Final GMP. After spending about $75 million in the last ten years to implement just some of the features of the GMP, the NPS estimates that an additional $145 million is needed to fund priority items remaining in the GMP. Thus the total government cost may be closer to $220 million for Yosemite's GMP. Clearly, more attention to the cost estimates associated with various features of the plan would have made the resulting GMP more realistic.

Given that funding has played a significant role in influencing the implementation of Yosemite National Park's GMP, more attention to economic factors may well have been warranted. Most important, information on the benefits of each feature of the plan relative to the cost might have alerted the NPS to certain features of the Final GMP that were not economically efficient and hence unlikely to pass scrutiny with Office of Management and Budget (OMB). This would have forced the NPS to search for more cost-effective management strategies to attain the same goals. Second, having solid information on the economic benefits of management features of the plan that were economically efficient would have helped YNP to compete for funding at the department level, with OMB and in Congress. An agency needs hard data to convince OMB and Congress to allocate millions of dollars to shuttle buses, remote parking, relocation of offices, and employee housing in a park. If estimates of the dollar benefits of reducing visitor congestion and improving air quality, which were associated with increased bus usage, and of improving scenic quality by relocating employee housing out of the valley had been part of the Final GMP, funding requests might well have fared better. To illustrate how such evaluations can be made for the GMP alternatives, we shall review an economic study of the Yosemite's GMP sponsored by the NPS and done by Richard Walsh (1980).

Improving GMP Planning in Yosemite National Park by Linking Recreation Benefits to Reduced Crowding

As part of a survey of Yosemite National Park visitors in 1975, Walsh (1980) asked visitors their current expenditures per trip. Then he asked how much higher those expenses would have to be before they would not visit the park. That difference is a consumer surplus or net willingness to pay (WTP) for Yosemite, which averaged $19.65 per visitor day. Then he asked people to assume that they saw one-half as many visitors and repeated the WTP question. The answer was an average value of $25 per visitor day. The increase in WTP of $5.35 between the ''see half as many'' and the current condition indicates the benefits of reducing congestion in Yosemite National Park. Across the different recreation activities, the increase in WTP per visitor (in 1980 dollars) for a 50% reduction in number of other parties seen ranged from $9.45 per day for backcountry users to $2 per day for people stayin in concessionaire lodging. Day users to the park would pay about $7 more per day, and NPS-developed campground users would pay about $4 more per day. This information can also be used to assist in determining optimal capacity as in figure 13.1, as well as to justify the budget needed to reduce visitor congestion.

To illustrate the application of these figures on WTP with changes in congestion, we shall calculate the change in benefits to day users associated with the shift in day visitor use from the valley to the rest of the park under the Proposed GMP. The Proposed GMP cuts day use in the valley by 36%. But benefits to valley day users only drop by 17%, since the reduced congestion increases the benefits per day to the remaining visitors. For the park as a whole, all but 940 of the displaced valley visitors are transferred to other areas of the park, so there is only a small net loss in visitation.. The WTP figures reported by Walsh (1980) for recreation activities suggest about $1.50 per visitor day differential to being in the valley over other areas in the park. Given this differential, then park-wide day-use visitor benefits increase by about $36,500 annually by shifting day-use visitors from the more congested valley to other, less congested areas of the park under the Proposed GMP. The reason for this is that the gain to the day-use visitors remaining in the valley is greater than the loss in benefits to people shifted to other (uncongested) areas of the park (in fact, as long as the WTP premium for the valley is less than $8, there is a net benefit to shifting some day use out of the congested valley to less congested areas in the park). Regulation increases benefits because the negative congestion externality is explicitly considered in setting use limits, whereas under existing "free entry" into the valley, individual choices result in overuse.

This relationship of increasing benefits park-wide with shifting use from the valley to the rest of the park is not a universal constant, as is illustrated in figure 13.4. The Proposed GMP, which reduces overnight use by 15%, results in a gain of $33,735 annually by shifting some of the current overnight use out of the valley. Using the type of marginal or incremental analysis performed here, Walsh concluded that an alternative with a 30% reduction in overnight use would have been the most beneficial (Walsh 1980:83). Unfortunately, the NPS really had no intermediate alternative between the draconian preservation Alternative 2, with its 75%

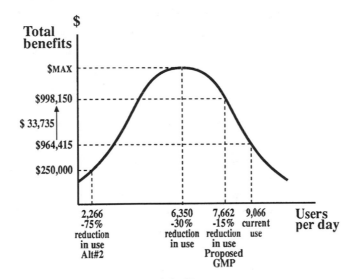

Figure 13.4.   Yosemite Valley-Overnight Use

reduction, and the proposed GMP, with its 15% reduction in overnight capacity. The added benefits of going from the 15% reduction in congestion in the Proposed GMP to 75% reduction is not worth the benefits forgone by the large displaced visitor use. That is, to attain a 75% reduction in congestion results in benefits forgone to a very large number of visitors that are no longer able to visit the valley. This loss is only partially offset by the higher benefits to the few remaining visitors in the Yosemite Valley. This relationship can be seen in figure 13.4.

The 30% use reduction becomes even more attractive when the operating and maintenance costs are considered. The alternatives that favored preservation and reduced use had substantially lower operating and maintenance costs. Walsh found that failure to include the operating and maintenance cost savings in the comparison of alternatives would have increased net benefits to the preservation alternative relative to the Proposed GMP.

Failure to compare carefully the incremental costs and benefits of different size reductions in visitor use and overnight facilities resulted in the NPS overlooking a better alternative. That is, the proposed GMP, at the margin, incurred greater additional costs by only cutting overnight use in the valley by 15%. In the range from a 15% to a 30% reduction in overnight visitors, the combined drop in operating, maintenance, and investment costs and the incremental benefits of less crowding (primarily among campers) resulted in this unstudied alternative having more benefits than the proposed GMP.

In addition, to help justify investment for out-of-valley parking and the operation costs of expanded shuttle bus service, the benefits of improved air quality and scenic quality could have been computed. Walsh (1980:85) estimates that a 30% improvement in air quality would be worth about $1 million a year in added visitor benefits. The NPS had a reasonably good visual evaluation of the intrusion of buildings and parking areas in the valley. However, no attempt was made to quantify the dollar benefits of removing these visual intrusions. Had this been done, it would have helped justify, to OMB and Congress, relocation of parking and buildings outside the valley.

As Walsh (1980) concludes, far too much effort went to obtaining the public's unconstrained preferences via the poster-sized map checklists, and not enough effort went into determining what management features people were willing to pay for. The number of completed public workbooks exceeded by a factor of 4 what NPS could actually use as data. Even this scaled-back data entry and analysis effort absorbed nearly all the funding available for other visitor surveys and analysis. In addition, since there was no attempt to get a random sample of the public or visitors in mailing out workbooks, the representativeness of the resulting information is certainly suspect. This unrepresentativeness is coupled with the limited usefulness of asking people to state what they would like in the park without having to consider the costs. Because of their realism and accompanying benefit-cost estimates, information on what visitors would be willing to pay for would have resulted in more alternatives with a greater chance of implementation. It is worth noting that the update to the *NPS Planning Procedures* (NPS-2), prepared in 1982, provides a more structured approach for obtaining cost estimates.

The Yosemite National Park example may represent the upper limit of complexity and public scrutiny in preparing a General Management Plan. Many other NPS units, such as Canyonlands National Park in Utah, have not required anywhere near the level of detailed effort that Yosemite involved. Chapter 11 stressed that it is important to tailor the planning effort to the resource values and issues at stake. Yosemite, with its three million visitors per year and significant public controversy, no doubt warrants such an intense planning effort. But we do not want to leave the impression that all GMPs must be this elaborate or take this long.

External Threats to Parks and the Need for Broader Multiagency Planning

So far this evaluation of NPS planning and management has focused on challenges to preserving National Park and Monument natural resources stemming from overuse by visitors and internal stresses on park resources (e.g., recreation development, livestock use, fire suppression, and so on). But along with these challenges come the rapidly growing threats to park resources that arise from public and private actions outside park boundaries. The magnitude of this problem has been growing as the once remote parks become surrounded by intensive human uses (e.g., hotels, vacation homes, mining, and so on) and the recognition that park boundaries do not match ecological units.

Some of the external threats are induced by the presence of the park. For example, the NPS emphasis on hotel, restaurant, and other visitor services being located outside the park has contributed to the large size of resort developments adjacent to many parks. Parks such as Yellowstone and Rocky Mountain literally have large towns abutting their boundaries. This problem would not be as critical as it is if park boundaries reflected natural ecological dividing zones. Unfortunately, most boundaries represent what was politically expedient at the time of park designation. As a result, water pollution, air pollution, pets, and exotic vegetation at the boundary of the park easily move into it.

In addition, many migratory animals, such as elk, are unable to find forage for their entire life cycles, even within the largest parks, such as Yellowstone and Rocky Mountain. Thus these animals move out of the parks during the winter and onto their historic winter range, which is now private ranchland or mountain subdivisions. As the animals move out of the parks, their populations are influenced more by state laws and concerns about controlling animal damage to fences, rangelands, and crops than by concerns about ecological management. Even when the species are endangered, such as grizzlies or wolves, protection of private property appears to supersede much else. This, of course, can completely undo efforts in a park to provide improved habitat for increased populations of these animals. In some cases, it has hampered NPS efforts to reintroduce native predators into park ecosystems.

Some of the threats to park resources arise from factors totally unrelated to the presence of the park. Adjoining National Forest or state lands often are managed under multiple-use mandates that are incompatible with park preservation. Timber harvesting, mining, and ORV use often have impacts that can be seen from the park and often directly influence park ecosystems. Once again, such external extractive uses of resources would not matter as much if the park boundaries enclosed ecological units, but they do not. Thus a uranium mine proposed outside Grand Canyon National Park could result in mine tailings washing into the park when frequent thunderstorms flood gullies that flow into the park. Air pollution from the coal-fired Navajo power plant in Page, Arizona, results in substantial haze in Grand Canyon National Park. Peak-power operations at Glen Canyon Dam result in erosion-producing water fluctuations in the Colorado River through Grand Canyon National Park.

Portions of Everglades National Park have dried up because of water diversions for flood control and water supply for lands adjacent to the park. Much of the water received in Everglades is polluted with pesticides from nearby agricultural runoff. Kentucky's underground Mammoth Cave National Park is being polluted by local sewage disposal. At one time, tapping geothermal energy outside of the boundary of Yellowstone National Park was considered by several agencies and companies, even though concerns existed about how these geothermal areas were linked to ones within the park (such as the Old Faithful geyser).

Recent ecological analysis has suggested that effective habitat in the park is reduced near the boundary when such activities as logging, mining, and so on, occur on adjacent lands. In some respects, activities outside the park can effectively reduce the size of what was legislatively protected.

As noted in chapter 3 on the rationale for public ownership, the problem in most of these external threats to National Parks is "negative externalities" stemming from incompatible uses of adjacent public or private lands. Recognizing these threats as externalities helps point the way to solutions for reducing them.

Potential Solutions to External Threats

Several solutions have been attempted for dealing with these many types of external threats. These include land acquisition by the NPS, preservation of viewsheds and integral vistas, and ecosystem management in conjunction with interagency integrated planning. All these potential solutions share the common theme of transforming land management decisions on adjacent lands from ones that ignored the spillover effects on park resources into ones that recognize the ecological interconnectedness of the lands in a common ecological unit (e.g., a watershed or life cycle habitat area). The goal is to optimize the total social value of the land area rather than make independent decisions that ignore the interconnectedness of the ecological unit and hence result in fewer benefits. There are three approaches to interconnected planning: land acquisition and boundary adjustments, integral vistas and viewsheds, and coordinated management of public lands.

Land Acquisition and Boundary Adjustments

When parks, monuments, or other NPS units abut other federal lands, boundary adjustments to conform to ecological boundaries are always possible. For example, the northern boundary of Rocky Mountain National Park was originally a straight line. But the ridgeline separating the park from the adjoining National Forest made a more obvious boundary from the point of view of visitors, watershed, and ecology. The agencies traded land so that this more logical boundary could be implemented.

Where state and private lands were intermingled or adjacent to National Parks, the federal government has allowed the state agency or private landowner to trade for other federal lands located elsewhere. This is particularly advantageous to a state that often was given scattered tracts as school sections when the public domain was established. Blocking up those lands into manageable units aids the state and reduces threats to park resources. Land trades and boundary adjustments result in the externalities being internalized by allowing the agency that will feel the effects of resource uses to control those resources.

In other cases, the National Park Service must either purchase the land outright (obtain "fee title") or purchase an easement to restrict certain landowner practices that threaten the park. In 1964, the Land and Water Conservation Fund was established whereby a portion of monies received by the federal government for offshore oil leasing would be used to buy private inholdings or new parkland. Depending on the amount and location of this land, acquisition can be quite expensive, often involving millions of dollars. In addition, there is often some local opposition to removing land from private ownership, since it no longer generates county property taxes. This objection has been partly overcome when the federal government provides "Payment in Lieu of Taxes" for federal lands to the county to cover a portion of what would have been collected had the land remained in private ownership.

Sometimes the NPS finds it worthwhile to "lease" conservation easements or development rights from adjacent landowners. In essence, the NPS pays the landowner to forgo some privately profitable activity on their land that is or would be detrimental to park resources. For example, the NPS may purchase scenic easements to prevent buildings. Another example might be the purchase of development rights to stop a subdivision likely to introduce exotic vegetation and pets into the park or result in consumptive water withdrawals detrimental to aquatic resources in the park. With conservation easements the landowner is compensated for forgoing some land use that, although privately profitable, reduces the total social benefits. In theory, such easements allow the NPS to attain its park protection objectives without paying the full price for the land or having to withdraw it from the county tax roles. In some cases, however, the administrative costs of negotiating and monitoring the easement coupled with the high percentage of the purchase price made up by these costs, makes this alternative less attractive in practice.

Integral Vistas and Viewsheds

During the late 1970s the NPS introduced the concept of maintaining viewsheds around parks. This strategy was in response to the concern that lands outside the parks provided much of the backdrop or background for viewing resources within the park. Of course, since vistas can extend for 10 to 50 miles from some parks, attempting to control unsightly land uses on lands surrounding many parks would result in substantial intrusion on the rights of private landowners or changes in congressionally mandated management of the surrounding public lands (which are often multiple-use lands). Thus this concept ran into much resistance and seems to have died a quiet death with the change in presidential administrations in 1980.

Coordinated Management of Public Lands

This approach can be carried out in two different (but not equally effective) ways. The existing mechanism is when state and federal agencies managing lands that are ecologically related across their many boundaries agree to coordinate planning and management of those lands. They recognize it is futile for one land-managing agency to plan for and manage its lands as wilderness while another opens up its adjacent lands for mining or off-road vehicles. To reconcile management direction on an ecologically related parcel of land, a cooperative management agreement may be signed. This often states what the acceptable uses are for this area that each agency will abide by (e.g., certain access restrictions, fire management, timber management, and so on). This cooperative agreement might even give the ''lead'' management responsibility to one agency or another. This agreement would then be cited in the management plan of each agency.

A more comprehensive form of integrated ecological management would be for all agencies holding land in one ecological unit simply to prepare one comprehensive management plan for that area. A separate plan would be developed for each ecological unit by all land-owning agencies and individuals. Thus the externalities are internalized to the new expanded group acting as one landowner. Since some members will gain and some will lose as commercial activities are restricted to optimize the value of the ecological unit, compensation may be necessary if private landowners are to participate voluntarily. In cases where there are conflicting public land use mandates, it may take congressional action to sanction the de facto reorganization of multiple public land units into an ecological unit and to clarify which of the conflicting mandates should apply. However, recent trends in Forest Service management toward maintaining biological diversity and sustainable forestry move its management philosophy closer to that of the National Park Service than ever before. The concluding chapter of this book addresses the potential for ecosystem management and provides a case study of the interagency efforts in planning in California and the Greater Yellowstone ecosystem.

14

Lessons From Current Public Land Management and Directions for the Future: Ecosystem Management

Importance of Budgetary Realism in Planning

One of the recurring themes from the case studies of public land planning is the importance of adequate funding for plan implementation. The most well-balanced plan can result in unbalanced management if some features are fully funded and others not funded at all. In the case of Yosemite National Park's General Management Plan, some problems were made even worse because of partial funding of plan features: Parking areas were removed from Yosemite Valley, but the associated bus shuttle system was not adequately expanded to fill the need for nonautomobile transportation. The result was a deterioration of the visitor experience in the park.

Another common theme in National Forest and BLM planning was an attempt to ''buy one's way out'' of trading off one resource for another. Agencies frequently attempted to appease all competing sides in a natural resource conflict by investing heavily in intense natural resource management. In essence, by adding more labor and capital the production possibilities curve could be shifted outward. As will be recalled from figure 8.4, this initially allows more of both resources to be produced from the given land area. However, such investments in range improvements, fencing, planting, fisheries improvements, and so on, are very expensive. Our review of implementation of these management plans has shown that the Executive Departments (the Department of Agriculture, for the Forest Service, and the Department of Interior, for BLM, NPS, and USFWS), the Office of Management and Budget, and sometimes Congress have been unwilling to fund the investments called for in the approved management plans. Consequently, more of both outputs cannot be produced, and the agency is left, as it was before the plan, trying to juggle excess demands for access to public land resources. The agency must deal with the fact that, given the available funds, it has promised more in the plan than it can deliver. One way to deal with this is to amend the plan. In the short run, the agency must often deal with the excess demand by scaling back all users of that resource or else adopt some decision rule about whose demands are satisfied. One can guess the potential for lobbying.

In the next round of forest and BLM planning there are two strategies for coping with potential funding shortfalls. First, planners and managers must be more realistic. Plans must represent cost-effective solutions to resolve problems and issues. It is often more efficient to trade one resource use for another than to engage in capital- and labor-intensive efforts to stimulate more production from a given amount of land.

If these investments in expanding the amount of multiple-use outputs available from a given area of public land are worthwhile, better economic analysis is needed to document this. All the Department of Interior public land agencies lag far behind the Forest Service in the quality and rigor of economic analysis performed on their management actions. More thorough economic analysis might go a long way in convincing the Office of Management and Budget as well as Congress that public land management investments provide competitive returns with competing social investments.

Importance of the Ecosystem Approach to Natural Resource Planning and Management

Inadequacy of Existing Public Land Laws and Management to Perform Ecosystem Planning

In some ways we now return to themes introduced early in this book. In chapter 1 we discussed the need for integrated resource management on public and private lands. Chapter 2 reviewed recent laws that required public land managers to abandon resource-by-resource plans in favor of one integrated plan that considered all the natural resources on their respective lands.

Although these laws emphasize integrated management of the agency's natural resources, they still focus "inward." That is, they concentrate only on the lands the agency has direct control over and give minimal attention to ecological interactions with adjoining lands. As we reviewed efforts by the Fish and Wildlife Service and the National Park Service to fulfill their congressional mandates to preserve wildlife and other natural resources, the inadequacy of this inward focus on management became obvious. Both National Wildlife Refuges and units of the National Park System are merely islands of habitat or small pieces in many species' overall habitat requirements. Many National Parks or Monuments do not include the entire watershed that influences the park environment. The reader will recall from maps in chapter 10 of the San Juan Resource Area, managed by BLM, that this area adjoins National Park lands (Canyonlands National Park and Glen Canyon National Recreation Area) as well as the Manti–LaSal National Forest. Optimal management in the San Juan Resource Area requires recognition that BLM shares the management of many ecosystems along the straight-line boundaries of the neighboring National Park Service and Forest Service lands.

Current Attempts to Coordinate Federal Land Management

The inward focus among neighboring landowners is most easily corrected when the adjacent landowner is the federal government. Some coordination of independent planning and management efforts has begun among the Forest Service, BLM, USFWS, and NPS. The Framework for Coordination in the Greater Yellowstone Area between the Forest Service and the NPS is a good example of emerging formal efforts to coordinate plans across administrative boundaries. This effort is examined in more detail as a case study in this chapter.

Greater planning coordination is possible within many existing public land laws. Under FLPMA, BLM is required to plan for consistency between its plans and those of neighboring local, state, and federal landowners. Unfortunately, this clause has not been fully utilized to coordinate planning to optimize land use among the various landowners; instead it has been used sometimes as a strategic device by local county commissioners to thwart BLM's efforts to provide for multiple uses other than mining and livestock grazing. However, BLM is moving toward blending its management with that of its surrounding federal neighbors. The case study of the San Juan Resource Area in southern Utah illustrated how it proposed a more resource-protected and nonmotorized-recreation-oriented management strategy where BLM lands bordered Canyonlands National Park.

The importance of federal and private-landowner resource interdependencies is also being recognized. In Montana, for example, the Forest Service has canceled some of its timber sales so as to compensate partially for the deterioration in water quality and elk habitat from timber harvesting on private lands.

These initial efforts at reconciling resource management objectives on adjoining public and private land represent important steps in the right direction. However, true ecosystem planning requires redefinition of planning areas beyond land ownership boundaries to include entire ecological units. It is to this type of ecosystem planning that we now turn.

The Scale of Ecosystem Planning

Integrated ecosystem planning for some natural resources may involve very small areas of land when a species or resource is very localized. However, for some species like grizzly bears, the necessary habitats may be significantly larger, such as the Greater Yellowstone Ecosystem, which spans Yellowstone National Park, several adjoining National Forests, and much private land. Private and public old-growth forests for the threatened northern spotted owl in Washington, Oregon, and California will require coordinated management. The North American Waterfowl Management Plan, with its goals and objectives organized by flyways that span public and private lands in Canada and the United States, is an example of international ecosystem management. Finally, some natural resource issues, such as ozone depletion in the atmosphere or global warming, may require that the earth be viewed as one ecosphere and managed accordingly.

Management Responses Along the Ecosystem Planning Continuum

Ecosystem planning puts the scale of the resource interaction first and artificial administrative boundaries second. A continuum of planning units ranges from rather small, isolated areas for some resources, such as a particular plant species, to nearly global planning units for atmospheric resources. The smaller the planning unit, the more likely that a single landowner will be able to manage the resource effectively. The larger the ecosystem interactions, the more likely that many landowners and agencies will be involved.

In dealing with resource management issues that involve large ecosystems, landowner coordination is only one of the first possible responses to this challenge. Unfortunately, a number of natural resource management agencies have not yet taken even this first step. Rather, we often have separate and largely uncoordinated public and private natural resource plans for the owner's respective lands and resources. Coordinating these resource management plans is a step toward ecosystem planning but may be inadequate in some cases. However, many other, more comprehensive choices are possible.

A Single Federal Ecosystem Management Plan

For those resources that are broad in scope, the next step may be one unified federal agency management plan. The management of all federal lands—including not only BLM, Forest Service, USFWS, and NPS but also military lands administered by Department of Defense and other federal agencies within the ecosystem—would follow this one mutually agreed plan. Each agency would have its own actions to implement on its own lands within this joint ecosystem, but what management actions to take and when to take them would have been decided jointly in one integrated ecosystem plan. Of course, each agency would follow its respective mandate on lands unrelated to the shared or joint ecosystem.

A Single State/Federal Ecosystem Management Plan

The next step in the continuum would be to develop one management plan for all federal and state lands within a particular ecosystem. The result would be a joint federal/state ecosystem plan that would guide the public agencies' resource management decisions. This would include management of state parks, state forests, and so on, along with federal lands. Such a joint state/federal plan may be appropriate when the ecosystem includes much intermixed federal and state lands, as is common in the western United States.

A Single Public Land Ecosystem Management Plan

A small extension of the preceding arrangement would be to include county and city lands that lie within the ecosystem into this one management plan. For example, when cities or counties own land in the ecosystem for protection of municipal water quality or as its own parkland, these areas would be included in the single ecosystem management plan. The management plan would likely not compromise the primary intent of these county and city lands but would blend their management into the overall goals for the ecosystem.

A Single Ecosystem Management Plan for All Landowners

Finally, there may be pervasive resource management issues that require that all land ownerships be linked in one ecosystem plan. For example, land management decisions by federal, state, county, city, and private landowners all influence the water quality of Lake Tahoe, and all are regulated. For natural resources such as this, what is required is the preparation and implementation of *one* ecosystem plan to which all parties must adhere. As will be discussed later, economic incentives must be provided to private landowners to ensure their participation, as enforcement authorities can never check on everyone.

An example of this single ecosystem management approach is provided by the Pinelands National Reserve in New Jersey. As described by Lilieholm (1990:10), "the 1.1 million acre Pinelands National Reserve forged a powerful alliance between private interests and federal, state and local governments to protect New Jersey's pinelands from encroaching development. The reserve is managed under a regional plan designed to channel development away from environmentally sensitive areas and into designated growth centers. Intergovernmental cooperation, landowner incentives and an innovative compensation package have resulted in widespread support and compliance."

Integrated Ecosystem Planning of Mixed Public and Private Lands

Evolving Definition of Private Property Rights

Integrated natural resource management may require continued evolution in the definition of private property rights for land, trees, and water. The extent of private property rights has been changing to recognize ecological interrelationships. For example, society has been redefining water rights to recognize the ecological benefits that water flows provide to such public trust resources as fish and wildlife (Casey 1984). Water rights in California are being revised so that water diversions that result in "unreasonable damage" to publicly held natural resources are no longer allowed. In setting minimum flows on hydropower projects, the Federal Energy Regulatory Commission is required to give "equal consideration" to fish and wildlife resources. Numerous states have Forest Practice Acts that control timber harvest practices on private lands. The goal is to internalize the negative externalities on fish and water quality associated with timber harvests on private lands. The owner has a right to harvest the timber but not to impose unreasonable costs on public waters and fisheries resources.

Combined ecological management of both private and publicly owned lands may take a significant step forward with the June 1990 decision to list the northern spotted owl as a threatened species under the Endangered Species Act. Protection of the owl involves millions of acres and numerous land ownerships (two federal agencies, three states, and in some areas private landowners). The timber harvest plans of these landowners must be coordinated to ensure that the owl does not become endangered or extinct. The remaining old growth can be viewed as a common nonrenewable oil or groundwater deposit with multiple owners. As has been demonstrated in many examinations of the problem of the commons, uncoordinated free-rider behavior associated with the "tragedy of the commons" (Hardin 1968) will continue unless the force of the Endangered Species Act is used to require coordinated management. The act appears to be forcing action in this case: a protection and recovery plan must be developed for the northern spotted owl. The state of California Department of Forestry is preparing a Habitat Conservation Plan for the Northern Spotted Owl and associated old-growth habitat on private and state lands so as to determine what logging can be approved on private lands in the state. This effort will serve as a focal point for the many landowners to develop ecologically and economically optimum harvest plans that allow for sharing the gains and losses associated with owl habitat protection.

Importance of Accurate Estimates of Costs and Benefits in Ecosystem Plans

Integrated natural resource planning and management across numerous landownerships requires better information on the costs and benefits of management than when each owner was concerned only about his or her own land. Where optimum ecological management requires restriction on private land use (because, for example, these private lands are often in critical winter range, valley bottoms, or riparian areas and therefore have high habitat value), the planners must be certain the public will compensate owners.

Thus integrated ecological planning must also be economic planning in the following sense. Private landowners and many state agencies will voluntarily engage in planning by ecological units only if they find it at least as beneficial as their current land management practices. Whatever the coordinated management plan suggests, there must be enough financial return to private landowners that they can pay their mortgages and taxes.

Potential Cost Savings of Including Private Lands as a Management Option

There is a potential mutually beneficial gain from coordinated ecological management in that a public agency may be able to meet its congressional mandate for resource preservation or multiple use more cheaply by including coordinated management on private lands as part of a management alternative. As mentioned earlier, this would work well when private landowners control habitats that may be the limiting factors for wildlife or when they control access to recreational resources, such as rivers, or their land has a major influence on water quality or quantity. The public agency might find that it gets a much larger return on its investment by putting a management practice on the more productive private lands than it would on an equivalent investment in public lands. For example, compensating a rancher for removing his or her cows from the river bottom may improve the fish habitat much more per dollar spent than would improving upland (publicly owned) tributary streams. But this cost savings must be documented to the agency heads, the Office of Management and Budget, and Congress if the funding for cattle removal is to be secured. One role of benefit-cost analysis in ecosystem planning is to identify these new cost savings in coordinated ecological management across public and private lands.

The Potential for the Federal Ecosystem Planning

The second round of National Forest planning required under NFMA provides a good opportunity to begin ecosystem planning on federal lands. The Forest Service should look outward to at least formally coordinate these new Forest Plans with surrounding federal and state landowners. Certainly, the BLM and the Forest Service should coordinate preparation of their respective second-generation Resource Management Plans and National Forest Plans when these areas are adjacent to one another or lie within the same ecosystem. These two agencies already share similar multiple-use mandates and some similar analysis techniques (the IMPLAN Input-Output model), and in many cases a wildlife species uses the lands of both agencies each year (e.g., high alpine areas managed by the Forest Service and mid-elevation lands under BLM management). As was illustrated on the map of the San Juan Resource Area (see figure 10.3), the proximity of these two landownerships would likely result in many cost savings when planning.

Where National Parks or Wildlife Refuges adjoin BLM lands, coordinating the production of a single ecosystem plan is particularly feasible because all these agencies are in the Department of Interior! A single plan could be developed that would meet the legal requirements of each agency yet optimize their resource values, rather than simply generate more interagency conflicts. Although the agency mandates are different, they have more in common than in conflict. The NPS is charged with preserving natural resources, including fish and wildlife, while allowing for their use by people, a responsibility similar in intent to that of the National Wildlife Refuges. Fish and wildlife constitute one of the multiple uses that BLM is charged with managing. Thus on lands within shared ecosystems, the similar components of the agencies' mandates would rationalize similar management actions. Where BLM and the Forest Service have multiple uses that are incompatible with preserving fish, wildlife, or primitive recreation, these other multiple uses need to occur in areas ecologically unrelated to the shared ecosystem or to be carried out so that they are much less incompatible with the shared ecosystem resources. Rather than stress the differences between agencies and mandates, it is time to emphasize similarities.

A pilot test of ecosystem planning could easily be performed where all federal land management agencies share adjacent lands within the same designated Wilderness ecosystem. On these lands Congress has given identical mandates to the different agencies. Unified plans for these Wilderness ecosystems would be a good place for federal land management agencies to begin.

In the end, entire ecosystems (including people) will likely be better off from greater integration of land management policy and practices across landownerships. The next sections present two current efforts at ecosystem planning. The first, the Greater Yellowstone Ecosystem, involves joint National Park Service and Forest Service efforts. The second, in California, involves federal, state, and local government efforts at ecosystem planning.

Case Study of Greater Yellowstone Area Coordinated Planning: The Beginnings of Ecosystem Planning Between USFS and NPS

The Greater Yellowstone Area (sometimes referred to as the Greater Yellowstone Ecosystem) is a 31,000-square-mile area that encompasses at its center Yellowstone National Park, designated in 1872 as the first U.S. National Park. The park is surrounded by portions of seven National Forests in three states. Intermingled in this Greater Yellowstone Ecosystem are substantial amounts of private lands, a moderate amount of state land, and Indian reservations. Finally, small amounts of BLM and U.S. Fish and Wildlife Service land are included in the Greater Yellowstone Ecosystem (GYE). Figure 14.1 illustrates the numerous landownerships and their many straight-line administrative boundaries.

The early history of public land management in GYE is typified by each agency managing its own lands, with little formal coordination. Timber sales, fire sup-

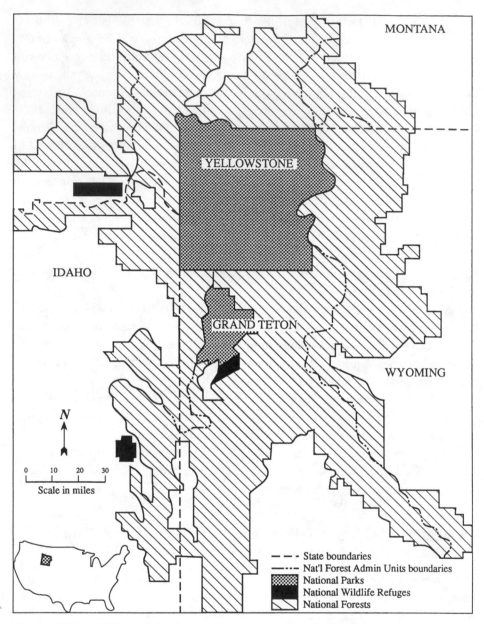

Figure 14.1.   A Map of the Greater Yellowstone Region Showing Major Administrative Units.

pression efforts, and oil and gas exploration on adjacent National Forests and BLM lands often resulted in spill-over impacts on Yellowstone National Park or wildlife species that spend their annual life cycles in and around Yellowstone National Park. In the early 1960s a Greater Yellowstone Coordinating Committee (GYCC) was created to foster communication between the federal agencies (GYCC 1990:5).

Management of grizzly bears, however, was what made the ecological interrelatedness of the federal lands in this area apparent. Wildlife biologist John Craighead introduced the term *Greater Yellowstone Ecosystem* in describing the "crossboundary" management strategy required for grizzly bears.

By 1986, a Congressional Research Service study concluded that there was a "relative lack of coordinated information for the entire GYE, and that this lack of coordination was harmful to GYE's fundamental values" (GYCC 1990:5). In September of 1986, the regional foresters for the three Forest Service Regions with National Forests abutting Yellowstone National Park signed a Memorandum of Understanding to begin to remedy this problem. In 1987, the GYCC compiled information from the NPS General Management Plan for Yellowstone with the data from the surrounding Forest Plans. The predicted trends over the next ten to fifteen years from implementing the existing plans were then analyzed. From this it was apparent that the Forest Plans and the NPS General Management Plan would need to be modified (i.e., formally amended) if GYE values were to be maintained. As a first step in that process, the GYCC developed a set of goals for management of the GYE. These were the shared "visions" of NPS and Forest Service managers for GYE. (BLM and USFWS are not formally participating agencies because they are relatively small public landholders and therefore have only liaison observers.)

The draft document contained three "fundamental philosophies," with numerous other specific subobjectives that have been drafted by the NPS and Forest Service (GYCC 1990). The three major goals were (1) to conserve the sense of naturalness and maintain ecosystem integrity; (2) to encourage opportunities that are biologically and economically sustainable; (3) to improve coordination among federal, state, and local governments as well as private landowners and users of the GYE. Thus explicit consideration of the effects of Forest Service management on Yellowstone National Park and vice versa would be part of the decision-making framework of these two agencies. This would be accomplished by making formal amendments to Forest Plans and revisions to Yellowstone National Park's General Management Plan (GYCC 1990: chapter 5:1). These amendments and revisions would help ensure that there are common goals, management standards, land practices and prescriptions for the two agencies' lands within the Greater Yellowstone Ecosystem.

The draft NPS/Forest Service joint report called Vision for the Future: A Framework for Coordination in the Greater Yellowstone Area, summarizes the difficult balancing act facing the agencies when Congress asked for more coordination in its oversight hearings: "In many instances, the coordinating criteria in the Vision clearly represent new ways of doing business, not only between the two agencies but also between the agencies and their many constituencies, contractors, and concessionaires. *The central element of the shift is toward ecosystem management,* but achieving that major shift will require many minor operational shifts. At the same time, the Forest Service and Park Service will not abandon their separate and often quite distinct mandates'' (GYCC 1990: chapter 4:1, emphasis added). The joint NPS/Forest Service report goes on to suggest that if Congress is serious about coordinated management and ecosystem planning, then additional funding and staffing will be required. Of course this is a two-way street, with the burden of proof on the two agencies to document the benefits of implementing ecosystem planning to warrant this additional cost. The draft *Vision for the Future* was seen by one of its coauthors, NPS Regional Director Lorraine Mintzmayer, as a model for interagency cooperation in this area and a model for other areas that would carry resource management well into the next century.

Unfortunately, such a visionary document that emphasized ecosystem concerns may have been ahead of its time in the northern Rocky Mountain region. Unlike the California case (to be discussed later), where the ecosystem planning effort was initiated by state agencies, the governors of Montana and Wyoming, along with their congressional delegations, were opposed to broad ecosystem planning, as described in the draft *Vision* document. These officials joined commodity groups in attacking the draft as limiting development on state and private lands and on National Forests.

These congressional delegations met with the Bush administration's respective assistant secretaries of agriculture (for Forest Service) and Fish, Wildlife, and Parks (Sewell in Interior) to urge major changes in the final document. After that meeting the Department of Interior rewrote the original statement. Neither Mintzmayer nor the Forest Service leader (Regional Forester John Mumma) was involved in the revision that became the final *Framework for Coordination of National Parks and Forests in the Greater Yellowstone Area.* In fact, both leaders were given forced reassignments away from offices and positions dealing with the Greater Yellowstone Area.

The final document retreated significantly from the draft objectives of ecosystem planning for protecting naturalness and biodiversity. Rather than seeking the common ground between the two agencies, the framework reinforces the separate missions of the two agencies with respect to the lands they manage. Although this may have satisfied the local politicians, no administrative rewrite can hide the management conflicts that gave rise to the need for ecosystem management in the Greater Yellowstone Ecosystem. The need for ecosystem planning is evident to most natural resource professionals in the area, but many intermountain politicians are still rooted in isolated agency planning. As the shortcomings of inward-focused agency planning become more and more obvious, the resistance to ecosystem planning in areas such as Yellowstone will end. Whether thrust upon this region by the Endangered Species Act or voluntarily accepted, more and more public lands management planning will become, if not explicitly at least implicitly, ecosystem planning by natural resource professionals.

A State and Federal Bioregional Approach to Resource Management in California

During the summer of 1991 state and federal resource management agencies within California signed a potentially historic agreement to implement a coordinated regional strategy to attain biological diversity on ten bioregional areas in California. The ten biogregions are shown in figure 14.2 and represent all landownerships within a bioregion. The public participants include California departments of Fish and Game, Forestry, and Parks as well as the federal Bureau of Land Management,

Figure 14.2. Generalized Bioregional Areas

Forest Service, Fish and Wildlife Service, and National Park Service (i.e., the major public landowners in California). To understand the significance of this accord we must understand both what they define as biological diversity and the means they propose to attain it.

The Goal: Biological Diversity

The participants define biological diversity broadly: from the genetic diversity of individual populations to ecosystems and finally large-scale regional landscapes. The goal is to maintain the full variety of life and its processes regardless of specific landownerships. In the case of ecosystems, biodiversity is indicated by the number of species, number of successional stages, variety of trophic levels, and interactions. At the landscape level there is recognition of the need for planned redundancy of widely separated but similar ecosystems as insurance against locally catastrophic events.

The Means to Attain Biodiversity

These agencies will rely on several institutional mechanisms to attain their goals. At the state level is an Executive Council composed of the heads of all the seven resource agency managers listed earlier, with the secretary of the California Resources Agency as the chair. "The Council will set statewide goals for the protection of biological diversity, recommend consistent statewide standards and guidelines, encourage cooperative projects and sharing of resources . . ." (California Resources Agency 1991:2). Of the several areas of specific cooperation listed, the most effective are likely to be land use planning, land acquisition, land exchange, private landowner assistance, and mitigation banks.

Each of the ten bioregions will have its own Bioregional Council. The composition of these councils will include the relevant land management agencies as well as local governments, environmental groups, and industry. The Bioregional Councils will tailor the state-level principles on biological diversity to the specific bioregion.

Within each bioregion, formation of watershed associations is encouraged. They will "develop specific cooperative projects" using a Coordinated Resource Management Plan. Figure 14.3 illustrates the relationships among the various institutions.

Experience with Coordinated Resource Management Plans in California

In some ways, the faith placed in local associations to develop a single Coordinate Resource Management Plan (CRMP) is well placed. Several resource management

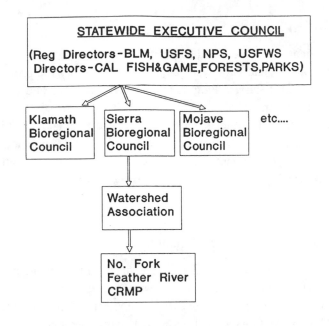

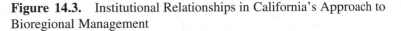

CRMP=Coordinated Resource Mgmt Plan

Figure 14.3. Institutional Relationships in California's Approach to Bioregional Management

success stories on multiple landownerships are due to CRMPs in California. One such CRMP is on a tributary of the North Fork Feather River, called Red Clover Creek. It involves three federal agencies, four state agencies, the county, a private utility company, and numerous private landowners. The first demonstration project on Red Clover Creek identified the riparian areas as being the critical factor on which soils, vegetation, water quality, fisheries, wildlife, and other resources depended. A jointly designed and funded rehabilitation effort on ten miles of stream on private land has been relatively successful. By constructing check dams, replanting willows, and restricting private livestock grazing, fisheries in the stream and waterfowl on reemerging wetlands have increased by 200% or more.

Importance of Coordination

The cornerstone at all three institutional levels of bioregional planning is the provision of a forum and focus for coordination among both private and public land managers. For any bioregion, the Bioregional Councils can establish consistent management practices that they will ask all participants to follow. Since no new legal authority is provided at any of the three institutional levels, only the spirit of consensus can bind the parties together.

Land Acquisition Strategies

However, coordinated easement and acquisition strategies are another mechanism open to Bioregional Councils or watershed associations. This approach has been used successfully in the Carrizo Plain area of San Luis Obispo and Kern Counties to increase public ownership of land from 30,000 acres to 180,000 acres. Cooperators included the federal agencies, such as BLM and Fish and Wildlife Service; state agencies, such as the California Department of Fish and Game and the Energy Commission; county governments; and private groups, such as the Nature Conservancy, the Audubon Society, and oil companies owning land and operating in the area.

Implications for Federal Land Management

The performance of this bioregional resource management effort will be worth watching over the next ten years. Its goal is a worthy one: maintenance of biological diversity by recognizing the importance of ecological rather than administrative boundaries. Unlike the Greater Yellowstone Ecosystem effort, which formally involves only two federal agencies, this effort in California includes nearly all the major state and federal resource management agencies. If federal land management agencies can successfully participate with state and local governments and private landowners in California in attaining biological diversity, then comprehensive ecosystem planning may replace the current inward-focused administrative planning we have reviewed in the earlier chapters. Bioregional planning and ecosystem management may well represent an integrated and proactive approach to public land management for the next century.

References

Adams, Darius and Richard Haynes. 1980. The 1980 softwood timber assessment market model. *Forest Science* 26, no. 3: supplement.

Allin, Craig. 1990. Agency values and wilderness management. In J. Hutcheson, Jr., F. Noe, and R. Snow, eds., *Outdoor Recreation Policy: Pleasure and Preservation*. New York: Greenwood Press.

Alward, Gregory and Charles Palmer. 1983. IMPLAN: An input-output analysis system for Forest Service planning. In *IMPLAN Training Notebook*. Land Management Planning, Rocky Mountain Forest and Range Experiment, U.S. Forest Service, Fort Collins, Colo.

Alward, G., H. Davis, K. Despotakis, and E. Lofting. 1985. Regional non-survey input-output analysis with IMPLAN. Paper presented at Southern Regional Science Association Conference. Washington, D.C.

Alward, Gregory, William Workman, and Wilbur Maki. 1992. Regional economic analysis for Alaskan wildlife resources. In G. Peterson, C. Swanson, D. McCollum, and M. Thomas, eds., *Valuing Wildlife Resources in Alaska*. Boulder, Colo. Westview Press.

Amend, Spencer. 1990. Letter from Spencer Amend, U.S. Fish and Wildlife Service, Fort Collins, Colo., to John Loomis. December 19.

Andrews, Karen. 1983. "Recreation benefits and costs of the proposed Deer Creek Reservoir." Wyoming Recreation Commission. Cheyenne, Wyo.

Andrews, Kent, Patrick Sousa, and Adrian Farmer. 1984. Final Report, Habitat Management Evaluation Model Project, Phase I-Feasibility. WELUT 85/W01, Western Energy and Land Use Team, Division of Biological Services, U.S. Fish and Wildlife Service, Fort Collins, Colo.

Bailey, Kenneth. 1987. *Methods of Social Research*. 3d ed. New York: The Free Press.

Baltic, Tony, John Hof, and Brian Kent. 1989. Review of Critiques of the USDA Forest Service Land Management Planning Process. General Technical Report RM-170, Rocky Mountain Forest and Range Experiment Station, U.S. Forest Service, Fort Collins, Colo.

Bangs, Edward, Theodore Bailey, Robert Delaney, Michael Hedrick, Richard Johnston, Norm Olson, and James Freidersdorff. 1986. Land use planning on the Kenai National Wildlife Refuge, Alaska. *Transactions of 51st North American Wildlife and Natural Resources Conference*. Washington, D.C.:Wildlife Management Institute.

Barlow, T., G. Helfand, T. Orr, and T. Stoel. 1980. *Giving Away the National Forests.* Washington, D.C.: Natural Resources Defense Council.

Bartlett, E. T. 1982. Valuing range forage on public rangelands. In F. J. Wagstaff, ed., *Proceedings: Range Economics Symposium and Workshop,* General Technical Report INT-149. Intermountain Forest and Range Experiment Station, U.S. Forest Service, Ogden, Utah.

Bartlett, E. T. 1984. Estimating benefits of range for wildland management and planning. In G. L. Peterson and A. Randall, eds., *Valuation of Wildland Resource Benefits.* Boulder, Colo.: Westview Press.

Baumol, William. 1978. On the social rate of discount. *American Economic Review* 68: 788–802.

Bean, Michael J. 1983. *The Evolution of National Wildlife Law.* Rev. ed. New York: Praeger.

Behan, Richard. 1992. Multi-resource Forest Management and National Forest Planning with EZ-IMPACT Simulation Models: How to Get by in Style Without a Linear Programming Fix. School of Forestry, Northern Arizona University, Flagstaff, Ariz.

Bishop, Dick. 1981. Alaska National Interest Lands Conservation Act: What Now? Alaska Department of Fish and Game, Juneau, Alaska.

Bishop, Richard and Thomas Heberlein. 1979. Measuring values of extra-market goods: Are indirect measures biased? *American Journal of Agricultural Economics* 61, no. 4: 926–930.

Bishop, Richard, Kevin Boyle, Michael Welsh, Robert Baumgartner, and Pamela Rathburn. 1986. Glen Canyon Dam Releases and Downstream Recreation: An Analysis of User Preferences and Economic Values. Final Report to the Bureau of Reclamation. HBRS, Madison, Wis.

Boadway, Robin. 1974. The welfare foundations of cost-benefit analysis. *Economic Journal* 84, no. 336: 926–939.

Boadway, Robin. 1979. *Public Sector Economics.* Cambridge, Mass.: Winthrop.

Bockstael, Nancy and Kenneth McConnell. 1981. Theory and estimation of household production function for wildlife recreation. *Journal of Environmental Economics and Management* 8 (September): 199–214.

Boskin, Michael, Marc Robinson, Terrance O'Reilly, and Praven Kumar. 1985. New estimates of the value of federal mineral rights and land. *American Economic Review* 75, no. 5: 923–936.

Bovee, Ken. 1982. A Guide to Stream Habitat Analysis Using the Instream Flow Incremental Methodology. Instream Information Paper No. 12. Washington, D.C.: U.S. Fish and Wildlife Service.

Bowes, Michael and John Krutilla. 1989. *Multiple Use Management: The Economics of Public Forestlands.* Washington, D.C.: Resources For the Future.

Bradley, Iver and James Gander. 1968. Utah Interindustry Study: An application of input-output analysis. *Utah Economic and Business Review* 28, no. 2.

Brookshire, David, Alan Randall, and John Stoll. 1980. Valuing increments and decrements in natural resource service flows. *American Journal of Agricultural Economics* 62: 478–488.

Brookshire, David, William Schulze, Mark Thayer, and Ralph d'Arge. 1982. Valuing public goods: A comparison of survey and hedonic approaches. *American Economic Review* 72, no. 1: 165–177.

Brookshire, David, Larry Eubanks, and Cindy Sorg. 1986. Existence values and normative economics: Implications for valuing water resources. *Water Resources Research* 22, no. 11:1509–1518.

Brown, Gardner and Robert Mendelshohn. 1984. The hedonic travel cost method. *Review of Economics and Statistics* 66 (August): 427–433.

Brown, Gardner and Michael Hay. 1987. Net economic recreation values for deer and waterfowl hunting and trout fishing. Working Paper No. 23, Division of Policy and Directives Management, U.S. Fish and Wildlife Service, Washington, D.C.

Bureau of Land Management. 1977. Final Environmental Impact Statement: Proposed Domestic Livestock Grazing Program for the Challis Planning Unit. Idaho State Office, Boise, Idaho.

Bureau of Land Management. 1983a. A Guide to Resource Management Planning on the Public Lands. No. 1983–574–232, Washington, D.C.: U.S. Government Printing Office.

Bureau of Land Management. 1983b. Planning, programming, budgeting; amendments to the planning regulations, elimination of unneeded provisions. *Federal Register* 48, no. 88: 20364–20375.

Bureau of Land Management. 1984a. Prescribed Resource Management Planning Action. BLM Manual Series, Subject 1616. Washington, D.C. April 6.

Bureau of Land Management. 1984b. Resource Management Plan, Approval, Use, and Modification. BLM Manual Series, Subject 1617. Washington, D.C. April 6.

Bureau of Land Management. 1985. SAGERAM Computer Program: Resource Investment Analysis. BLM Manual, Handbook H-1743–1. Washington, D.C.

Bureau of Land Management. 1986. Supplemental Program Guidance, BLM Manual Series, Subject 1620. Washington, D.C. November 14.

Bureau of Land Management. 1989. *Public Land Statistics 1988,* vol. 173. Washington, D.C.: U.S. Department of Interior, March.

Burt, Oscar and Durwood Brewer. 1971. Evaluation of net social benefits from outdoor recreation. *Econometrica* 39 (September): 813–827.

California Resources Agency. 1991. Memorandum of Understanding: California's Coordinated Regional Strategy to Conserve Biological Diversity. Sacramento.

Casey, E. 1984. Water Law–Public Trust Doctrine. *Natural Resources Journal* 24, no. 3: 809–825.

Cesario, Frank. 1976. Value of time in recreation benefit studies. *Land Economics* 52: 32–41.

Cherfas, Jeremy. 1990. Software for hard choices. *Science* 250: 367–368. October.

Clawson, Marion and Jack Knetsch. 1966. *Economics of Outdoor Recreation.* Baltimore, Md.: Johns Hopkins University Press.

Clawson, Marion. 1975. *Forests for Whom and for What?* Washington, D.C.: Resources for the Future.

Clawson, Marion. 1978. The concept of multiple use forestry. *Environmental Law* 8: 281–308.

Clawson, Marion. 1983. *The Federal Lands Revisited.* Washington, D.C.: Resources for the Future

Colby, Bonnie. 1987. *Water Markets in Theory and Practice.* Boulder, Colo.: Westview Press.

Coppedge, Robert. 1977. Income Multipliers in Economic Impact Analysis, No. 400 X-5. Cooperative Extension Service, New Mexico State University, Las Cruces.

Cortner, Hanna and Dennis Schweitzer. 1981. Institutional limits to national public planning for forest resources: The Resources Planning Act. *Natural Resources Journal* 21: 203–222.

Cortner, Hanna and Dennis Schweitzer. 1983. Limits to hierachical planning and budgeting systems: The case of public forestry. *Journal of Environmental Management* 17: 191–205.

Cory, Dennis and William Martin. 1985. Valuing wildlife for efficient multiple use: Elk versus cattle. *Western Journal of Agricultural Economics* 10: 282–293.

Crawford, John. 1986. Land use planning in the Bureau of Land Management. *Transactions of 51st North American Wildlife and Natural Resources Conference.* Washington, D.C.: Wildlife Management Institute.

Crowe, Douglas. 1984. Comprehensive Planning for Wildlife Resources. Wyoming Game and Fish Commission. Cheyenne, Wyo.

Cummings, Ronald, David S. Brookshire, and William Schulze, eds. 1986. *Valuing Environmental Goods: An Assessment of the Contingent Valuation Method.* Totowa, N.J.: Rowman and Allanheld.

Culhane, Paul. 1981. *Public Land Politics: Interest Group Influence on the Forest Service and the Bureau of Land Management.* Washington, D.C.: Resources for the Future.

Dana, Samuel and Sally Fairfax. 1980. *Forest and Range Policy.* 2d ed. New York: McGraw-Hill.

Danzig, G. 1963. *Linear Programming and Extension.* Princeton, N.J.: Princeton University Press.

Davis, Lawrence and Norman Johnson. 1987. *Forest Management.* 3d ed. New York: McGraw-Hill.

Davis, O. and A. Whinston. 1965. Welfare economics and the theory of second best. *Review of Economic Studies* 32: 1–14.

Davis, Robert, K. 1963. The Value of Outdoor Recreation: An Economic Analysis of the Maine Woods. Ph.D. dissertation, Harvard University, Cambridge, Mass.

deFranceaux, Cynthia. 1987. National Park Service planning. *Trends* 24, no. 2: 13–19.

Donnelly, Dennis and Louis Nelson. Net Economic Value of Deer Hunting in Idaho. Resource Bulletin, Rocky Mountain Forest and Range Experiment Station, 1985.

Duffield, John. 1984. Travel Cost and Contingent Valuation: A Comparative Analysis. In V. Smith and A. Witte, eds., *Advances in Applied Micro-Economics.* Greenwich, Conn.: JAI Press.

Duffield, John, John Loomis, and Rob Brooks. 1987. The Net Economic Value of Fishing in Montana. Technical Report, Montana Department of Fish, Wildlife, and Parks. Helena, Mont.

Dwyer, John, John Kelly, and Michael Bowes. 1977. Improved Procedures for Valuation of the Contribution of Recreation to National Economic Development. Water Resources Center Report 128. University of Illinois at Urbana-Champaign.

Dyer, A. A. and E. T. Bartlett. 1974. Decision making systems in resource management. Unpublished paper, Colorado State University, Fort Collins, Colo.

Dyer, A. A. 1984. Public natural resource management and valuation of nonmarket outputs. In National Academy of Sciences, National Research Council, *Developing Strategies for Rangeland Management.* Boulder, Colo.: Westview Press.

Dykstra, Dennis. 1984. *Mathematical Programming for Natural Resource Management.* New York: McGraw-Hill.

Everhart, William. 1983. *The National Park Service.* Boulder, Colo.: Westview Press.

Farmer, Adrian, Michael Armbruster, James Terrell, and Richard Schroeder. 1982. Habitat Models for Land Use Planning: Assumptions and Strategies for Development. *Transactions of the 47th North American Wildlife and Natural Resources Conference.* Washington, D.C.: Wildlife Management Institute, pp. 47–56.

Farmer, Adrian, Patrick Sousa, Rodney Olson, and Jacy Hoekenstrom. 1989. A Microcomputer Program for Use in Designing Cost Effective Habitat Management Plans. In *Freshwater Wetlands and Wildlife,* R. Sharitz and J. Gibbons, eds. DOE Symposium Series No. 61, USDOE, Oak Ridge, TN.

Feldstein, Martin. 1985. International trade, budget deficits and the interest rate. *Journal of Economic Education* 16: 189–193.

Fish and Wildlife Service. 1980. Habitat Evaluation Procedures (HEP). Ecological Services Manual 102. Division of Ecological Services. Washington, D.C.

Fish and Wildlife Service. 1982. Refuge Manual. Washington, D.C.: U.S. Fish and Wildlife Service.

Fish and Wildlife Service. 1984. Draft Environmental Impact Statement: Management of Charles M. Russell National Wildlife Refuge. Regional Office. Denver, Colo.

Fish and Wildlife Service. 1985. Final Environmental Impact Statement: Management of Charles M. Russell National Wildlife Refuge. Regional Office. Denver, Colo.

Fish and Wildlife Service. 1988a. Arctic National Wildlife Refuge: Comprehensive Conservation Plan, Environmental Impact Statement, Wilderness Review and Wild River Plan. Draft Summary. Region 7, Anchorage, Alaska. January.

Fish and Wildlife Service. 1988b. Arctic National Wildlife Refuge: Comprehensive Conservation Plan, Environmental Impact Statement, Wilderness Review and Wild River Plan. Final. Region 7, Anchorage, Alaska. September.

Fish and Wildlife Service. 1989. *Waterfowl 2000: News from the North American Waterfowl Management Plan* 1, no. 5. Twin Cities, Minn. September.

Flaters, Curtis and Thomas Hoekstra. 1989. An Analysis of the Wildlife and Fish Situation in the United States: 1989–2040. General Technical Report RM 178. Rocky Mountain Forest and Range Experiment Station, U.S. Forest Service, Fort Collins, Colo.

Foresta, Ronald. 1984. *America's National Parks and Their Keepers.* Washington, D.C.: Resources for the Future.

Forest Service. 1981. An Assessment of the Forest and Range Land Situation in the United States. Forest Resource Report No. 22. Washington, D.C. October.

Forest Service. 1982. National forest system land and resource management planning. *Federal Register* 47, no. 190: 43026–43052. September 30.

Forest Service. 1985. Hyalite-Porcupine Buffalo Horn Wilderness Study Report. Gallatin National Forest, Bozeman, Mont.

Forest Service. 1986. Siuslaw National Forest Plan, Draft Environmental Impact Statement. Siuslaw National Forest, Corvallis, Oreg.

Forest Service. 1989. An Analysis of the Land Base Situation in the United States: 1989–2040. General Technical Report RM 181, Rocky Mountain Forest and Range Experiment Station. Fort Collins, Colo.

Forest Service. 1989. RPA Assessment of the Forest and Rangeland Situation in the United States, 1989. Forestry Research Report No. 26. Washington, D.C.

Forest Service. 1990a. Recommended 1990 RPA Program: A Long-Term Strategic Plan. Washington, D.C. May.

Forest Service. 1990b. Shasta Costa: From a New Perspective. Siskiyou National Forest, Gold Beach Ranger District, Gold Beach, Oreg.

Forest Service. 1991. Advance notice of proposed rulemaking: National Forest System land and resource management planning, 36 CFR Part 219. *Federal Register* 56 (February 15): 6507–6538.

Foss, Phillip. 1958. *The Battle for Soldier Creek.* The Inter-University Case Program Publication No. 65. University, Ala.: University of Alabama Press.

Foss, Phillip. 1960. *Politics and Grass.* Seattle, Wash.: University of Washington Press.

Foster, H. and Bruce Beattie. 1979. Urban residential demand for water in the United States. *Land Economics* 55, no. 1: 43–58.

Freeman, Myrick. 1979. *The Benefits of Environmental Improvement.* Baltimore, Md.: Johns Hopkins University Press.

Garcia, Margot. 1987. FORPLAN and land management planning. In Peter Dress and Richard Fields, eds., *The 1985 Symposium on Systems Analysis in Forest Resources.* Athens, Ga.: Georgia Center for Continuing Education.

Gee, Kerry. 1981. Estimating Economic Impacts of Adjustments in Grazing on Federal Lands and Estimating Federal Rangeland Forage Values. Technical Bulletin 143. Colorado State University Experiment Station, Fort Collins, Colo.

Gibbons, Diana. 1986. *The Economic Value of Water.* Washington, D.C.: Resources for the Future.

Gibbs, Deborah. 1982. A Comparison of Travel Cost Models Which Estimate Recreation Demand for Water Resource Development. Master's Thesis, Department of Mineral Economics, Colorado School of Mines, Golden, Colo.

Godfrey, Bruce. 1982. Economics and multiple use management on federal rangelands, In F. J. Wagstaff, ed., *Proceedings: Range Economics Symposium and Workshop.* General Technical Report INT-149. Intermountain Forest and Range Experiment Station, U.S. Forest Service, Ogden, Utah.

Gray, Lee and Robert Young. 1984. Valuation of water on wildlands. In G. Peterson and A. Randall, eds. *Valuation of Wildland Resource Benefits.* Boulder, Colo.: Westview Press.

Greater Yellowstone Coordinating Committee. 1990. Vision for the Future: A Framework for

Coordination in the Greater Yellowstone Area, Draft. Custer National Forest, Billings, Mont. August.

Hahn, W., T. Crawford, K. Nelson, and R. Bowe. 1989. Estimated Forage Values for Grazing National Forest Lands. USDA Economic Research Service Staff Report 89–51. Washington, D.C.

Hanemann, Michael. 1984. Welfare evaluations in contingent valuation experiments with discrete responses. *American Journal of Agricultural Economics* 66, no. 3: 332–341.

Hardin, Garrett. 1968. Tragedy of the commons. *Science* 162: 1243–1248.

Harmon, Mark, William Ferrell and Jerry Franklin. 1990. Effects on Carbon Storage of Conversion of Old-Growth Forests to Young Forests. *Science* 247 (February 9):699–701.

Heller, David, James Maxwell, and Mit Parsons. 1983. Modelling the Effects of Forest Management on Salmonid Habitat. Siuslaw National Forest, Corvallis, Oreg.

Hendee, John, George Stankey, and Robert Lucas. 1978. Wilderness Management. Miscellaneous Publication No. 1365. U.S. Forest Service. Washington, D.C.

Hoekstra, Thomas, Greg Alward, A. Dyer, John Hof, Daniel Jones, Linda Joyce, Brian Kent, Robert Lee, Randall Scheffield, and Robert Williams. 1990. *Critique of Land Management Planning, Vol. 4: Analytical Tools and Information.* FS-455. Policy Analysis Staff, U.S. Forest Service, Washington, D.C.

Hoehn, John and Alan Randall. 1987. A satisfactory benefit cost indicator from contingent valuation. *Journal of Environmental Economics and Management* 14: 226–247.

Hof, John and John Loomis. 1986. A note on the marginal valuation of underpriced facilities. *Public Finance Quarterly* 14, no. 4: 489–498.

Irland, Lloyd. *Wilderness Economics and Policy.* Lexigton, Mass.: Lexington Books.

Iverson, David and Richard Alston. 1986. The Genesis of FORPLAN: A Historical and Analytical Review of Forest Service Planning Models. General Technical Report INT-214. Intermountain Forest and Range Experiment Station, U.S. Forest Service. Ogden, Utah.

Jameson, Donald, Mary Ann Dohn Moore, and Pamela Case. 1982. *Principles of Land and Resource Management Planning.* Washington, D.C.: Land Management Planning, U.S. Forest Service.

Johnson, K. N., T. W. Stuart, and S. A. Crim. 1986. *FORPLAN2: An Overview.* Washington, D.C.: Land Management Planning, U.S. Forest Service.

Joyce, Linda. 1989. An Analysis of the Range Forage Situation in the United States: 1989–2040. General Technical Report RM-180. Rocky Mountain Forest and Range Experiment Station, U.S. Forest Service. Fort Collins, Colo.

Joyce, Linda, Bruce McKinnon, John Hof, and Thomas Hoekstra. 1983. Analysis of Multiresource Production for National Assessments and Appraisals. General Technical Report RM-101. Rocky Mountain Forest and Range Experiment Station, U.S. Forest Service, Fort Collins, Colo.

Just, Richard, Darrell Hueth, and Andrew Schmitz. 1982. *Applied Welfare Economics and Public Policy.* Englewood Cliffs, N.J.: Prentice-Hall.

Keeney, Ralph and H. Raiffa. 1976. *Decisions with Multiple Objectives.* New York: John Wiley.

Keith, J. and K. Lyon. 1985. Valuing wildlife management: A Utah deer herd. *Western Journal of Agricultural Economics* 10: 216–222.

Kent, Brian. 1989. Forest Service Land Management Planners' Introduction to Linear Programming. General Technical Report RM-173, Rocky Mountain Forest and Range Experiment Station, U.S. Forest Service, Fort Collins, Colo.

Knetsch, J., R. Brown, and W. Hansen. 1976. Estimating expected use and value of recreation sites. In C. Gearing, W. Swart, and T. Var, eds., *Planning for Tourism Development: Quantitative Approaches.* New York: Praeger.

Kriesel, Warren and Alan Randall. 1986. Evaluating national policy by contingent valuation. Paper presented at the Annual Meetings of the American Agricultural Economics Association, Reno, Nevada. July.

Krutilla, John. 1967. Conservation reconsidered. *American Economic Review* 57: 787–796.

Krutilla, John and Jon Haigh. 1978. An integrated approach to national forest management. *Environmental Law* 8: 373–415.

Laffin, Curt. 1990. Letter from Curt Laffin, U.S. Fish and Wildlife Service, Regional Office, Newton Corner, Mass., to John Loomis. December 28.

Larsen, Gary, Arnold Holden, Dave Kapaldo, John Leasure, Jerry Mason, Hal Salwasser, Susan Yonts-Shepard, and William Shands. 1990. *Critique of Land Management Planning, Vol. 1: Synthesis of the Critique of Land Management Planning.* FS-454. Washington, D.C.: Policy Analysis Staff. U.S. Forest Service.

Leman, Christopher. 1984. Formal vs. de facto systems of multiple use planning in the Bureau of Land Management: Integrated comprehensive and focused approaches. In National Research Council, National Academy of Sciences, *Developing Strategies for Rangeland Management.* Boulder, Colo.: Westview Press.

Leshy, John. 1987. *The Mining Law: A Study in Perpetual Motion.* Washington, D.C.: Resources for the Future.

Lilieholm, Robert. 1990. Alternatives in regional land use. *Journal of Forestry* April: 10–11.

Lind, Robert. 1990. Reassessing the government's discount rate policy in light of new theory and data in a world economy with a high degree of capital mobility. *Journal of Environmental Economics and Management* 18, no. 2: S-8–S-28.

Lofting, Edward. 1985. Use-Make Matrices and Gross Regional Trade Flows, Report EEA 85-02. Engineering Economics Associates, Berkeley, Calif.

Loomis, John. 1982. Use of travel cost models for evaluating lottery rationed recreation: Application to big game hunting. *Journal of Leisure Research* 14, no. 2: 117–124.

Loomis, John. 1987a. The economic value of instream flow: Methodology and benefit estimates for optimum flows. *Journal of Environmental Management* 24, no. 2: 169–179.

Loomis, John. 1987b. Economic efficiency analysis, bureaucrats and budgets: A test of hypotheses. *Western Journal of Agricultural Economics* 12, no. 1: 27–34.

Loomis, John. 1987c. Balancing public trust resources of Mono Lake and Los Angeles's water right: An economic approach. *Water Resources Research* 23, no. 8: 1449–1456.

Loomis, John. 1988a. *Economic Benefits of Pristine Watersheds.* Englewood, Colo.: American Wilderness Alliance.

Loomis, John. 1988b. The bioeconomic effect of timber harvesting on recreational and commercial salmon and steelhead fishing: A case study of the Siuslaw National Forest. *Marine Resource Economics* 5, no. 1: 43–60.

Loomis, John. 1989a. A more complete accounting of costs and benefits of timber sales. *Journal of Forestry* 87, no. 3: 19–23.

Loomis, John. 1989b. Test-retest reliability of the contingent valuation method: A comparison of general population and visitor responses. *American Journal of Agricultural Economics* 71, no. 1: 76–84.

Loomis, John. 1990. Comparative reliability of the dichotomous choice and open-ended contingent valuation techniques. *Journal of Environmental Economics and Management* 18: 78–85.

Loomis, John and William Brown. 1984. The use of regional travel cost models to estimate the net economic value of recreational fishing at new and existing sites. In *Making Economic Information More Useful for Salmon and Steelhead Production Decisions.* NOAA Technical Memorandum NMFS F/NBWR-8, Portland, Oregon.

Loomis, John, Joseph Cooper, and Stewart Allen. 1988. The Montana Elk Hunting Experience: A Contingent Valuation Assessment of Economic Benefits to Hunters. Montana Department of Fish, Wildlife, and Parks, Helena, Mont.

Loomis, John, Dennis Donnelly, and Cindy Sorg-Swanson. 1989. Comparing the economic value of forage on public lands for wildlife and livestock. *Journal of Range Management* 42, no. 2: 134–138.

Lyon, Jack. 1985. Elk and cattle on the National Forests: A simple question of allocation or a complex management problem? *Western Wildlands* 11: 16–19.

MacCleary, Douglas. 1985. Memo from Deputy Assistant Secretary to Chief, U.S. Forest Service, on USDA Decision on Review of Administrative Decision by the Chief of the Forest Service Related to Administrative Appeals of the Forest Plans and EIS's for the San Juan National Forest and the Grand Mesa, Uncompahgre and Gunnison National Forest. Office of the Secretary, U.S. Department of Agriculture, Washington, D.C. July 31.

Maler, K. G. 1974. *Environmental Economics: A Theoretical Inquiry.* Baltimore, Md.: Johns Hopkins University Press.

Manning, Robert. 1985. *Studies in Outdoor Recreation.* Corvallis, Oreg.: Oregon State University Press.

Markoff, John, Gilbert Shapiro, and Sasha Weitman. 1974. Toward the integration of content analysis and general methodology. In D. Heise, ed., *Social Methodology: 1975.* San Francisco: Jossey-Bass.

Martin, W., J. Tinney, and R. Gum. 1978. A welfare economic analysis of the potential competition between hunting and cattle ranching. *Western Journal of Agricultural Economics* 3: 87–97.

Matulich, Scott, Jeffrey Hanson, Ivan Lines, and Adrian Farmer. 1982. HEP as a planning tool: Application to waterfowl enhancement. *Transactions of 47th North American Wildlife and Natural Resources Management Conference.* Washington, D.C.: Wildlife Management Institute, pp. 111–127.

McConnell, Kenneth. 1977. Congestion and willingness to pay: A study of beach use. *Land Economics* 53, no. 2: 185–195.

McConnell, K. and I. Strand. 1981. Measuring the cost of time in recreation demand analysis: An application to sport fishing. *American Journal of Agricultural Economics* 63: 153–156.

McConnell, Kenneth. 1985. The Economics of outdoor recreation. In A. V. Kneese and J. L. Sweeney, eds., *Handbook of Natural Resource and Energy Economics,* Vol. II. New York: North-Holland.

McKean, John. 1981. An Input-Output Analysis of Sportsman Expenditures in Colorado. Colorado Water Resources Research Institute Technical Report No. 26. Colorado State University, Fort Collins, Colo. January.

McKean, John. 1983. The economic base model. Unpublished mimeo. Department of Economics, Colorado State University, Fort Collins, Colo. Spring.

Mendelsohn, R. 1984. An application of the hedonic framework for recreation modeling to the valuation of deer. In V. K. Smith and D. Witte, ed., *Advances in Applied Microeconomics,* Greenwich, Conn.: JAI Press, pp. 89–101.

Mendelsohn, Robert and Gardner Brown. 1983. Revealed preference approaches to valuing outdoor recreation. *Natural Resources Journal* 23: 607–618.

Menz, Fredrick and Donald Wilton. 1983. Alternative ways to measure recreational values by the travel cost method. *American Journal of Agricultural Economics* 61: 153–156.

Merkhofer, Miley and Ralph Keeney. 1987. A multiattribute utility analysis of alternative sites for the disposal of nuclear waste. *Risk Analysis* 7, no. 2: 173–194.

Milhous, Robert, David Wegner and Thomas Waddle. 1984. A User's Guide to the Physical Habit Simulation System (PHABSIM) Instream Flower Information Paper No. 11. Washington, D.C.: U.S. Fish and Wildlife Service.

Miller, Joseph. 1973. Congress and the Origins of Conservation: Natural Resources Policy, 1865–1900. Ph.D. dissertation, Minneapolis, University of Minnesota.

Mishan, E. J. 1981. *Introduction to Normative Economics.* New York: Oxford University Press.

Mitchell, Robert and Richard Carson. 1989. *Using Surveys to Value Public Goods: The Contingent Valuation Method.* Washington, D.C.: Resources for the Future.

Moab District. 1985. San Juan Resource Management Plan: Management Situation Analysis. Bureau of Land Management, Moab, Utah.

Moab District. 1986. Draft San Juan Resource Management Plan and Environmental Impact Statement. Bureau of Land Management, Moab, Utah. May.

Moab District. 1987. Proposed San Juan Resource Management Plan and Final Environmental Impact Statement. Bureau of Land Management, Moab, Utah. September.

Muhn, James and Hanson Stuart. 1988. *Opportunity and Challenge: The Story of the BLM.* Washington, D.C.: Bureau of Land Management, U.S. Department of Interior.

Mumy, Gene and Steve Hanke. 1975. Public investment criteria for underpriced public products. *American Economic Review* 65: 712–720.

Musgrave, Richard. 1959. *The Theory of Public Finance.* New York: McGraw-Hill.

National Park Service. 1975. The Workbook for Yosemite Master Plan. Yosemite Planning Team, San Francisco, Calif. October.

National Park Service. 1978. Draft Environmental Impact Statement: General Management Plan, Yosemite National Park. Prepared by Denver Service Center, Denver, Colo. August.

National Park Service. 1980. Final Environmental Impact Statement: General Management Plan, Yosemite National Park. Prepared by Denver Service Center, Denver, Colo. October.

National Park Service. 1987. *The National Parks: Index 1987.* Washington, D.C.: Office of Public Affairs, U.S. Department of the Interior.

National Park Service. 1988. *Management Policies.* Washington, D.C.

National Park Service. 1989. Draft Yosemite GMP Examination Report: A Review of the 1980 General Management Plan. Western Regional Office, San Francisco, Calif. August.

Natural Resources Defense Council vs. Morton. 1974. 388 Federal Supplement, 840.

Nelson, Jack. 1984. A modeling approach to large herbivore competition. In National Research Council, National Academy of Sciences, *Developing Strategies for Rangeland Management.* Boulder, Colo.

Nelson, Robert. 1980. The new range wars: Environmentalists versus cattlemen for the public rangelands. Unpublished manuscript, Economics Staff, Office of Policy Analysis, U.S. Dept. of Interior, Washington, D.C.

Nelson, Robert. 1981. Making sense of the Sagebrush Rebellion: A long-term strategy for the public lands. Paper presented at the Third Annual Conference of the Association for Public Policy Analysis and Management, Washington, D.C.

Olsen, Norm. 1991. Personal communication between Norm Olsen, U.S. Fish and Wildlife Service, Washington, D.C., and John Loomis. January 4.

O'Toole, Randal. 1988. *Reforming the Forest Service.* Washington, D.C.: Island Press.

Palmer, Charles. 1983. I-O study team contributions (originally published by Natural Resource Economics Division, USDA Economic Research Service, January 1976). Reprinted in *IMPLAN Training Notebook.* Land Management Planning, Rocky Mountain Forest and Range Experiment, U.S. Forest Service, Fort Collins, Colo.

Peterson, R. D. 1982. An introduction to input-output economics. *Colorado Economic Issues* 6, no. 5. Cooperative Extension Service, Colorado State University, Fort Collins, Colo. January.

Radtke, Hans, Stan Detering, and Ray Brokken. 1985. A comparison of economic impact estimates for changes in the federal grazing fee: secondary vs primary data I/O models. *Western Journal of Agricultural Economics* 10, no. 2: 382–390.

Randall, Alan and John Stoll. 1980. Consumer surplus in commodity space. *American Economic Review* 71, no. 3: 449–457.

Randall, Alan and John Stoll. 1983. Existence value in a total valuation framework. In R. Rowe and L. Chestnut, eds., *Managing Air Quality and Scenic Resources at National Parks and Wilderness Areas.* Boulder, Colo.: Westview Press.

Reed, Nathaniel and Dennis Drabelle. 1984. *The United States Fish and Wildlife Service.* Boulder, Colo.: Westview Press.

Richardson, Harry. 1972. *Input-Output and Regional Economics.* New York: John Wiley.

Robinson, Bob, Lamar White, Geoffrey Johnson, and Michelle Gambone. 1989. National Wildlife Refuges: Continuing Problems with Incompatible Uses Call for Bold Action. RCED-89–196. U.S. General Accounting Office, Washington, D.C.

Rosenthal, Donald. 1987. The necessity for substitute prices in recreation demand analysis. *American Journal of Agricultural Economics* 69, no. 4: 828–837.

Rosenthal, D., J. Loomis, and G. Peterson. 1984. The Travel Cost Model: Concepts and Applications. General Technical Report RM 109. Rocky Mountain Forest and Range Experiment Station, U.S. Forest Service, Fort Collins, Colo.

Row, Clark, Fred Kaiser, and John Sessions. 1981. Discount rate for long-term Forest Service investments. *Journal of Forestry* 79: 367–369.

Sabatier, Paul, John Loomis, and Catherine McCarthy. 1990. Professional norms, external constituencies and hierarchical controls: An analysis of U.S. Forest Service planning decisions. Paper presented at the 1990 Annual Meeting of the Midwest Political Science Association. Chicago, Ill.

Sassone, P. and W. Schaffer. 1978. *Cost Benefit Analysis: A Handbook.* New York: Academic Press.

Schulze, William, Ralph D'Arge, and David Brookshire. 1981. Valuing environmental commodities: Some recent experiments. *Land Economics* 57, no. 2: 151–172.

Schulze, William, David Brookshire, Eric Walther, Karen MacFarland, Mark Thayer, Regan Whitworth, Shaul Ben-David, William Malm, and John Molenar. 1983. The economic benefits of preserving visibility in the national parklands of the Southwest. *Natural Resources Journal* 23: 149–173.

Sellar, Christine, Jean-Paul Chavas, and John Stoll. 1986. Specification of the logit model: The case of valuation of nonmarket goods. *Journal of Environmental Economics and Management* 13, no. 4: 382–390.

Sellars, Richard. 1989. Scenery vs. ecology in the national parks. *Courier* 34, no 7: 4–7. National Park Service, Washington, D.C.

Shands, William, Thomas Waddell, and German Reyes. 1988. *Below Cost Timber Sales in the Broad Context of National Forest Management.* Washington, D.C.: The Conservation Foundation.

Shands, William, Alaric Sample, and Dennis LeMaster. 1990. *Critique of Land Management Planning, Vol. 2. National Forest Planning: Searching for a Common Vision.* FS-453. Washington, D.C.: Policy Analysis Staff, U.S. Forest Service.

Shrank, Barney. 1990. Personal communication. Regional Office, U.S. Fish and Wildlife Service, Denver, Colo. July 3.

Smith, K., W. Desvousges, and M. McGivney. 1983. The opportunity cost of travel time in recreation demand models. *Land Economics* 59: 259–278.

Smith, V. Kerry, William Desvousges, and Ann Fisher. 1986. A comparison of direct and indirect methods for estimating environmental benefits. *American Journal of Agricultural Economics* 68, no. 2: 280–290.

Smith, V. Kerry and Yoshiaki Kaoru. 1990. Signals or noise? Explaining the variation in recreation benefit estimates. *American Journal of Agricultural Economics* 72, no. 2: 419–434.

Sorg, Cindy and John Loomis. 1984. Empirical Estimates of Amenity Forest Values: A Comparative Review. General Technical Report RM-107. Rocky Mountain Forest and Range Experiment Station, U.S. Forest Service, Fort Collins, Colo.

Sorg, Cindy and Lou Nelson. 1986. The Net Economic Value of Elk Hunting in Idaho. Resource Bulletin RM-12. Rocky Mountain Forest and Range Experiment Station, U.S. Forest Service, Ft. Collins, Colo.

Sousa, Patrick. 1987. Habitat Management Models for Selected Wildlife Management Practices in the Northern Great Plains. REC-ERC 87–11, Denver Federal Center, Office of Environmental Technical Services, U.S. Bureau of Reclamation, Denver, Colo.

Stroup, R. and J. Baden. 1983. *Natural Resources: Bureaucratic Myths and Environmental Management.* San Francisco, Calif.: Pacific Institute.

Sullivan, Jay. 1983. Determining changes in final demand for IMPLAN economic impact analysis. In *IMPLAN Training Notebook.* Land Management Planning, Rocky Mountain Forest and Range Experiment, U.S. Forest Service, Fort Collins, Colo.

Teeguarden, Dennis. 1990. *Critique of Land Management Planning, Vol. 11. National Forest Planning Under RPA/NFMA: What Needs Fixing?* FS-462. Policy Analysis Staff, U.S. Forest Service. Washington, D.C.

Thomas, J. W. 1984. Fee hunting on the public lands—an Appraisal. *Transactions of North American Wildlife and Natural Resources Conference* 49: 455–468.

Thomas, J. W. and D. E. Toweill. 1982. *Elk of North America.* Harrisburg, Penn.: Stackpole Books.

Tittman, Paul and Clifton Brownell. 1984. Appraisal Report: Estimating Market Rental Value of Public Rangelands in the Western United States Administered by USDA–Forest Service and USDI–Bureau of Land Management, Vol. 1. PB84–242205. Washington, D.C.

U.S. Agricultural Research Service. 1982. Proceedings of a Workshop on Estimating Erosion and Sediment Yields on Rangelands. Western Series No. 26. U.S. Department of Agriculture, Oakland, Calif.

U.S. Army Corps of Engineers. 1990. Recreation Task Force—Final Report. Headquarters, U.S. Army Corps of Engineers, Washington, D.C.

U.S. Bureau of Reclamation. 1987. Report on Refuge Water Supply Investigations, Central Valley, Hydrologic Basin, California, Volume 1, Mid-Pacific Region, Sacremento, CA.

U.S. Department of Agriculture. 1978. Predicting Rainfall Erosion Losses. Agricultural Handbook No. 357. Washington, D.C.

U.S. Department of Commerce. 1986. *Regional Multipliers: A User Handbook for the Regional Input-Output Modelling System.* RIMS II. Washington, D.C.: Bureau of Economic Analysis, U.S. Department of Commerce.

U.S. Department of Interior. 1986. Natural resource damage assessments; final rule. 43 CFR Part 11. *Federal Register* 51 (August 1): 27674–27753.

U.S. Department of Interior. 1990. A Management Plan for Agricultural Subsurface Drainage and Related Problems on the Westside Sam Joaquin Valley. Final Report of the San Joaquin Valley Drainage Program, Sacramento, CA.

U.S. General Accounting Office. 1986. Land Management: Forest Planning Costs at the Boise and Clearwater National Forests in Idaho. Report GAO/RCED 87–28FS. Washington, D.C.

U.S. General Accounting Office. 1989. Federal Land Management: The Mining Law of 1872 Needs Revision. Report RCED-89–72. General Accounting Office. Washington, D.C.

U.S. Water Resources Council. 1979. Procedures for Evaluation of National Economic Development (NED) Benefits and Costs in Water Resources Planning (Level C). *Federal Register* 44, 23): 72892–72976.

U.S. Water Resources Council. 1983. *Economic and Environmental Principles for Water and Related Land Resources Implementation Studies.* Washington, D.C.: U.S. Government Printing Office.

Varan, Hal, 1990. *Intermediate Microeconomics: A Modern Approach.* New York: Norton.

Vaughan, William and Clifford Russell. 1982. Valuing a fishing day: An application of a systematic varying parameter model. *Land Economics* 58: 450–463.

Verburg, Edwin and Richard Coon. 1987. Planning in the United States Fish and Wildlife Service. *Trends* 24, no. 2: 20–26.

Vogely, William. 1984. Valuation of mineral resources. In G. Peterson and A. Randall, eds., *Valuation of Wildland Resource Benefits.* Boulder, Colo.: Westview Press.

Walker, John. Tall trees, people and politics: The opportunity costs of Redwood National Park. *Contemporary Policy Issues* 2, no. 5: 22–29.

Wallensnoin, Rolf. 1984. Division of Environmental Coordination's Review of Draft Comprehensive Management Plan and Environmental Impact Statement, Kenai, National Wildlife Refuge. U.S. Fish and Wildlife Service, Washington, D.C. May 10.

Walsh, Richard. 1980. An Economic Evaluation of the General Management Plan for Yosemite National Park. Colorado Water Resources Research Institute, Colorado State University, Fort Collins, Colo.

Walsh, Richard. 1986. *Recreation Economic Decisions.* State College, Penn.: Venture Press.

Walsh, Richard, Ray Ericson, Daniel Arosteguy and Michael Hansen. 1980a. An Empirical Application of a Model for Estimating the Recreation Value of Instream Flow. Colorado Water Resources Research Institute, A-036, Colorado State University, Fort Collins, Colo.

Walsh, Richard, Robert Aukerman, Robert Milton. 1980b. Measuring the Benefits and the Economic Value of Water in High Country Reservoirs. Colorado Water Resources Research Institute, B-175, Colorado State University, Fort Collins, Colo.

Walsh, Richard and John Olienyk. 1981. Recreation demand effects of mountain pine beetle damage to the quality of forest recreation resources in the Colorado Front Range. Unpublished paper, Department of Economics, Colorado State University, Fort Collins, Colo.

Walsh, Richard and Lynde Gilliam. 1982. Benefits of wilderness expansion with excess demand for Indian peaks. *Western Journal of Agricultural Economics* 7: 1–12.

Walsh, Richard, Nicole Miller, and Lynde Gilliam. 1983. Congestion and willingness to pay for expansion of skiing capacity. *Land Economics* 59, no. 2: 195–210.

Walsh, Richard, John Loomis, and Richard Gillman. 1984. Valuing option, existence and bequest demands for wilderness. *Land Economics* 60, no. 1: 14–29.

Walsh, Richard, Larry Sanders, and John Loomis. 1985. *Wild and Scenic River Economics: Recreation Use and Preservation Values.* Denver, Colo.: American Wilderness Alliance.

Walsh, Richard, Donn Johnson, and John McKean. 1989. Issues in nonmarket valuation and policy application: A retrospective glance. *Western Journal of Agricultural Economics* 14, no. 1: 178–188.

Walsh, Richard, Larry Sanders, and John McKean. 1990. The consumptive value of travel time on recreation trips. *Journal of Travel Leisure* 29, no. 1: 17–24.

Ward, Frank. 1985. Economics of optimally managing wild rivers for instream benefits. Unpublished paper, Department of Agricultural Economics, New Mexico State University, Las Cruces, N. Mex.

Ward, Frank and John Loomis. 1986. The travel cost method as an environmental policy assessment tool: A review of the literature. *Western Journal of Agricultural Economics* 11, no. 2: 164–178.

Weisbrod, Burton. Collective consumption services of individual consumption goods. *Quarterly Journal of Economics* 78, no. 3: 471–477.

Welsh, Michael. 1986. Exploring the Accuracy of the Contingent Valuation Method: Comparisons with Simulated Markets. Ph. D. dissertation, Department of Agricultural Economics, University of Wisconsin.

Wilkinson, Charles and H. Michael Anderson. 1987. *Land and Resource Planning in the National Forests.* Washington, D.C.: Island Press.

Williams, David. 1987. Planning approaches in the Bureau of Land Management. *Trends* 24, no. 2: 27–35.

Wilson, J., G. Marousek, and K. Gee. 1985. Economic impacts of BLM grazing policies on Idaho cattle ranchers. Research Bulletin No. 136. Agricultural Experiment Station, University of Idaho, Moscow, Idaho.

Young, Robert. 1973. The price elasticity of demand for water: A case study of Tucson, Arizona. *Water Resources Research* 9, no. 6: 1068–1072.

Index

Designer: Susan Clark
Text: Caslon 540
Compositor: Impressions, *a division of* Edwards Brothers
Printer: Edwards Brothers
Binder: Edwards Brothers